AF680095

# Blockchain applications in agriculture

## Revolutionizing the food supply chain

**Duong Hoai An**
*Thai Nguyen University of Agriculture and Forestry, Vietnam*

Series in Business and Finance
Vernon Press

Copyright © 2024 Vernon Press, an imprint of Vernon Art and Science Inc, on behalf of the author.

All rights reserved. No part of this publication may be reproduced, stored in a retrieval system, or transmitted in any form or by any means, electronic, mechanical, photocopying, recording, or otherwise, without the prior permission of Vernon Art and Science Inc.

www.vernonpress.com

*In the Americas:*
Vernon Press
1000 N West Street, Suite 1200,
Wilmington, Delaware 19801
United States

*In the rest of the world:*
Vernon Press
C/Sancti Espiritu 17,
Malaga, 29006
Spain

Series in Business and Finance

Library of Congress Control Number: 2024937246

ISBN: 978-1-64889-949-2

Product and company names mentioned in this work are the trademarks of their respective owners. While every care has been taken in preparing this work, neither the authors nor Vernon Art and Science Inc. may be held responsible for any loss or damage caused or alleged to be caused directly or indirectly by the information contained in it.

Every effort has been made to trace all copyright holders, but if any have been inadvertently overlooked the publisher will be pleased to include any necessary credits in any subsequent reprint or edition.

Cover design by Vernon Press with elements from Freepik.

# Contents

# Foreword

Son Nghiem

*The Australian National University*

In the intricate web of modern living, the age-old adage "you are what you eat" reverberates with newfound significance. The quality, safety, and nutritional content of our food are paramount to our health and wellbeing. However, the complexities of contemporary food supply chains often shroud these critical factors in obscurity. Dr. An Duong, an agricultural economist with a profound interest in information technology, has dedicated years to bridging the gap between agriculture and technology. His latest endeavour, "Blockchain Applications in Agriculture: Revolutionizing the Food Supply Chain," epitomizes his relentless pursuit of leveraging technology for the betterment of food systems.

Dr. Duong's expertise lies at the nexus of agriculture and technology, where he has meticulously researched the transformative potential of blockchain technology within the food supply chain. With a deep understanding of the challenges plaguing the agricultural sector, Dr. Duong illuminates how blockchain offers a beacon of hope, promising transparency, traceability, and accountability. This book represents the culmination of his endeavours, offering a comprehensive resource for policymakers, food suppliers, and consumers alike.

Spanning ten chapters with more than 400 pages, this book serves as a veritable guide through the labyrinth of blockchain applications in agriculture. From demystifying blockchain technology to elucidating its manifold applications within the agricultural domain, Dr. Duong leaves no stone unturned. Each chapter unfolds a new facet of blockchain's role in revolutionizing the food supply chain, providing readers with actionable insights and practical solutions. Moreover, the book ventures into the realm of future trends, offering a glimpse into the evolving landscape of agricultural technology.

What sets this book apart is its reader-friendly structure. At the outset of each chapter, readers will find a neat summary accompanied by a memorable quote, setting the stage for the ensuing exploration. Such clarity and self-containment render the book conducive to non-linear reading, allowing readers to delve into chapters of interest at their discretion.

Through a blend of theoretical discourse, practical examples, and real-world case studies, Dr. Duong navigates the complex terrain of blockchain applications in agriculture with finesse. Readers will emerge equipped with a nuanced understanding of how blockchain technology can drive positive change within the agricultural sector, paving the way for a more transparent, efficient, and sustainable food supply chain.

Son Nghiem
Associate Professor of Health Economics
Department of Health Economics, Wellbeing and Society
The Australian National University

# Preface

In an era of unprecedented technological advancements, few innovations have captured the world's attention and potential as profoundly as blockchain technology. With its decentralized and transparent nature, blockchain has the power to transform industries across the globe, and agriculture stands at the forefront of this revolution. The convergence of blockchain and agriculture has opened up new possibilities for securing the global food supply chain, promoting sustainability, enhancing productivity, and ensuring food safety.

"Blockchain Applications in Agriculture: Revolutionizing the Food Supply Chain" is a comprehensive exploration of the transformative potential of blockchain technology in the agricultural sector. This book is designed to provide a holistic understanding of blockchain's applications, its underlying principles, and how it can reshape every aspect of the agriculture industry.

Chapter 1, "Introduction to Blockchain Technology," serves as the foundation of our journey. We delve into the fundamental concepts of blockchain, exploring its decentralized architecture, immutability, and cryptographic security. We also acknowledge the blockchain disadvantages and limitations. Through this chapter, readers will gain a solid grasp of the technology that underpins the subsequent chapters.

Chapter 2, "Overview of the Agriculture Industry," takes us deep into the workings of the agriculture industry. By understanding the challenges faced by the agricultural sector such as farmers, producers, and consumers, we can appreciate the potential impact of blockchain in addressing these issues. From small-scale farming to large-scale agricultural operations, this chapter examines the nuances of the industry and sets the stage for the subsequent exploration of blockchain-based solutions.

Chapter 3, "Blockchain-based Supply Chain Management," showcases the transformative power of blockchain in securing and optimizing the food supply chain. We delve into the potential of distributed ledgers to enhance transparency, traceability, and efficiency in the movement of agricultural products from farm to table. By leveraging smart contracts and immutable records, blockchain has the capacity to revolutionize supply chain management, ensuring fair trade, reducing waste, and eliminating fraud.

Chapter 4, "Farming and Crop Management," uncovers how blockchain technology can revolutionize traditional farming practices. From precision agriculture to smart contracts for land tenure, blockchain's decentralized nature can enhance productivity, improve resource allocation, and empower farmers with data-driven insights. By leveraging blockchain's capabilities, farmers can make informed decisions, optimize crop yields, and promote sustainable farming practices.

Chapter 5, "Livestock Management and Animal Welfare," delves into the potential of blockchain in transforming livestock management practices and ensuring animal welfare. From tracking the origin and health records of livestock to certifying ethical practices in animal farming, blockchain-based solutions offer unprecedented transparency and accountability. Through this chapter, readers will explore how blockchain technology can enhance the integrity and trustworthiness of the livestock industry.

Chapter 6, "Food Safety and Quality Assurance," emphasizes the critical role that blockchain plays in ensuring the safety and quality of our food. We explore how blockchain's immutable and transparent nature can facilitate real-time monitoring, traceability, and verification of food products. By reducing the risk of contamination, preventing counterfeit goods, and enabling swift recalls, blockchain technology holds tremendous promise for safeguarding public health and improving consumer confidence.

Chapter 7, "Agricultural Finance and Insurance," highlights the transformative potential of blockchain in reshaping financial and insurance systems within the agriculture industry. By facilitating secure and transparent transactions, blockchain can streamline financial processes, improve access to capital for farmers, and enhance risk management through smart contracts and parametric insurance. This chapter explores the possibilities of financial inclusion and resilience that blockchain technology brings to the agriculture sector.

Chapter 8, "Sustainable Agriculture and Supply Chain Sustainability," underscores the role of blockchain in promoting sustainable agriculture and ensuring supply chain sustainability. We examine how blockchain's transparency and traceability can incentivize environmentally friendly practices, promote fair trade, and enable consumers to make informed choices. From reducing carbon footprints to supporting ethical sourcing, blockchain has the potential to usher in a new era of sustainability in the agriculture industry.

Chapter 9, "Case Studies and Real-world Examples," presents a collection of inspiring case studies and real-world examples where blockchain technology has been successfully implemented in agriculture. From small-scale initiatives to large-scale projects, these cases showcase the tangible benefits and transformative power of blockchain in addressing industry-specific challenges. Through these stories, readers will gain insights into the practical applications of blockchain and draw inspiration for their own endeavors.

Chapter 10, "Future Trends and Challenges," explores the exciting possibilities and potential challenges that lie ahead in the intersection of blockchain and agriculture. From the emergence of new technologies such as the Internet of Things (IoT) and artificial intelligence (AI) to regulatory considerations and scalability issues, this chapter examines the evolving landscape and offers reflections on what the future holds for blockchain in agriculture.

"Blockchain Applications in Agriculture: Revolutionizing the Food Supply Chain" aims to be a comprehensive resource for researchers, policymakers, farmers, and industry professionals seeking to understand and harness the transformative power of blockchain technology in the agriculture sector. By providing a balanced blend of theoretical concepts, practical insights, and real-world examples, this book equips readers with the knowledge and inspiration to navigate the blockchain revolution and drive innovation in agriculture.

As we embark on this journey together, I invite you to explore the possibilities, envision the future, and join the movement to revolutionize the food supply chain through blockchain technology.

# Chapter 1
# Introduction to blockchain technology

*Unveiling the Digital Revolution: Exploring the Building Blocks of Blockchain Technology* - Unknown


## Abstract

This chapter serves as an introduction to the concept of blockchain technology and its potential applications in various industries. Specifically focusing on agriculture, the chapter explores the fundamental principles and components of blockchain technology, highlighting its relevance and potential benefits in farming and crop management.

The chapter begins with a concise overview of blockchain technology, explaining its underlying principles and core components. Readers will gain a solid understanding of how blockchain operates as a decentralized and transparent ledger system, ensuring data integrity and security. Key concepts such as distributed ledgers, consensus mechanisms, and immutability are explored in detail.

Furthermore, the chapter delves into the unique features of blockchain technology that make it particularly suitable for the agricultural sector. It discusses the inherent benefits of transparency, traceability, and accountability offered by blockchain, and how these features can enhance various aspects of agriculture.



**Keywords**: Blockchain technology, Agriculture, Farming, Crop management, Decentralized ledger system, Data integrity, Security, Distributed ledgers, Consensus mechanisms, Transparency


### Chapter contents

- Understanding the fundamentals of blockchain technology
- Distributed ledger technology and its benefits
- Blockchain consensus mechanisms

***

## 1.1. Understanding the fundamentals of blockchain technology

In recent years, blockchain technology has garnered significant attention for its transformative potential across various industries. But what exactly constitutes the fundamentals of blockchain technology? Through an exploration guided by Gayvoronskaya and Meinel (2021), Ghiro et al. (2021), Lafourcade and Lombard-Platet (2020), Pilkington (2016), Radziwill (2018), Singhal, Dhameja, and Panda (2018) and Zheng, Xie, Dai, Chen, and Wang (2018), we uncover the core principles encompassing the definition of blockchain, the intricate concept of blocks and chains, the pivotal role of decentralization and trust, the application of cryptography, and the immutable and transparent nature inherent to blockchain technology.

### 1.1.1. Definition of blockchain

*a. Defining blockchain*

According to Gayvoronskaya and Meinel (2021), Ghiro et al. (2021), Lafourcade and Lombard-Platet (2020), Pilkington (2016) and Yaga, Mell, Roby, and Scarfone (2019), blockchain and its components can be defined as follows:

- Distributed Ledger Technology: At its core, blockchain is a type of distributed ledger technology (DLT). We will discuss how a distributed ledger differs from a traditional centralized ledger and the advantages it offers in terms of transparency, security, and immutability.
- Blocks and Chains: Blockchain gets its name from the way data is organized and secured. We will explore the concept of blocks, which contain batches of transactions, and how they are linked together in a chain through cryptographic hashing.
- Decentralization: A key characteristic of blockchain is its decentralized nature. We will explain how blockchain operates on a network of multiple participants or nodes, with no central authority or control. We will discuss the benefits of decentralization, such as increased security, resilience, and censorship resistance.

*b. Key components of blockchain*

- Transactions: Transactions are the fundamental building blocks of the blockchain. We will examine what constitutes a transaction, including the sender and receiver addresses, the transaction amount, timestamps, and any additional data associated with it.
- Cryptographic Hash Functions: Cryptographic hash functions play a vital role in blockchain technology. We will explore how hash functions generate

unique digital fingerprints for blocks and transactions, ensuring data integrity and security.

- Consensus Mechanisms: Consensus mechanisms are essential to establish agreement among participants in a decentralized network. We will discuss popular consensus algorithms such as Proof of Work (PoW) and Proof of Stake (PoS) and their role in validating transactions and maintaining the integrity of the blockchain.
- Smart Contracts: Smart contracts are self-executing contracts with predefined conditions encoded on the blockchain. We will explain how smart contracts automate and enforce the terms of agreements, eliminating the need for intermediaries and enhancing efficiency and trust.

*c. Features and advantages of blockchain*

- Transparency: Blockchain provides transparency by allowing all participants to view and verify transactions. We will discuss how transparency contributes to trust, accountability, and the prevention of fraud in various sectors.
- Security and Immutability: Blockchain ensures the security and immutability of data through cryptographic mechanisms. We will explore how cryptographic hashes, distributed consensus, and decentralized storage prevent tampering and unauthorized modifications, making blockchain highly secure and resistant to fraud.
- Trust and Disintermediation: Blockchain technology reduces the need for intermediaries by enabling direct peer-to-peer transactions. We will discuss how blockchain enhances trust among participants by removing reliance on third parties and creating a more efficient and cost-effective system.
- Potential Applications: We will provide examples of real-world applications where blockchain is being implemented, such as finance, supply chain management, healthcare, voting systems, and more. We will discuss the transformative potential of blockchain in these domains and the benefits it offers.

### 1.1.2. The concept of blocks and chains

*a. Anatomy of a block*

- Block Structure: We will examine the structure of a typical block in a blockchain. This includes components such as a header, a list of transactions, and a cryptographic hash. We will discuss the purpose and function of each component and how they contribute to the overall integrity and security of the block.

- Transaction Data: Transactions are the core elements stored within a block. We will explore the types of information included in transactions, such as sender and receiver addresses, transaction amounts, timestamps, and additional dates. We will discuss how transactions are validated and added to a block.
- Nonce and Proof of Work: The nonce is a crucial element in the block header that miners manipulate to find a valid solution to the proof-of-work problem. We will discuss the role of the nonce and how it is used to ensure the security and immutability of the blockchain.

*b. Linking blocks: hashes and pointers*

- Cryptographic Hash Functions: Cryptographic hashes play a vital role in blockchain technology. We will explore the properties of cryptographic hash functions, such as collision resistance and determinism. We will discuss how hash functions are used to create a unique digital fingerprint for each block and ensure the integrity of the blockchain.
- Merkle Trees: Merkle trees provide an efficient way to verify the integrity of a large number of transactions within a block. We will examine how Merkle trees are constructed using cryptographic hashes and how they enable efficient verification of transaction data.
- Block Pointers: Block pointers are the links that connect blocks in a blockchain. We will discuss how block pointers utilize cryptographic hashes to reference the previous block, creating a chain of blocks. We will explore the significance of these pointers in establishing the immutability and security of the blockchain.

*c. The power of the chain: immutable and distributed ledger*

- Immutability: The arrangement of blocks in a chain, coupled with cryptographic hashing, ensures the immutability of the blockchain. We will discuss how any modification to a single block alters the hash of that block and subsequent blocks, making it computationally infeasible to tamper with the blockchain's historical records.
- Distributed Ledger: The chain of blocks is maintained across a decentralized network of nodes. We will explore the role of consensus algorithms and network participation in ensuring the consistency and synchronisation of the blockchain. We will discuss how every participant in the network maintains a copy of the blockchain, providing redundancy and enhancing the system's robustness.

- Forks and Longest Chain Rule: Forks are temporary divergences in the blockchain where multiple valid chains emerge. We will discuss the role of the longest chain rule in resolving forks and establishing a single canonical chain, ensuring consensus and the integrity of the blockchain.

### 1.1.3. Decentralization and trust in blockchain

*a. The essence of decentralization*

Decentralization lies at the heart of blockchain technology. It fundamentally challenges the traditional model of centralized authority and places control in the hands of a distributed network of participants. We will explore the key aspects of decentralization, including:

- Distributed Ledger: Blockchain's decentralized nature stems from its distributed ledger. We will examine how this ledger is maintained across multiple nodes, ensuring that no single entity has exclusive control over the data.
- Peer-to-peer Network: A peer-to-peer network architecture forms the backbone of decentralized blockchains. We will discuss how nodes communicate and validate transactions directly with each other, eliminating the need for intermediaries.
- Removal of Central Points of Failure: Decentralization mitigates the risks associated with a single point of failure. We will explore how blockchain's resilience is derived from the redundancy and robustness of its decentralized architecture.

*b. Trust in blockchain*

Trust is a fundamental aspect of any financial or societal system, and blockchain technology introduces innovative mechanisms to establish and enhance trust. We will examine how blockchain fosters trust through the following elements:

- Immutability and Transparency: The immutability of blockchain records ensures that once a transaction is recorded, it cannot be altered. We will discuss how this immutability, coupled with transparency, creates a system where participants can trust the integrity of the data and transactions.
- Consensus Mechanisms: Consensus mechanisms play a crucial role in establishing trust in blockchain networks. We will explore popular mechanisms such as Proof of Work (PoW) and Proof of Stake (PoS) and discuss how they ensure agreement and prevent malicious activities.
- Smart Contracts: Smart contracts enable automated, trustless execution of agreements on the blockchain. We will examine how these self-executing

contracts eliminate the need for intermediaries, enhancing trust and reducing the potential for fraud or manipulation.

- Community Governance: Blockchain often involves decentralized governance models where decisions are made collectively by the community. We will discuss how this participatory governance model instills trust and ensures that the network evolves in a fair and transparent manner.

*c. The implications of decentralization and trust*

The combination of decentralization and trust in blockchain technology has far-reaching implications across various domains. We will explore the following areas:

- Financial Systems: Blockchain's decentralized nature has disrupted traditional financial systems, providing opportunities for financial inclusion, reducing transaction costs, and improving security. We will discuss how decentralized finance (DeFi) applications are transforming the financial landscape.
- Supply Chain and Logistics: Blockchain's ability to establish trust and transparency is particularly valuable in supply chain and logistics management. We will examine how decentralized solutions enhance traceability, authenticity verification, and overall efficiency in these industries.
- Governance and Voting Systems: Decentralized governance mechanisms powered by blockchain enable transparent and tamper-resistant voting systems. We will discuss the potential of blockchain-based governance to enhance democracy, reduce corruption, and empower individuals.
- Data Security and Privacy: Blockchain's decentralized architecture offers enhanced data security and privacy by reducing reliance on centralized entities and minimizing the risk of data breaches. We will explore the implications of decentralized data storage and authentication.

### 1.1.4. Cryptography in blockchain

*a. Fundamentals of cryptography*

The foundational aspects of cryptography are elucidated by El Ioini and Pahl (2018), and Lubbe (1998) as follows:

- Encryption: Encryption forms the basis of secure communication in the blockchain. We will discuss symmetric and asymmetric encryption techniques and their applications in protecting data confidentiality.
- Hash Functions: Hash functions are critical cryptographic tools used in blockchain to ensure data integrity. We will explore the properties of hash

functions, such as collision resistance and determinism, and how they are employed in blockchain systems.

- Digital Signatures: Digital signatures provide authentication and non-repudiation in blockchain transactions. We will examine the concepts of public key cryptography, key pairs, and the process of generating and verifying digital signatures.
- Key Management: Effective key management is essential for maintaining the security of blockchain systems. We will discuss key generation, storage, and distribution techniques, as well as the concept of key pairs and their role in cryptographic operations.

*b. Securing transactions and data in blockchain*

- Secure Transactions: Cryptography plays a crucial role in securing transactions on the blockchain. We will explore how digital signatures are used to verify the authenticity and integrity of transaction messages, preventing tampering and ensuring trust.
- Merkle Trees: Merkle trees are cryptographic data structures used to efficiently verify the integrity of large sets of data. We will examine how Merkle trees are employed in blockchain to validate the consistency of transactions and prevent data manipulation.
- Privacy-enhancing Techniques: Cryptography also enables privacy-enhancing techniques in blockchain. We will discuss concepts such as zero-knowledge proofs and ring signatures, which allow for anonymous transactions and preserve the privacy of participants.

*c. Consensus and cryptographic algorithms*

- Proof of Work (PoW): PoW is a consensus algorithm used in blockchain, and it relies on cryptographic puzzles to secure the network. We will delve into the cryptographic computations involved in PoW and the role they play in achieving consensus and preventing malicious activities.
- Proof of Stake (PoS): PoS is an alternative consensus algorithm that leverages participants' stake in the network. We will explore how cryptography is used in PoS to select validators and ensure the integrity of the consensus process.
- Byzantine Fault Tolerance (BFT): BFT algorithms ensure the resilience of blockchain networks against malicious or faulty nodes. We will examine how cryptographic techniques, such as digital signatures and threshold cryptography, are employed in BFT consensus protocols.

*d. Future developments in cryptography for blockchain*

- Post-quantum Cryptography: The advent of quantum computers poses a potential threat to traditional cryptographic algorithms. We will discuss the emergence of post-quantum cryptography and its implications for blockchain security.
- Homomorphic Encryption: Homomorphic encryption enables computations on encrypted data without revealing the underlying information. We will explore the potential applications of homomorphic encryption in blockchain, such as secure data sharing and privacy-preserving smart contracts.
- Multi-party Computation (MPC): MPC allows multiple participants to jointly compute a function while keeping their individual inputs private. We will discuss the relevance of MPC in blockchain settings, enabling secure collaborative operations without compromising data privacy.

### 1.1.5. Immutable and transparent nature of blockchain

*a. Understanding immutability*

Immutability lies at the core of blockchain technology. In a traditional centralized system, data can be altered or manipulated by a single entity with sufficient privileges. This poses significant risks, such as fraud, data breaches, and disputes over the integrity of information. However, blockchain tackles this problem by introducing immutability.

In a blockchain network, information is stored in blocks, which are cryptographically linked to each other in a chain. Once a block is added to the chain, it becomes nearly impossible to alter or delete the data it contains. This is achieved through the use of cryptographic hashing algorithms, which generate unique digital signatures for each block. Even a small change in the data would result in a completely different hash value, making it evident that tampering has occurred.

The immutability of blockchain enhances trust and security in various ways. Financial transactions recorded on a blockchain become verifiable and auditable by anyone on the network, eliminating the need for intermediaries. In industries such as supply chain management, where trust and provenance are crucial, the immutability of blockchain ensures that the information about the origin and movement of goods remains tamper-proof.

*b. Unveiling transparency*

Transparency is another fundamental characteristic of blockchain. Traditional systems often suffer from opacity, as data is controlled and managed by centralized entities. This lack of transparency breeds mistrust and creates information asymmetry between parties involved in transactions. Blockchain technology addresses this issue by introducing a transparent and decentralized network.

In a blockchain, all participants have access to the entire transaction history, stored in a distributed ledger. Every transaction is visible to all network participants, ensuring that no single entity can manipulate or obscure the information. This transparency leads to increased accountability and trust among participants.

Transparency in blockchain can be categorised into two levels: public and private. Public blockchains, such as Bitcoin and Ethereum, allow anyone to participate and view the entire transaction history. Private blockchains, on the other hand, restrict access to a select group of participants, ensuring confidentiality within the network.

The transparency provided by blockchain has far-reaching implications. In financial systems, it allows for real-time auditing, reducing the risk of fraudulent activities. In governmental applications, blockchain can enable transparent voting systems, ensuring the integrity of elections. Moreover, transparency in supply chain management enables consumers to make informed decisions about the products they purchase, promoting ethical and sustainable practices.

### *c. Challenges and considerations*

While the immutable and transparent nature of blockchain brings significant benefits, it also poses challenges and considerations. One key challenge is striking a balance between transparency and privacy. While transparency is desirable in many cases, certain sensitive data may need to be protected. Solutions such as zero-knowledge proofs and privacy-focused blockchains are being developed to address this challenge.

Another consideration is the scalability of blockchain networks. As the number of transactions and participants increases, the size of the blockchain grows, leading to longer validation times and increased storage requirements. Innovative approaches like sharding and layer-two solutions are being explored to enhance scalability without compromising security and transparency.

Furthermore, the governance of blockchain networks must be carefully designed to ensure transparency and inclusivity. Decisions regarding protocol upgrades and network modifications should involve the consensus of the network participants, avoiding centralized control and maintaining the integrity of the system.

## 1.2. Distributed ledger technology and its benefits

### 1.2.1. Introduction to distributed ledger technology (DLT)

Liu, Farahani, and Firouzi (2020) underscored the pivotal role of distributed ledger technology (DLT) within the realm of blockchain. Thus, comprehending its nuances, distinguishing characteristics, diverse applications, and navigating potential challenges becomes imperative when implementing this transformative technology.

*a. Understanding distributed ledger technology*

At its core, DLT is a decentralized and distributed database that enables the recording, sharing, and verification of data across multiple participants or nodes in a network. Unlike traditional centralized databases, where data is stored and managed by a single authority, DLT distributes the responsibility of data management among the network participants. This distributed nature brings several advantages, including increased security, transparency, and resilience.

DLT utilizes cryptographic techniques to ensure the integrity and security of data. The data is organized into blocks, which are cryptographically linked to form a chain of transactions or events. Each participant in the network maintains a copy of the ledger, and consensus mechanisms are employed to validate and agree upon the state of the ledger across the network.

*b. Characteristics of distributed ledger technology*

DLT possesses several key characteristics that differentiate it from traditional databases:

- Decentralization: DLT operates in a decentralized manner, with no central authority controlling the entire network. Instead, consensus algorithms are used to achieve agreement among participants on the validity of transactions and the state of the ledger.
- Transparency: DLT promotes transparency by allowing all participants to have visibility into the transactions and the current state of the ledger. This transparency enhances trust, accountability, and auditability, as participants can verify the integrity of the data.
- Immutability: Similar to blockchain, DLT ensures immutability by making it extremely difficult to alter or delete data once it is recorded in the ledger. The use of cryptographic hashing and consensus mechanisms makes tampering with the data highly impractical.

- Security: DLT incorporates strong cryptographic measures to secure the data and protect it from unauthorized access or manipulation. The distributed nature of the ledger also enhances security by reducing the risk of a single point of failure or attack.

*c. Applications of distributed ledger technology*

Distributed Ledger Technology (DLT) has shown versatility across diverse industries (Li & Kassem, 2021; Wang, Han, & Beynon-Davies, 2019). Its applicability spans various sectors, including finance, healthcare, supply chain management, and beyond. These studies underscore the wide-ranging potential of DLT to revolutionize traditional processes and enhance efficiency, transparency, and security in numerous domains.

- Financial Services: DLT has gained significant traction in the financial sector, enabling efficient and secure transactions, reducing settlement times, and eliminating the need for intermediaries. It has given rise to cryptocurrencies, such as Bitcoin, and has the potential to revolutionize areas like cross-border payments, remittances, and smart contracts.
- Supply Chain Management: DLT provides transparency and traceability in supply chain processes, ensuring the authenticity and provenance of goods. It enables real-time tracking of products, enhances efficiency in logistics, and helps combat counterfeiting and fraud.
- Healthcare: DLT has the potential to transform healthcare systems by securely storing and sharing patient records, improving interoperability between healthcare providers, and enhancing data privacy and consent management.
- Governance: DLT can enable transparent and secure voting systems, ensuring the integrity of elections and reducing fraud. It also has applications in areas such as land registries, identity management, and public administration.
- Internet of Things (IoT): DLT can facilitate secure and decentralized communication and data exchange among IoT devices. It can enable autonomous machine-to-machine transactions, improve data integrity, and enhance the security and privacy of IoT networks.

*d. Considerations and challenges*

While Distributed Ledger Technology (DLT) holds immense promise, it also presents notable considerations and challenges, as highlighted by Wang, Han, et al. (2019), including:

- Scalability: DLT networks face scalability challenges as the number of participants and transactions increase. The processing and storage requirements can become significant, and efforts are being made to develop scalable solutions without compromising the security and decentralization aspects.
- Interoperability: As different DLT platforms and networks emerge, interoperability becomes crucial to enable seamless data exchange and collaboration between them. Standardization efforts and interoperability protocols are being developed to address this challenge.
- Regulatory and Legal Frameworks: The regulatory and legal frameworks surrounding DLT are still evolving. Issues such as data privacy, cross-border transactions, smart contract enforceability, and liability require careful consideration and adaptation of existing laws.

### 1.2.2. Comparison of centralized and decentralized systems

*a. Centralized systems*

Centralized systems are characterized by a single point of control or authority over data, decision-making, and operations. In this model, a central entity or organization has the power to manage and control the entire system. Some key features of centralized systems include:

- Control and Authority: In a centralized system, decision-making power and control over resources lie in the hands of a central authority. This authority has the ability to dictate rules, policies, and access privileges.
- Efficiency and Simplicity: Centralized systems often offer streamlined processes and efficient resource allocation since decision-making is centralized. They can be easier to manage and maintain due to the concentration of power.

*b. Decentralized Systems*

Decentralized systems, on the other hand, distribute control, decision-making, and operations across multiple participants or nodes (Blokdyk, 2021). These systems aim to eliminate the need for a central authority and promote autonomy, transparency, and collaboration. Some key features of decentralized systems include:

- Autonomy and Empowerment: Decentralized systems empower individual participants by giving them control over their own resources, data, and decision-making. Participants can act autonomously within the predefined rules and protocols of the system.

- Transparency and Trust: Decentralized systems promote transparency by making information accessible to all participants. This transparency enhances trust, as the actions and decisions within the system can be verified and audited by multiple participants.

*c. Comparative analysis*

To better understand the advantages and disadvantages of centralized and decentralized systems, let's compare them across different aspects:

- Control and Authority: Centralized systems offer a clear chain of command and decision-making hierarchy. This centralized control can lead to efficient operations and quick decision-making. However, it also concentrates power, making the system vulnerable to corruption, single points of failure, and abuse of authority. In contrast, decentralized systems distribute control among participants, reducing the risk of abuse and promoting autonomy. However, decision-making can be slower due to the need for consensus among participants.
- Efficiency and Scalability: Centralized systems often excel in efficiency and scalability. With a single authority managing resources, they can optimize processes, allocate resources effectively, and quickly adapt to changing demands. However, their scalability is limited by the capacity and capabilities of the central authority. Decentralized systems may face challenges in terms of efficiency and scalability due to the need for consensus and coordination among multiple participants. However, advancements in technology and protocols are addressing these limitations.
- Transparency and Trust: Centralized systems may lack transparency, as access to information is controlled by the central authority. This opacity can breed mistrust and raise concerns about data integrity and security. In contrast, decentralized systems provide transparency by making information visible to all participants. This transparency enhances trust, accountability, and the verifiability of actions within the system.
- Security and Resilience: Centralized systems can be vulnerable to security breaches and single points of failure. If the central authority is compromised, the entire system is at risk. In decentralized systems, the distribution of data and decision-making reduces the vulnerability to attacks and single points of failure. The immutability and cryptographic measures employed in decentralized systems, such as blockchain, enhance security and resilience.

*d. Considerations and context*

When comparing centralized and decentralized systems, it's important to consider the context and specific requirements of the application. Some factors to consider include:

- Data Privacy: Centralized systems may provide better control over data privacy, as access is managed by the central authority. Decentralized systems, especially public blockchains, prioritize transparency over privacy. However, privacy-focused decentralized systems, such as private blockchains or privacy-enhancing technologies, can address privacy concerns.
- Governance and Decision-making: Centralized systems offer streamlined decision-making but at the expense of autonomy and inclusivity. Decentralized systems enable participants to have a voice in decision-making, but reaching a consensus can be more time-consuming.
- Regulatory Compliance: Centralized systems may have advantages in terms of regulatory compliance, as the central authority can enforce rules and regulations. Decentralized systems may face challenges in adhering to existing regulations, but they also offer opportunities for self-regulation and innovation in regulatory frameworks.

### 1.2.3. Advantages of DLT in agriculture

*a. Supply chain traceability*

DLT offers unprecedented traceability in agricultural supply chains. By recording every transaction and movement of goods on a decentralized ledger, DLT enables end-to-end visibility of the entire supply chain. From seed to shelf, stakeholders can track the origin, production methods, storage conditions, and transportation details of agricultural products (Antonucci et al., 2019; Cottrill, DeCovny, & Harris, 2018; Detwiler, 2020; FAO, 2019; Tomu, 2020; Wang, Han, et al., 2019). This enhanced traceability enables consumers to make informed choices about the food they purchase, promoting transparency and trust. It also assists in identifying and mitigating potential issues such as contamination, fraud, and counterfeit products. In the event of a food safety or quality concern, DLT allows for swift and accurate recalls, minimizing the impact on public health and preserving consumer confidence.

*b. Improved inventory management*

DLT provides real-time and accurate information about inventory levels and product availability (Cottrill et al., 2018). With a shared and decentralized ledger, stakeholders such as farmers, distributors, and retailers can have up-to-date visibility into inventory levels, reducing inefficiencies and optimizing supply and demand. Farmers can better plan their production cycles based on market demand, reducing wastage and improving resource allocation. Distributors and retailers can optimize their inventory management, ensuring timely restocking and reducing out-of-stock situations. This improved coordination and transparency in inventory management contribute to cost savings, streamlined operations, and enhanced customer satisfaction.

*c. Efficient payments and financial inclusion*

DLT has the potential to revolutionize financial transactions in the agricultural sector. Traditional payment systems often pose challenges, especially in remote and underbanked regions. DLT-based solutions, such as digital tokens or cryptocurrencies, can facilitate secure and efficient peer-to-peer transactions, eliminating the need for intermediaries. By leveraging DLT, farmers can receive fair and transparent payments directly from buyers or consumers. This reduces the dependency on middlemen, minimizes transaction costs, and enhances financial inclusion for small-scale farmers. DLT-based financial systems can also enable access to credit and insurance services, empowering farmers to invest in their operations and mitigate risks.

*d. Smart contracts and automation*

DLT enables the use of smart contracts, which are self-executing agreements with predefined conditions and actions. In the agricultural sector, smart contracts can automate various processes, such as land leases, crop insurance, and commodity trading (Hamilton, 2020; Kuo, Kim, & Ohno-Machado, 2017; Li & Kassem, 2021; Ølnes, Ubacht, & Janssen, 2017). For example, a smart contract can automatically release insurance payouts to farmers in the event of a predefined weather event, eliminating the need for lengthy claims processes. Land lease agreements can be encoded into smart contracts, ensuring transparent and automated payments based on agreed-upon terms. This automation reduces administrative burdens, minimizes disputes, and improves efficiency throughout the agricultural value chain.

*e. Data sharing and collaborative research*

DLT facilitates secure and efficient data sharing among stakeholders in the agricultural ecosystem. Researchers, farmers, agronomists, and policymakers can collaborate and share data on crop yields, soil conditions, weather patterns, and pest management techniques (Bermeo-Almeida et al., 2018; Parmentola, Petrillo, Tutore, & De Felice, 2022). By securely sharing data on a decentralized ledger, stakeholders can gain insights, make data-driven decisions, and collectively address challenges in the agricultural sector. This collaborative approach fosters innovation, encourages best practices, and accelerates research and development efforts, ultimately leading to sustainable and resilient agricultural systems.

### 1.2.4. Increased security and data integrity

Implementing blockchain technology in agricultural supply chains has been shown to enhance security and ensure data integrity. By utilizing blockchain's decentralized and immutable ledger, stakeholders can securely record and track various stages of the agricultural supply chain, from production to distribution. This heightened level of security and data integrity helps to mitigate risks such as fraud, tampering, and inaccuracies, ultimately fostering trust and transparency within the agricultural industry (Hang, Ullah, & Kim, 2020; Ølnes et al., 2017).

*a. Blockchain technology*

Blockchain, the underlying technology behind cryptocurrencies, has gained significant attention due to its potential to enhance security and data integrity (Staples et al., 2017). By design, blockchain is a decentralized and immutable ledger that records transactions in a transparent and tamper-resistant manner. The key features of blockchain that contribute to increased security and data integrity are:

- Decentralization: Blockchain operates on a decentralized network of computers, known as nodes. Each node maintains a copy of the entire blockchain, ensuring that no single entity has control over the entire system. This decentralized structure eliminates single points of failure and makes the system more resistant to hacking or data manipulation.
- Cryptographic Security: Blockchain employs advanced cryptographic algorithms to secure the data. Each transaction is cryptographically linked to the previous one, forming a chain of blocks. The use of cryptographic hash functions ensures the integrity of the data within each block, making it highly resistant to tampering.
- Consensus Mechanisms: Blockchain networks utilize consensus mechanisms to validate and agree on the state of the ledger. Through mechanisms such as Proof of Work (PoW) or Proof of Stake (PoS), participants in the network reach a consensus on the validity of transactions. This consensus process ensures that only legitimate transactions are added to the blockchain, enhancing security and data integrity.

*b. Cryptographic techniques*

Beyond blockchain, various cryptographic techniques contribute to increased security and data integrity in digital systems (Sayeed, Marco-Gisbert, & Caira, 2020). These techniques include:

- Encryption: Encryption transforms data into an unreadable form using cryptographic algorithms. Only authorized parties with the encryption key can decrypt and access the original data. Encryption protects sensitive information from unauthorized access and ensures confidentiality.
- Digital Signatures: Digital signatures provide a means to verify the authenticity and integrity of digital documents or transactions. They use public-key cryptography to create a unique digital signature that is linked to the signer's identity. Digital signatures enable recipients to verify the origin of a document and ensure that it hasn't been tampered with.
- Hash Functions: Hash functions are cryptographic algorithms that transform data into fixed-length hash values. These hash values are unique to the input data and act as digital fingerprints. Any slight modification in the input data results in a completely different hash value. Hash functions ensure data integrity by providing a reliable means to verify the integrity of data without revealing the original content.

*c. Advantages of increased security and data integrity*

The increased security and data integrity provided by blockchain technology and cryptographic techniques offer several advantages:

- Trust and Transparency: The use of blockchain ensures transparency in data transactions, enabling participants to verify and audit the integrity of information. This transparency fosters trust among parties and reduces the reliance on intermediaries for data validation.
- Protection Against Tampering: The immutability of blockchain and the use of cryptographic techniques make it extremely difficult for malicious actors to tamper with data. Any attempt to modify data in a blockchain leaves a trace and requires the consensus of the network to be accepted, enhancing data integrity.
- Confidentiality and Privacy: Cryptographic techniques, such as encryption, protect sensitive data by ensuring that only authorized parties can access it. This promotes confidentiality and privacy, especially when handling personal or sensitive information.
- Data Auditability and Compliance: The transparent and tamper-resistant nature of blockchain enables efficient data auditability, making it easier to

comply with regulations and standards. Organizations can demonstrate compliance by providing verifiable and auditable records of data transactions.

- Enhanced Cybersecurity: The use of cryptographic techniques, combined with the decentralized nature of blockchain, strengthens cybersecurity defenses. It mitigates the risk of data breaches, unauthorized access, and identity theft, protecting sensitive information from malicious activities.

### 1.2.5. Enhanced transparency and trust in transactions

*a. Blockchain technology*

Blockchain technology has gained recognition for its ability to provide transparency and trust in transactions. It operates as a decentralized and immutable ledger, where transactions are recorded in a transparent and tamper-resistant manner. The key aspects of blockchain that enhance transparency and trust are:

- Decentralization: Blockchain operates on a network of decentralized nodes, each maintaining a copy of the entire ledger. This decentralized structure eliminates the need for a central authority, making transactions transparent and reducing the risk of manipulation or fraud by a single entity.
- Immutable and Transparent Records: Every transaction recorded on the blockchain is stored in a block that is linked to the previous one, forming an unchangeable chain of records. Once a transaction is recorded, it cannot be altered retroactively, ensuring the integrity and transparency of the transaction history.
- Public Visibility: In public blockchains, anyone can access and view the transaction history. This transparency allows participants to verify and validate transactions, fostering trust among parties. It also serves as a deterrent to fraudulent activities, as any attempts to tamper with the data can be easily identified.

*b. Smart contracts*

Smart contracts, which are self-executing agreements with predefined conditions and actions, further enhance transparency and trust in transactions. By automating contract execution, smart contracts eliminate the need for intermediaries and ensure that transactions are executed as agreed upon. The benefits of smart contracts include:

- Transparency: Smart contracts are stored on the blockchain, making them transparent and accessible to all relevant parties. This transparency allows participants to view the terms and conditions of the contract, ensuring a shared understanding of the transaction.

- Immutability: Once deployed on the blockchain, smart contracts cannot be modified or tampered with. The terms and conditions defined in the contract are enforced automatically, ensuring trust and eliminating the risk of one party changing the agreement without the consent of others.
- Automated Execution: Smart contracts eliminate the need for intermediaries or third parties to execute and enforce the terms of the agreement. The self-executing nature of smart contracts ensures that transactions occur as intended, reducing the risk of human error or manipulation.

*c. Decentralized systems*

Beyond blockchain, decentralized systems contribute to enhanced transparency and trust in transactions. These systems distribute control and decision-making among multiple participants, reducing the reliance on a central authority. The advantages of decentralized systems include:

- Trust among Peers: Decentralized systems promote trust among participants by allowing them to have direct interactions and share information without relying on a central authority. This trust is built on the consensus mechanisms and protocols established within the system.
- Verifiability and Auditing: In decentralized systems, transactions and actions can be verified and audited by multiple participants. This verifiability enhances transparency and ensures that transactions are conducted in a trustworthy manner. Participants can access and review the transaction history, promoting accountability.
- Inclusive Decision-making: Decentralized systems provide opportunities for participants to contribute to decision-making processes. Consensus mechanisms allow participants to collectively agree on the validity of transactions and the rules governing the system. This inclusive approach fosters trust and promotes transparency in decision-making.

*d. Advantages of enhanced transparency and trust*

The enhanced transparency and trust facilitated by blockchain technology, smart contracts, and decentralized systems offer several advantages:

- Reduced Fraud and Manipulation: The transparent nature of transactions and the immutability of records reduce the opportunities for fraud and manipulation. Participants can verify and validate transactions independently, ensuring that the agreed-upon terms are followed.

- Increased Efficiency: The elimination of intermediaries and the automation of transactions through smart contracts streamline the transaction process. This increased efficiency reduces delays, eliminates paperwork, and minimizes the chances of errors, enhancing trust and satisfaction among participants.
- Improved Accountability: The transparency and verifiability of transactions foster accountability among participants. Each transaction is traceable and auditable, allowing for easy identification of responsible parties in case of disputes or non-compliance.
- Enhanced Market Access and Fairness: By providing transparent and trustworthy transactions, these technologies facilitate access to markets for individuals and businesses that may have faced barriers in traditional systems. This increased market access promotes fairness and equal opportunities for participation.
- Reinforced Consumer Confidence: The transparency and trust established through these technologies build consumer confidence. Customers can trace the origin and history of products, verify authenticity, and ensure fair practices, ultimately promoting trust and loyalty.

## 1.3. Blockchain consensus mechanisms

This section of the book provides a comprehensive overview of various consensus mechanisms in blockchain networks, including Proof of Work (PoW), Proof of Stake (PoS), Practical Byzantine Fault Tolerance (PBFT), Delegated Proof of Stake (DPoS), and other mechanisms. The discussion examines the functionality and applicability of each mechanism, particularly in the context of agriculture. Insights from Lashkari and Musilek (2021) and Li, Jiang, Chen, Luo, and Wen (2020) contribute to the analysis presented in this section.

### 1.3.1. Consensus in blockchain networks

*a. Understanding consensus*

Consensus refers to the collective agreement among participants in a blockchain network on the order and validity of transactions. In traditional centralized systems, a central authority is responsible for maintaining and validating the ledger. However, in a decentralized blockchain network, consensus is achieved through a distributed consensus mechanism that allows participants to reach an agreement without relying on a central authority (Li, Li, et al., 2020).

The primary goals of achieving consensus in a blockchain network are:

- Validating Transactions: Consensus ensures that only valid transactions are added to the blockchain. By collectively agreeing on the validity of transactions, participants prevent the inclusion of fraudulent or unauthorized transactions.
- Establishing Order: Consensus determines the order in which transactions are added to the blockchain. This order ensures that the entire network reaches a consistent state and prevents conflicts or double-spending.
- Maintaining Integrity: Consensus mechanisms ensure the integrity of the distributed ledger. By agreeing on the state of the ledger, participants prevent tampering, data manipulation, and unauthorized changes to the blockchain.

*b. Consensus mechanisms*

Different consensus mechanisms are employed in blockchain networks to achieve agreement among participants (Buterin, 2014, 2023; Sanka & Cheung, 2021; Wang, Hoang, et al., 2019). Let's explore some of the commonly used consensus mechanisms:

- Proof of Work (PoW): PoW is the consensus mechanism popularised by Bitcoin. In this mechanism, participants, known as miners, compete to solve complex mathematical puzzles to validate transactions and create new blocks. The first miner to solve the puzzle is rewarded and their proposed block is added to the blockchain. PoW ensures security by making it computationally expensive to alter the blockchain's history.
- Proof of Stake (PoS): PoS is an alternative consensus mechanism that relies on the participants' ownership or stake in the network. Instead of competing through computational power, validators are chosen to create new blocks based on the number of tokens they hold and are willing to "stake" as collateral. PoS consumes less energy compared to PoW and provides incentives for participants to hold and support the network.
- Delegated Proof of Stake (DPoS): DPoS is a variation of PoS where participants vote to select a limited number of delegates or "witnesses" who are responsible for validating transactions and creating blocks. These witnesses take turns in securing the network, reducing the computational and energy requirements of block production. DPoS offers fast transaction confirmation times and efficient governance structures.
- Practical Byzantine Fault Tolerance (PBFT): PBFT is a consensus mechanism suitable for permissioned blockchains, where participants are known and trusted. PBFT relies on a leader-based system, where a leader node proposes a

block, and other nodes in the network validate and agree on its validity. PBFT can tolerate a certain number of faulty or malicious nodes and provides fast consensus, making it suitable for enterprise applications.

- Proof of Authority (PoA): PoA is a consensus mechanism that relies on a predetermined set of approved validators who are trusted to maintain the blockchain. Validators are typically known entities with a reputation to uphold. PoA provides fast transaction finality and is commonly used in private or consortium blockchains.

*c. Consensus and blockchain network design*

The choice of consensus mechanism depends on the specific requirements, use case, and characteristics of the blockchain network. Factors such as scalability, energy efficiency, security, decentralization, and trust among participants play a role in selecting an appropriate consensus mechanism. Furthermore, hybrid consensus models are emerging, combining the strengths of different mechanisms to achieve a balance between scalability, security, and decentralization.

*d. The importance of consensus*

Consensus mechanisms are vital for the functioning and success of blockchain networks. They ensure the integrity, security, and reliability of the distributed ledger. By establishing agreement among participants, consensus mechanisms enable trust and transparency in transactions, eliminate the need for intermediaries, and facilitate decentralized governance. Consensus mechanisms continue to evolve as researchers and developers explore new approaches to address scalability, energy consumption, and other limitations associated with existing mechanisms. The selection and implementation of an appropriate consensus mechanism are critical considerations when designing and deploying blockchain networks.

### 1.3.2. Proof of Work (PoW) consensus mechanism

*a. Understanding proof of work*

Proof of Work is a consensus mechanism introduced by Bitcoin's creator, Satoshi Nakamoto. It is designed to address the challenge of reaching consensus in a decentralized network without relying on a central authority. PoW requires participants, known as miners, to perform computationally intensive tasks to validate transactions and secure the network. The core principle of PoW is as follows: miners compete to solve complex mathematical puzzles, known as hash functions, in order to add new blocks to the blockchain. The puzzles require significant computational power and energy expenditure to solve. The miner who successfully solves the puzzle first is granted the right to create the next block and is rewarded with cryptocurrency tokens.

*b. How PoW works*

The PoW process involves the following steps:

- Transaction Verification: Miners collect and verify transactions by confirming their validity and ensuring that the participants have sufficient funds to complete the transaction.
- Hash Calculation: Once a group of transactions is verified, miners package them into a block. The miner's task is to find a specific hash value for the block that meets certain criteria. This process involves repeatedly applying a hash function to the block data with different inputs until the desired hash value is found.
- Difficulty Adjustment: The difficulty of the puzzle is adjusted dynamically based on the total computational power of the network. The goal is to maintain a consistent block creation rate, typically by adjusting the complexity of the hash function. This ensures that new blocks are added to the blockchain at a predictable rate.
- Mining Competition: Miners compete with each other to solve the puzzle by performing massive computational calculations. They use specialized hardware, such as ASICs (Application-specific Integrated Circuits), to gain an advantage in the competition.
- Proof Verification: Once a miner successfully solves the puzzle and finds a valid hash value, they broadcast the solution to the network. Other participants can easily verify the solution by reapplying the hash function to the block data and confirming that it meets the required criteria.
- Block Reward: The miner who solves the puzzle and creates a new block is rewarded with cryptocurrency tokens. This incentivizes miners to contribute their computational power to secure the network.

*c. Advantages of PoW*

The PoW consensus mechanism offers several advantages:

- Security: PoW ensures the security of the blockchain network by making it computationally expensive to alter the transaction history. Changing the content of a block would require recalculating the hash value for that block and all subsequent blocks, which is highly resource-intensive.

- Decentralization: PoW promotes decentralization by allowing any participant with sufficient computational power to become a miner. No central authority controls the validation process, ensuring that no single entity has undue influence over the network.
- Sybil Attack Resistance: PoW mitigates the risk of Sybil attacks, where an adversary creates multiple identities to gain control over a network. In PoW, an attacker would need an enormous amount of computational power, making such attacks economically infeasible.
- Fairness: PoW provides a fair method for distributing newly created cryptocurrency tokens. Miners are rewarded based on their contributions to securing the network, incentivizing participation and maintaining the network's integrity.

*d. Limitations and considerations*

Despite its advantages, PoW has some limitations and considerations:

- Energy Consumption: The computational requirements of PoW result in significant energy consumption. As the network grows and mining becomes more competitive, the energy consumption of PoW-based blockchains has raised concerns about sustainability and environmental impact.
- Scalability: PoW can be resource-intensive and limits the transaction processing speed of a blockchain. The time required to solve the puzzles and reach a consensus can result in slower transaction confirmations and potential network congestion.
- Centralization of Mining Power: Over time, mining has become highly specialized, with specialized hardware and large-scale mining operations dominating the network. This concentration of mining power in the hands of a few entities raises concerns about centralization.
- Potential 51% Attack: A 51% attack occurs when an entity or a group controls more than 50% of the total mining power in the network. This level of control could allow malicious actors to manipulate the blockchain's history and compromise its security.

*e. Evolving beyond PoW*

To address the limitations of PoW, alternative consensus mechanisms, such as Proof of Stake (PoS) and Delegated Proof of Stake (DPoS), have emerged. These mechanisms offer energy efficiency, faster transaction confirmation times, and improved scalability. Many blockchain projects are actively exploring these alternatives or hybrid models to enhance their networks' efficiency and sustainability.

### 1.3.3. Proof of Stake (PoS) consensus mechanism

*a. Understanding proof of stake*

Proof of Stake is a consensus mechanism that selects validators to create new blocks based on their stake or ownership of cryptocurrency tokens in the network. Unlike PoW, which relies on computational power, PoS determines block validators by considering their economic stake in the system. The higher the stake held by a participant, the higher the chance of being chosen to validate transactions and create new blocks. The primary goal of PoS is to ensure the security and integrity of the blockchain network while addressing some of the limitations associated with PoW, such as energy consumption and scalability (Buterin, 2023; Nguyen et al., 2019).

*b. How PoS works*

The PoS consensus mechanism operates based on the following principles:

- Stake and Validator Selection: Participants who hold a certain amount of cryptocurrency tokens in their wallet can "stake" them as collateral. The selection of validators to create new blocks is typically based on a random or deterministic algorithm that takes into account the participants' stake.
- Block Validation and Creation: Chosen validators are responsible for validating transactions and creating new blocks. The validator's chance of being selected for block creation is proportional to the stake they hold. Validators are incentivized to act honestly, as they risk losing their stake if they validate fraudulent transactions.
- Transaction Verification: Validators verify the authenticity and validity of transactions based on predefined consensus rules. This includes confirming that the sender has sufficient funds and that the transaction adheres to network rules.
- Block Finality: In PoS, block finality can be achieved more quickly compared to PoW. Once a block is created and added to the blockchain, it is considered final and cannot be easily reversed or changed.
- Rewards and Incentives: Validators who successfully validate transactions and create blocks are rewarded with transaction fees and newly minted cryptocurrency tokens. These rewards serve as an incentive for participants to hold and stake tokens, contributing to the network's security and stability.

*c. Advantages of PoS*

The PoS consensus mechanism offers several advantages:

- Energy Efficiency: PoS consumes significantly less energy compared to PoW since it eliminates the need for miners to perform resource-intensive computations. Validators are selected based on their stake, reducing the energy consumption associated with traditional mining processes.
- Scalability: PoS enables faster transaction confirmations and higher throughput compared to PoW. The block creation time is typically shorter, allowing for quicker settlement and increasing the network's capacity to handle a larger number of transactions.
- Security: PoS promotes network security by aligning the interests of validators with the overall stability of the blockchain. Validators have a financial stake in the network and risk losing their stake if they act maliciously or validate fraudulent transactions, incentivizing them to act honestly.
- Decentralization: PoS can potentially enhance decentralization by allowing a wider range of participants to become validators. The barrier to entry is lowered as validators are selected based on their stake, rather than requiring expensive mining hardware.
- Reduced Environmental Impact: The energy efficiency of PoS contributes to a smaller carbon footprint, making it a more environmentally friendly consensus mechanism compared to PoW.

*d. Considerations and challenges*

While PoS offers several advantages, there are considerations and challenges to be aware of:

- Potential Centralization: PoS systems may still face challenges related to centralization if a small group of participants accumulates a significant amount of tokens, giving them disproportionate control over the network.
- Long-term Security: The long-term security of PoS networks depends on the distribution and decentralization of token ownership. Ensuring a diverse and widespread distribution of tokens is crucial to prevent potential attacks and maintain the network's resilience.
- Nothing-at-stake Problem: The "nothing-at-stake" problem refers to the theoretical scenario where validators have nothing to lose by staking conflicting blocks during a network fork. Various mechanisms, such as penalties and slashing, have been proposed to mitigate this risk and discourage malicious behavior.

- Initial Token Distribution: Fair and equitable token distribution at the inception of a PoS network is essential to prevent undue influence and concentration of power.

*e. Evolving PoS variants*

Since the introduction of PoS, various PoS variants and hybrid models have emerged, each with its own characteristics and benefits. These variants include Delegated Proof of Stake (DPoS), Practical Byzantine Fault Tolerance (PBFT), and Byzantine Fault Tolerant Proof of Stake (BFT-PoS). These variations aim to enhance scalability, governance, and consensus efficiency while maintaining the security and decentralization of the blockchain network.

### 1.3.4. Practical Byzantine Fault Tolerance (PBFT)

*a. Understanding Byzantine Fault Tolerance*

Before delving into PBFT, let's first understand the concept of Byzantine Fault Tolerance (BFT). In a distributed system, Byzantine faults refer to arbitrary and potentially malicious behavior exhibited by nodes or participants. These faults can include sending incorrect or conflicting information, delaying or dropping messages, or even impersonating other nodes.

Byzantine Fault Tolerance aims to address the challenge of reaching a consensus in the presence of such malicious behavior. The goal is to achieve agreement among a set of nodes, even when a certain number of nodes behave in a Byzantine manner.

*b. How PBFT works*

The PBFT consensus algorithm operates based on the following principles:

- Replica Model: PBFT assumes a model where a set of nodes, known as replicas, work together to achieve consensus. These replicas communicate with each other to reach an agreement on the order and validity of transactions.
- Phases of Consensus: PBFT divides the consensus process into multiple phases: a) Request: A client sends a request to the replicas to execute a specific operation. b) Pre-Prepare: Replicas agree on the proposed request and its order. c) Prepare: Replicas reach a state where they agree on the ordering of requests. d) Commit: Replicas finalize the consensus by agreeing on the validity of the request. e) Reply: Replicas send the response to the client.

- Message Exchange: During each phase, replicas exchange messages to gather votes and confirm the agreement. These messages include the proposed request, prepared messages, and committed messages. The communication enables replicas to determine the most agreed-upon request and achieve consensus.
- Fault Tolerance: PBFT can tolerate up to (n-1)/3 Byzantine faulty replicas in a network of n replicas. This fault tolerance property ensures that even if a subset of replicas behaves maliciously, the consensus can still be reached as long as the majority of replicas are honest.
- View Change: PBFT includes a mechanism for handling faulty primary replicas. If the primary replica behaves maliciously or becomes unresponsive, a new primary replica is selected through a view change process, ensuring the progress of the consensus algorithm.

*c. Advantages of PBFT*

The PBFT consensus algorithm offers several advantages:

- Fast Consensus: PBFT achieves consensus in a relatively short time compared to Proof of Work (PoW) or Proof of Stake (PoS) algorithms. The number of message exchanges and the deterministic nature of the algorithm contribute to faster confirmation of transactions.
- Finality: PBFT provides immediate finality, meaning that once consensus is reached, the agreed-upon transaction order is considered definitive and cannot be altered or reversed without the consensus of the replicas.
- Byzantine Fault Tolerance: PBFT is designed to handle Byzantine faults, ensuring the security and integrity of the distributed system. It can tolerate a certain number of malicious replicas while maintaining consensus.
- Low Energy Consumption: PBFT does not rely on resource-intensive computational tasks like PoW. As a result, it consumes significantly less energy, making it an efficient consensus algorithm for environments where energy efficiency is a priority.

*d. Considerations and challenges*

While PBFT offers advantages in terms of speed and fault tolerance, there are considerations and challenges to be aware of:

- Network Size: PBFT is well-suited for networks with a small to medium number of replicas. As the network size increases, the communication overhead and message complexity can become a challenge, potentially impacting the algorithm's performance.

- Trusted Setup: PBFT assumes that a certain number of replicas are honest and behave correctly. Therefore, a trusted setup is required to establish the initial set of replicas and their identities. The trust assumption can be a limitation in some decentralized environments.
- Centralization of Power: In PBFT, replicas take turns acting as primary nodes to propose requests. This rotation can result in a certain level of centralization of power and influence among replicas, which may introduce vulnerabilities if replicas collude or act maliciously.

*e. Applications of PBFT*

PBFT has found applications in various domains, including blockchain technology, distributed databases, and consensus protocols. Its fast consensus and fault tolerance properties make it suitable for applications that require immediate finality and Byzantine fault tolerance guarantees, such as financial systems, payment networks, and enterprise blockchains. By leveraging PBFT, these systems can achieve robust consensus mechanisms that ensure reliability, security, and efficiency in processing transactions and maintaining system integrity (Cachin, 2016; Driscoll, Hall, Sivencrona, & Zumsteg, 2003; Luu et al., 2016).

### 1.3.5. Delegated Proof of Stake (DPoS)

A comprehensive understanding of Delegated Proof of Stake (DPoS) and its implications in blockchain technology is essential for navigating the complexities of decentralized systems. DPoS, as a consensus mechanism, delegates decision-making power to selected network participants, impacting the security, scalability, and governance of blockchain networks. Thus, gaining insight into DPoS enables stakeholders to make informed decisions and contribute effectively to the advancement of blockchain technology (Zhang, Wu, & Wang, 2020).

*a. Understanding DPoS*

DPoS is designed to address the limitations of both traditional Proof of Work (PoW) and Proof of Stake (PoS) mechanisms. It introduces a delegated governance model where token holders vote to elect a set of trusted nodes, known as delegates or witnesses, who are responsible for validating transactions and creating blocks on behalf of the network.

The key idea behind DPoS is to achieve consensus through a smaller set of trusted delegates, enhancing scalability, transaction throughput, and energy efficiency while maintaining decentralization and security.

*b. How DPoS works*

The DPoS consensus mechanism operates based on the following principles:

- Delegates and Voting: Token holders in the network have the power to vote and elect delegates. The number of delegates can vary depending on the blockchain implementation. Typically, the top-ranked delegates with the highest number of votes become active delegates responsible for block validation and creation.
- Block Validation and Creation: Active delegates take turns to validate transactions and create new blocks in a predefined order. The rotation of delegates ensures that different participants have the opportunity to contribute to the consensus process.
- Voting Weight and Influence: Voting power is typically proportional to the number of tokens held by a participant. The more tokens a participant holds, the more influence they have in electing delegates and shaping the governance of the blockchain network.
- Block Production Time: DPoS networks usually have shorter block production times compared to PoW or PoS systems. This allows for faster transaction confirmations and improves the overall throughput of the network.
- Incentives and Rewards: Delegates who actively participate in block validation and creation are rewarded with transaction fees and newly minted cryptocurrency tokens. This incentivizes delegates to act honestly and perform their duties diligently.

*c. Advantages of DPoS*

DPoS offers several advantages over traditional consensus mechanisms:

- Scalability and Throughput: By limiting the number of active delegates responsible for block validation, DPoS improves the scalability and transaction throughput of the blockchain network. Shorter block production times ensure quicker settlement and faster confirmation of transactions.
- Energy Efficiency: DPoS consumes significantly less energy compared to PoW-based systems since it eliminates the need for resource-intensive computations. The block validation process relies on the selected delegates, reducing the overall energy footprint of the consensus mechanism.
- Governance and Flexibility: DPoS incorporates a delegated governance model, allowing token holders to participate in decision-making and shaping the direction of the blockchain network. The voting mechanism

provides stakeholders with a voice in electing delegates and proposing network upgrades.

- Decentralization: While DPoS relies on a smaller set of trusted delegates, it still maintains decentralization by allowing token holders to participate in the election process. The rotation of delegates ensures that different participants have the opportunity to contribute to the consensus process.
- Resistance to Sybil Attacks: Delegated Proof of Stake (DPoS) exhibits greater resilience against Sybil attacks in contrast to Proof of Work (PoW) or Proof of Stake (PoS) systems. This is because voting power within DPoS is contingent on the quantity of tokens possessed, thereby necessitating an attacker to obtain a substantial number of tokens to exert influence, rendering such attempts economically unfeasible.

*d. Considerations and challenges*

While DPoS offers several advantages, there are considerations and challenges to be aware of:

- Centralization of Power: The DPoS model relies on delegates to validate transactions and create blocks. If a small group of delegates gains significant voting power or colludes, it can potentially lead to centralization and concentration of power, raising concerns about censorship or manipulation.
- Voter Apathy: In DPoS, voter participation is crucial for maintaining a healthy and decentralized network. However, achieving high participation rates can be challenging, as token holders may exhibit voter apathy or lack interest in the governance process.
- Trust in Delegates: The election of trustworthy delegates is essential for the security and integrity of the network. Token holders need to carefully assess the qualifications, track record, and reliability of delegates before voting.

*e. Applications of DPoS*

DPoS has found applications in various blockchain networks and projects. It is particularly well suited for applications that require high transaction throughput, low latency, and efficient governance. DPoS-based networks have been utilized in decentralized exchanges, gaming platforms, and social media platforms, among others.

### 1.3.6. Other consensus mechanisms and their applicability in agriculture

*a. Practical Byzantine Fault Tolerance (PBFT)*

We have already discussed PBFT in a previous chapter, highlighting its features and benefits. PBFT's ability to handle Byzantine faults and its fast consensus mechanism makes it suitable for applications where immediate finality and Byzantine fault tolerance are critical. In the agricultural sector, PBFT could be applied in supply chain management systems to ensure the integrity and transparency of data related to the production, processing, and distribution of agricultural products.

*b. Proof of authority (PoA)*

Proof of Authority (PoA) is a consensus mechanism that relies on a set of approved authorities or validators to validate and confirm transactions. These authorities are typically known and trusted entities within the network. PoA offers faster transaction times and lower energy consumption compared to PoW or PoS, as it does not require resource-intensive computational tasks or stake-based selection. In agriculture, PoA can be used in applications that require a trusted network of participants, such as tracking the provenance of organic or fair trade products, where designated authorities can validate and authenticate the compliance of products with specific standards (Shamieh, Wang, & Hussein, 2020).

*c. Proof of elapsed time (PoET)*

Proof of Elapsed Time (PoET) is a consensus mechanism that relies on a random selection process, similar to PoW, but without computational competition. In PoET, each participant requests a random wait time, and the participant with the shortest wait time is selected as the leader to propose the next block. PoET is designed to be energy-efficient, as it avoids the resource-intensive nature of PoW. In agriculture, PoET could be applicable in scenarios where participants need to take turns in executing certain tasks, such as water irrigation management, pest control scheduling, or crop rotation planning, ensuring fairness and preventing conflicts of interest (Xu, Zhang, et al., 2022).

*d. Proof of capacity (PoC)*

Proof of Capacity (PoC), also known as Proof of Space, is a consensus mechanism that leverages the storage capacity of participants' devices to determine their mining power. Participants precompute a large amount of storage space and provide proofs that demonstrate their commitment to storing and validating the blockchain. PoC is energy-efficient, as it does not rely on computational power. In agriculture, PoC could find applications in resource management systems, where participants contribute their available storage capacity to store and share data related to soil quality, weather patterns, or irrigation schedules (Shamieh et al., 2020; Xu, Zhang, et al., 2022).

*e. Hybrid consensus mechanisms*

Hybrid consensus mechanisms combine different elements from multiple consensus mechanisms to achieve a balance between scalability, security, and decentralization. For example, combining PoW and PoS can result in a hybrid mechanism where PoW is used for block creation, and PoS is used for block validation and selection of validators. Hybrid mechanisms provide flexibility and allow for customization based on the specific requirements of an application. In agriculture, hybrid consensus mechanisms can be utilized in complex supply chain networks, where different stages of the supply chain may require different consensus mechanisms based on their specific needs and constraints.

*f. Considerations for agricultural applications*

When considering the applicability of different consensus mechanisms in agriculture, several factors should be taken into account:

- Scalability: The ability of the consensus mechanism to handle the transactional demands of agricultural applications, such as supply chain management or data sharing.
- Energy Efficiency: The energy consumption associated with the consensus mechanism, as agriculture promotes sustainability and environmental responsibility.
- Governance and Trust: The level of decentralization, governance models, and trust assumptions inherent in the consensus mechanism.
- Security and Integrity: The ability of the consensus mechanism to ensure data integrity, prevent tampering, and protect against malicious attacks.
- Adaptability: The consensus mechanism's ability to accommodate changing requirements and the evolving nature of the agricultural industry.

## 1.4. Disadvantages and limitations of blockchain

Amidst fervent enthusiasm and high expectations, blockchain technology has emerged as a transformative force across numerous industries. Its decentralized and immutable nature promises to reshape traditional systems, yet, akin to any technological advancement, it is not without its constraints and drawbacks. In this exploration, a critical examination of the key limitations and disadvantages inherent in blockchain technology is undertaken.

### 1.4.1. Disadvantages of blockchain

*a. Transaction Throughput*

Transaction throughput, the measure of how many transactions a blockchain can process per unit of time, emerges as a critical bottleneck in the scalability of blockchain networks. This limitation arises from the foundational principles of public blockchains like Bitcoin and Ethereum, which prioritize decentralization and cryptographic security over transaction speed. While these principles uphold the integrity and trustworthiness of blockchain transactions, they also impose constraints on the network's capacity to handle high volumes of transactions efficiently. The inherent design of public blockchains dictates that every transaction must be validated and recorded by a distributed network of nodes, a process that entails significant computational overhead and time. Consequently, these networks often struggle to scale beyond a few dozen transactions per second, a rate that pales in comparison to the demands of large-scale applications serving millions of users concurrently. This disparity between transaction demand and network capacity introduces delays, congestion, and unpredictability, impairing the user experience and impeding the seamless functioning of blockchain-based systems. The limited transaction throughput of blockchain networks poses formidable challenges for applications requiring real-time transaction processing and high-volume data throughput. Industries such as payment processing, supply chain management, and decentralized finance (DeFi) demand rapid transaction settlement and scalability to accommodate a diverse array of users and transactions. However, the current transaction throughput constraints of blockchain networks hinder the widespread adoption of blockchain technology for such use cases, limiting its utility and applicability in practical settings. Addressing the transaction throughput limitations of blockchain networks requires innovative scaling solutions that strike a balance between decentralization, security, and efficiency. Initiatives exploring off-chain scaling techniques, such as layer-two protocols and sidechains, aim to alleviate congestion on the main blockchain while enhancing transaction throughput and scalability. Additionally, advancements in consensus algorithms, such as proof of stake (PoS) and delegated proof of stake (DPoS), offer promising avenues for improving transaction throughput without compromising network security or decentralization. By overcoming the transaction throughput challenges, blockchain technology can realize its full potential as a transformative force across diverse industries and applications (Chauhan, Malviya, Verma, & Mor, 2018; Kaur & Gandhi, 2020).

*b. Smart Contract Risks*

While heralded for their promise of automation and self-execution of agreements, smart contracts introduce a new frontier of risks and vulnerabilities within the blockchain ecosystem. Smart contract bugs, coding errors, and unanticipated interactions can precipitate dire consequences, ranging from substantial financial losses to compromising the security and integrity of blockchain-based systems. The immutable nature of blockchain amplifies these risks, as once a smart contract is deployed onto the blockchain, it becomes resistant to modification or revocation, magnifying the impact of any vulnerabilities present within its codebase. The DAO hack of 2016 stands as a poignant testament to the inherent dangers associated with smart contract vulnerabilities. In this watershed event, a flaw in the code was ruthlessly exploited, resulting in the illicit siphoning of millions of dollars' worth of cryptocurrency. The ramifications of this exploit reverberated throughout the blockchain community, serving as a sobering reminder of the critical importance of vigilance and robust security measures in the realm of smart contracts. Addressing the multifaceted risks posed by smart contracts necessitates a comprehensive approach encompassing rigorous testing, meticulous auditing, and ongoing monitoring of smart contract codebases. Thorough testing protocols are essential to identify and rectify potential vulnerabilities and coding errors before deployment, mitigating the risk of exploitation by malicious actors. Furthermore, independent audits by qualified professionals can provide an additional layer of assurance regarding the security and reliability of smart contract implementations. Beyond initial deployment, continuous monitoring and auditing of smart contract code are imperative to detect and respond promptly to emerging threats and vulnerabilities. Automated monitoring tools and vulnerability scanners can help identify anomalous behavior and potential exploits, enabling proactive intervention to safeguard the integrity of blockchain-based applications. In conjunction with technical safeguards, fostering a culture of security awareness and education within the blockchain community is paramount to mitigate smart contract risks effectively. Promoting best practices in smart contract development, facilitating knowledge sharing, and encouraging collaboration among developers and security experts can fortify defenses against evolving threats and vulnerabilities. By prioritizing security and diligence in smart contract development and deployment, the blockchain community can cultivate a resilient ecosystem that inspires trust and confidence in blockchain-based applications (Mezquita, Valdeolmillos, González-Briones, Prieto, & Corchado, 2019).

*c. Compliance and Privacy Concerns*

While blockchain's transparency and immutability offer unparalleled integrity and trust in data transactions, they also introduce complexities in navigating privacy regulations and compliance standards, notably exemplified by the General Data Protection Regulation (GDPR). The fundamental principles of blockchain, including the public accessibility of data and the immutability of recorded information, pose inherent challenges in safeguarding personal and sensitive data. Storing such information on a public blockchain exposes it to perpetual public scrutiny, potentially leading to privacy infringements and violations of stringent regulatory frameworks. Moreover, the immutable nature of blockchain means that once data is recorded, it cannot be altered or erased, even to comply with regulatory mandates such as the 'right to be forgotten' under GDPR. This lack of flexibility in data management conflicts with the core tenets of privacy regulations, complicating efforts to maintain compliance while utilizing blockchain technology. Furthermore, the global nature of blockchain networks presents additional hurdles, as regulatory requirements vary across jurisdictions, leading to uncertainty and inconsistency in compliance efforts. To address these challenges, organizations leveraging blockchain must carefully navigate the intersection of technology and regulatory compliance. Implementing robust data protection measures, such as encryption and permissioned access, can mitigate privacy risks while ensuring adherence to regulatory mandates. Additionally, collaborative efforts between policymakers, industry stakeholders, and technologists are essential to develop adaptable frameworks that reconcile the innovative potential of blockchain with the imperative of protecting individual privacy rights (Akanfe, Lawong, & Rao, 2024; Finck, 2018; Tatar, Gokce, & Nussbaum, 2020; Walters, 2019; Wirth & Kolain, 2018; Zyskind & Nathan, 2015).

*d. Carbon Footprint*

The environmental impact of blockchain technology, particularly exemplified by Bitcoin mining, has become a subject of intense scrutiny and debate. The energy-intensive process of mining cryptocurrencies consumes a substantial amount of electricity, with some estimates indicating that Bitcoin mining alone surpasses the energy consumption of entire nations. This voracious appetite for energy has significant implications for carbon emissions and exacerbates concerns about climate change and environmental sustainability. The carbon footprint of blockchain networks has drawn sharp criticism from environmental advocates and policymakers alike, who highlight the paradoxical nature of a technology designed to decentralize power while centralizing energy consumption. The proliferation of energy-intensive consensus mechanisms, such as proof-of-work (PoW), has become a focal point of contention, as they prioritize computational power and electricity consumption over environmental considerations. In response to mounting pressure, there have been calls for the adoption of more energy-efficient consensus mechanisms and sustainable mining practices within the blockchain ecosystem. Initiatives exploring alternative consensus algorithms, such as proof-of-stake (PoS) and proof-of-authority (PoA), aim to reduce the carbon footprint of blockchain networks by minimizing energy consumption and incentivizing more eco-friendly mining practices. However, transitioning to these alternative consensus mechanisms poses its own set of challenges, including security concerns and resistance from vested interests within the blockchain community. Moreover, achieving widespread adoption of sustainable mining practices requires concerted efforts from industry stakeholders, regulatory bodies, and environmental advocates to foster a culture of environmental responsibility and accountability within the blockchain ecosystem. Ultimately, addressing the carbon footprint of blockchain technology requires a multifaceted approach that balances innovation with environmental stewardship, ensuring that the promise of decentralized technologies does not come at the expense of planetary well-being (Badea & Mungiu-Pupăzan, 2021; Parmentola et al., 2022; Shi et al., 2023; Wendl, Doan, & Sassen, 2023).

*e. Navigating the Intersection of GDPR and Blockchain: Balancing Privacy Rights with Decentralized Principles*

Regulations such as the General Data Protection Regulation (GDPR) can clash with certain principles inherent in blockchain technology, particularly regarding the control and management of personal data. The GDPR, enacted by the European Union (EU), aims to safeguard individuals' privacy rights and regulate the processing of personal data by organizations. While the GDPR embodies important protections for data subjects, its requirements can pose challenges for blockchain applications that prioritize decentralization and immutability (Belen-Saglam, Altuncu, Lu, & Li, 2023; Dutta, Das, Dey, & Bhattacharya, 2020).

- Right to Erasure (Right to Be Forgotten): One of the fundamental principles of the GDPR is the right to erasure, also known as the right to be forgotten. This right grants individuals the power to request the deletion or removal of their personal data from databases and systems. However, the immutable nature of blockchain technology, wherein data once recorded cannot be altered or deleted, directly conflicts with this principle. Once personal data is recorded on a blockchain, it becomes practically impossible to comply with requests for erasure without compromising the integrity and immutability of the blockchain.

- Data Minimization and Purpose Limitation: The GDPR also mandates that organizations collect and process only the minimum amount of personal data necessary for specific purposes. Additionally, organizations must not retain personal data for longer than is necessary for the purposes for which it was collected. However, blockchain systems typically store data indefinitely, as each new block added to the chain contains a permanent record of all transactions. This perpetual storage of data on the blockchain may contravene the GDPR's principles of data minimization and purpose limitation.
- Data Controller and Processor Responsibilities: Under the GDPR, organizations are classified as either data controllers or data processors and are subject to specific obligations and responsibilities regarding the processing of personal data. However, the decentralized nature of blockchain networks blurs the lines of traditional data controller and processor roles. In a blockchain network, multiple participants (nodes) collectively validate and record transactions, making it challenging to identify a single entity as the data controller or processor. This decentralized architecture complicates compliance with GDPR requirements related to data controller and processor responsibilities.
- Cross-border Data Transfers: The GDPR imposes restrictions on the transfer of personal data outside the European Economic Area (EEA) to countries that do not provide an adequate level of data protection. However, blockchain networks operate on a global scale, with data replicated and stored across multiple nodes located in various jurisdictions. Ensuring compliance with GDPR restrictions on cross-border data transfers presents significant challenges for blockchain applications, as data may traverse international borders without explicit consent from data subjects.

### 1.4.2. Limitations of Blockchain

*a. Scalability*

Scalability stands out as one of the most pressing challenges confronting blockchain technology. As blockchain networks expand in size and scope, they encounter significant hurdles in maintaining efficient transaction processing capabilities. The growth of the blockchain inevitably entails an increase in the time and computational resources necessary to validate and record transactions. This augmentation in processing demands inevitably translates to slower transaction speeds and heightened operational costs, undermining the feasibility of blockchain solutions for applications requiring rapid, high-volume transaction throughput. The decentralized architecture inherent in blockchain exacerbates the scalability conundrum. Unlike centralized systems where a single entity can oversee transaction validation, blockchain networks rely on consensus mechanisms that mandate every network node to independently validate and record each transaction. While this decentralization confers robustness and resilience to the network, it simultaneously introduces inefficiencies and bottlenecks in transaction processing. The need for consensus among network participants introduces latency and overhead, further impeding the scalability of blockchain systems. Moreover, as blockchain networks expand to accommodate a growing user base and increasing transaction volumes, the storage and bandwidth requirements escalate exponentially. This amplifies the challenges associated with scalability, as the infrastructure necessary to sustain the burgeoning demands of the network becomes increasingly prohibitive. Consequently, blockchain scalability emerges as a critical barrier inhibiting the widespread adoption of blockchain technology across diverse industries and applications. Efforts to address this limitation entail exploring innovative scaling solutions, such as sharding, layer-two protocols, and off-chain scaling approaches, to enhance the efficiency and scalability of blockchain networks without compromising decentralization and security (Khan, Jung, & Hashmani, 2021; Lashkari & Musilek, 2021; Sanka & Cheung, 2021).

*b. Security Vulnerabilities*

Despite its reputation for robust security, blockchain technology remains susceptible to a range of vulnerabilities and threats. While blockchain's decentralized architecture and cryptographic principles provide a foundation for security, they do not render it impervious to attack. Indeed, various factors contribute to the landscape of security concerns confronting blockchain ecosystems. One notable vulnerability arises from the code execution capabilities embedded within smart contracts, self-executing agreements encoded on the blockchain. Smart contract bugs, arising from coding errors or vulnerabilities in the underlying software, can lead to unforeseen consequences, including financial losses and system disruptions. High-profile incidents, such as the DAO hack of 2016, underscore the risks associated with smart contract vulnerabilities and the potential for exploitation by malicious actors. Furthermore, blockchain networks face the looming threat of 51% attacks, wherein a single entity or coalition of entities gains majority control over the network's computing power, enabling them to manipulate transactions and undermine the integrity of the blockchain. While such attacks remain relatively rare, they highlight the inherent vulnerabilities of proof-of-work consensus mechanisms and the concentration of mining power among a select few entities. Additionally, the advent of quantum computing poses a looming threat to the cryptographic algorithms underpinning blockchain security. Quantum computers, with their exponentially greater computational power, could theoretically break existing cryptographic schemes, compromising the confidentiality and integrity of blockchain transactions. While quantum-resistant cryptographic techniques are under development, the eventual emergence of quantum computing necessitates proactive measures to fortify blockchain security against this emerging threat. Moreover, the immutable nature of blockchain, while heralded for its tamper-resistant properties, presents challenges in rectifying errors or addressing fraudulent activities once transactions are recorded. The inability to alter or erase data poses significant implications for dispute resolution, compliance, and regulatory compliance, particularly in cases of erroneous or malicious transactions. In light of these security vulnerabilities, the blockchain community must remain vigilant and proactive in implementing robust security measures and best practices. This entails rigorous code auditing, vulnerability assessments, and the adoption of resilient consensus mechanisms and cryptographic protocols to mitigate risks and uphold the integrity and trustworthiness of blockchain ecosystems. Additionally, ongoing research and collaboration are essential to stay ahead of emerging threats and ensure the continued security and resilience of blockchain technology in an evolving threat landscape (Kushwaha, Joshi, Singh, Kaur, & Lee, 2022; Lee, 2019; Li, Jiang, et al., 2020; Sayeed et al., 2020; Staples et al., 2017; Yamashita, Nomura, Zhou, Pi, & Jun, 2019; Zamani, He, & Phillips, 2020).

*c. Regulatory Uncertainty*

The regulatory landscape surrounding blockchain technology is characterized by a pervasive sense of ambiguity and uncertainty across many jurisdictions worldwide. This regulatory ambiguity stems from the rapid pace of technological advancement outpacing legislative frameworks, resulting in a regulatory "gray area" that leaves businesses and individuals navigating uncertain terrain. The absence of clear regulations and guidelines poses formidable challenges for enterprises and innovators seeking to leverage blockchain solutions. Without defined parameters or regulatory standards, businesses face uncertainty regarding legal compliance, licensing requirements, and the legality of blockchain-based activities. This regulatory uncertainty undermines confidence in the legal framework surrounding blockchain technology, stifling innovation and inhibiting investment in the burgeoning blockchain space. Moreover, the lack of regulatory clarity introduces operational complexities and legal risks for organizations operating within the blockchain ecosystem. The absence of clear guidelines leaves businesses vulnerable to potential legal disputes, regulatory enforcement actions, and compliance issues. Furthermore, the uncertainty surrounding the legal status of blockchain-based assets, such as cryptocurrencies and tokenized securities, creates challenges for market participants seeking to navigate regulatory compliance obligations and investor protection measures.

In addition to stifling innovation and investment, regulatory uncertainty engenders a climate of hesitancy and caution among stakeholders in the blockchain industry. Enterprises may adopt a conservative approach to blockchain adoption, opting to delay or scale back investments in blockchain initiatives due to regulatory uncertainties and associated risks. This risk aversion impedes the development and maturation of the blockchain ecosystem, hampering its potential to drive transformative change across industries. Addressing regulatory uncertainty requires proactive engagement and collaboration between industry stakeholders, policymakers, and regulatory authorities. Establishing clear and comprehensive regulatory frameworks tailored to the unique characteristics of blockchain technology is paramount to fostering innovation, protecting consumers, and ensuring the integrity of financial markets. By providing certainty and clarity, regulatory frameworks can catalyze responsible innovation and investment in blockchain technology, unlocking its full potential to drive economic growth and societal progress (Clohessy, Acton, & Rogers, 2019; Cumming, Johan, & Pant, 2019).

*d. Environmental Impact*

The environmental ramifications of blockchain technology have emerged as a focal point of discourse, with particular attention directed towards the energy consumption inherent in blockchain consensus mechanisms, notably exemplified by proof of work (PoW). The mining process, integral to maintaining the integrity and security of blockchain networks, demands significant computational power, resulting in a staggering consumption of electricity and consequential carbon emissions. This energy-intensive nature of blockchain operations has sparked widespread concern regarding its environmental sustainability and raised pertinent questions about the balance between technological innovation and ecological responsibility. Critics contend that the environmental toll exacted by blockchain technology, particularly through PoW consensus mechanisms, may overshadow the purported benefits offered by the technology. The substantial energy requirements of blockchain mining activities often surpass the annual electricity consumption of entire nations, underscoring the magnitude of its environmental footprint. Furthermore, the carbon emissions stemming from blockchain operations contribute to climate change and exacerbate global environmental challenges, prompting apprehensions about the net ecological impact of blockchain technology. The discord between the decentralized ethos underpinning blockchain technology and its environmental consequences has prompted calls for greater accountability and sustainability within the blockchain ecosystem. Advocates stress the imperative of reconciling technological advancement with environmental stewardship, emphasizing the need for innovative solutions to mitigate the environmental impact of blockchain operations. In response to mounting concerns, the blockchain community is actively exploring alternative consensus mechanisms and sustainable mining practices to address environmental challenges. Efforts to transition towards more energy-efficient consensus algorithms, such as proof of stake (PoS) and delegated proof of stake (DPoS), seek to reduce energy consumption while maintaining network security and decentralization. Furthermore, initiatives promoting transparency and accountability in energy consumption data within the blockchain ecosystem are gaining momentum, enabling stakeholders to make informed decisions regarding sustainable blockchain practices. By prioritizing environmental sustainability and embracing responsible innovation, the blockchain community can play a pivotal role in advancing a greener, more sustainable future while harnessing the transformative potential of blockchain technology (Polemis & Tsionas, 2023; Rana, Giungato, Tarabella, & Tricase, 2019).

# Chapter 2

# Overview of the agriculture industry

*"Seeding a Resilient Future: Nurturing Growth, Accountability, and Trust in Agriculture"* - Unknown

**Abstract**

This chapter provides an in-depth overview of the agriculture industry and examines the potential benefits of integrating blockchain technology into its operations. It explores the unique challenges faced by the agriculture sector and discusses how blockchain can address these challenges to enhance efficiency, transparency, and sustainability. The chapter begins by establishing a comprehensive understanding of the agriculture industry, highlighting its intricacies, complexities, and specific challenges. Next, the chapter examines the various benefits that blockchain can offer to the agriculture sector. It explores how blockchain's decentralized nature, immutability, and transparency can address issues such as traceability, food safety, fraud prevention, and supply chain inefficiencies. Furthermore, the chapter provides an overview of existing applications and initiatives that have utilized blockchain in the agriculture field. It showcases real-world examples of how blockchain has been implemented to improve traceability, enable fair trade, enhance quality control, and promote sustainable practices.

**Keywords**: Lack of transparency, Data Dilemma in Agriculture, The Plight of Small-scale Farmers, Stumbling Blocks, Agricultural Advancement, Dilemma of Market Access, Fair Prices, Complexity of Certification, Agriculture industry, Blockchain technology, Efficiency, Transparency, Sustainability, Challenges, Traceability, Food safety, Fraud prevention, Supply chain inefficiencies

**Chapter contents**

- Current challenges in the agriculture sector and the role of blockchain
- Importance of traceability and transparency in the food supply chain
- Potential applications of blockchain technology in agriculture

***

## 2.1. Current challenges in the agriculture sector and the role of blockchain

The contemporary hurdles within the agricultural sector, along with the potential solutions offered by blockchain technology, have been scrutinized by notable authorities such as FAO (2019), Iansiti and Lakhani (2017), Niknejad, Ismail, Bahari, Hendradi, and Salleh (2021) and Tomu (2020). The forthcoming Section 2.1 will delve deeper into the insights provided by these sources, shedding light on their implications for addressing the industry's challenges. Through the analysis of these authoritative works, valuable perspectives and strategies will be uncovered, paving the way for informed decision-making and innovative approaches to overcome agricultural sector obstacles.

### 2.1.1. Lack of Transparency

The intricate web of agricultural supply chains involves numerous participants, from farmers and producers to distributors and retailers. As products traverse through this complex network, the lack of transparency becomes a breeding ground for various challenges. Instances of food fraud, where products are misrepresented, and contamination, which poses significant risks to public health, are exacerbated by the current inadequacies in traceability. Additionally, inefficient logistics can lead to delays, waste, and increased costs (Blokdyk, 2021; Kamilaris, Fonts, & Prenafeta-Boldú, 2019).

Blockchain's Role in Agricultural Transparency Includes:

- A Transparent and Immutable Ledger: Blockchain technology, best known as the underlying framework for cryptocurrencies, holds the promise of revolutionizing supply chain transparency in agriculture. At its core, blockchain is a decentralized and immutable ledger that records transactions in a transparent and secure manner. This inherent nature of blockchain makes it an ideal candidate to address the lack of visibility plaguing agricultural supply chains (Ahmed et al., 2022).
- Tracing the Journey from Farm to Table: The application of blockchain in agriculture involves creating a comprehensive and unchangeable record of every transaction and movement of products throughout the supply chain. This ledger, accessible to all authorized participants, enables stakeholders to trace the entire journey of agricultural products from the farm to the consumer's table. Each step, be it planting, harvesting, processing, or distribution, is meticulously recorded in a secure and transparent manner.
- Rapid Issue Identification and Resolution: The real-time and transparent nature of blockchain empowers stakeholders to identify and address issues swiftly. In the event of contamination or fraud, the source can be traced back with precision, allowing for targeted recalls and interventions. This

capability is instrumental in safeguarding food safety and quality, thereby bolstering consumer trust in the agricultural supply chain (Ahmed et al., 2022).

- Reducing Food Fraud and Contamination: By implementing blockchain, the agricultural sector can significantly reduce instances of food fraud. The ability to verify the authenticity of products at every stage of the supply chain acts as a deterrent to fraudulent activities. Moreover, the transparent and traceable nature of blockchain helps in identifying and isolating contaminated products promptly, preventing widespread health risks.
- Streamlining Logistics for Efficiency: Inefficiencies in logistics often result in delays, waste, and increased costs. Blockchain's real-time tracking and automated smart contracts can streamline logistics processes, optimizing the movement of products and reducing operational inefficiencies. This not only enhances the overall efficiency of the supply chain but also contributes to the sustainability of agricultural practices.

### 2.1.2. The Data Dilemma in Agriculture

Agriculture, once rooted in age-old practices, has undergone a digital transformation, with data emerging as a linchpin for success. Crop information, ranging from soil health to yield projections, weather patterns dictating planting and harvesting seasons, and market data influencing pricing and distribution strategies — all these facets rely on an intricate web of data. However, this digital evolution brings forth a new challenge — the need to ensure the security and integrity of this wealth of agricultural information (Coble, Mishra, Ferrell, & Griffin, 2018; Li, Jiang, et al., 2020; McBratney, Whelan, Ancev, & Bouma, 2005).

Blockchain's Role in Data Security Includes:

- Decentralization – A Pillar of Security: Blockchain technology, renowned for its decentralized architecture, steps in as a formidable solution to the challenges posed by the dependence on agricultural data. Unlike centralized databases that are vulnerable to single points of failure and unauthorized access, a decentralized blockchain distributes data across a network of nodes. This inherently reduces the risk of data breaches and enhances overall security.
- Cryptographic Safeguards: The cryptographic foundations of blockchain contribute significantly to data security. Data stored on a blockchain is encrypted, rendering it resistant to tampering and unauthorized alterations. This cryptographic layer ensures that only authorized parties with the requisite cryptographic keys can access and modify the data, reinforcing the integrity of the information.

- Streamlining Verification Processes: Smart contracts, the self-executing contracts with coded terms and conditions, play a pivotal role in automating data verification processes. In the agricultural context, smart contracts can be programmed to validate data automatically, ensuring that it adheres to predefined criteria. This automation not only expedites the verification process but also minimizes the potential for human error, enhancing the accuracy and reliability of the data.
- Reducing the Risk of Tampering: One of the primary concerns in the realm of agricultural data is the risk of tampering, whether intentional or accidental. Smart contracts on the blockchain act as guardians, validating the authenticity of data at each step. Any attempt to manipulate the data triggers alerts, preserving the integrity of the information and mitigating the risk of fraudulent practices.

### 2.1.3. The Plight of Small-scale Farmers

Small-scale farmers, despite being the backbone of global food production, find themselves grappling with a significant impediment — limited access to credit. Traditional banking systems often demand collateral and rely heavily on credit histories, criteria that many small-scale farmers are unable to fulfill. This financial barrier stifles growth, restricts investments in essential resources, and hampers the overall economic resilience of these farmers (Awazi, Tchamba, & Avana, 2019; Kelly, Adesina, & Gordon, 2003; Ruml & Qaim, 2021).

Blockchain's Financial Revolution Includes:

- Decentralized Financial Systems: Blockchain technology emerges as a beacon of hope, offering a transformative solution to the financial challenges faced by small-scale farmers. At its core, blockchain enables the creation of decentralized financial systems, cutting through the bureaucratic layers of traditional banking and fostering a more inclusive financial ecosystem.
- Smart Contracts – The Architects of Transparent and Automated Loans: Central to this financial revolution are smart contracts, self-executing contracts with predefined terms and conditions written into code. These contracts enable the establishment and execution of transparent and automated loan agreements. By leveraging blockchain's transparency and security, smart contracts mitigate the need for intermediaries and traditional collateral, making the lending process more accessible and equitable for small-scale farmers.
- Breaking Down Traditional Barriers: Blockchain's decentralized financial systems and smart contracts dismantle the traditional barriers that have long hindered financial inclusion for small-scale farmers. The reliance on

collateral and credit history is replaced by transparent and automated processes, offering farmers a fair and accessible avenue for securing the financial support they need to invest in their operations.

- Democratizing Finance for Agricultural Growth: The democratization of finance through blockchain not only addresses the immediate financial challenges but also catalyzes broader economic growth within the agricultural sector. Small-scale farmers, armed with newfound access to credit, can make essential investments in seeds, equipment, and technology, ultimately enhancing productivity and contributing to the overall resilience of the agricultural ecosystem.

### 2.1.4. Stumbling Blocks to Agricultural Advancement

The ambiguity surrounding land titles and disputes over ownership has long been a stumbling block to agricultural development. In many regions, the lack of a clear and transparent land registry system opens the door to legal complications, disputes, and impediments to investment. Agricultural progress often falters when the foundation of land ownership is uncertain, deterring potential investors and hindering sustainable development (Heard-Lauréote, 2007).

Blockchain's Role in Land Registries Includes:

- A Secure and Transparent Ledger: Enter blockchain technology, offering a transformative solution to the age-old challenges of land titles and ownership disputes. At its core, blockchain provides a secure, decentralized, and transparent ledger. When applied to land registries, it becomes an immutable record of ownership, minimizing the risk of fraud, disputes, and other impediments to agricultural development.
- Mitigating Fraud and Ensuring Transparency: Utilizing blockchain for land registries means that every transaction related to land ownership is securely recorded and time-stamped. The decentralized nature of blockchain ensures that this ledger is not controlled by a single entity, reducing the risk of fraudulent activities. Each entry is transparent, traceable, and resistant to tampering, thereby providing a clear and trustworthy record of ownership.
- Enhancing Efficiency and Trust: The application of blockchain in land registries streamlines property transactions, making them more efficient and trustworthy. Smart contracts, a cornerstone of blockchain technology, can automate the execution of property transactions based on predefined rules. This not only reduces the administrative burden but also minimizes the scope for errors, ensuring a seamless and transparent process for transferring land ownership.

- Empowering Agricultural Investment: A transparent and secure land registry system powered by blockchain technology has profound implications for agricultural investment. Investors and stakeholders gain confidence when they can rely on an unambiguous record of land ownership. The reduction in disputes and legal complexities encourages a more favorable environment for investments in agricultural projects, thereby fostering sustainable development.

### 2.1.5. The Dilemma of Market Access and Fair Prices

Farmers, often at the mercy of intermediaries and market fluctuations, face a dual challenge — accessing markets and ensuring fair compensation for their hard work. Traditional market structures can introduce inefficiencies, leading to a lack of transparency, unequal bargaining power, and delayed payments. The resulting impact on the livelihoods of farmers underscores the urgent need for innovative solutions (Barrett, 2010; Gouët & Van Paassen, 2012; Shiferaw, Hellin, & Muricho, 2011).

Blockchain's Market Revolution Includes:

- Decentralized Marketplaces – A Paradigm Shift: Blockchain technology introduces a paradigm shift by enabling the creation of decentralized marketplaces. These digital ecosystems empower farmers to directly connect with buyers, eliminating the need for multiple layers of intermediaries. Through transparent and secure transactions facilitated by blockchain, farmers gain greater control over the sale of their produce and access to a broader market landscape.
- Smart Contracts: Automating Fair Transactions: At the heart of this market revolution are smart contracts, self-executing agreements with predefined rules encoded into the blockchain. Smart contracts enable the automation of payment processes, ensuring that farmers receive fair compensation in a transparent and timely manner. By reducing the reliance on intermediaries, smart contracts contribute to a more equitable distribution of profits along the agricultural supply chain.
- Direct Farmer-Buyer Connections: Blockchain-powered decentralized marketplaces foster direct connections between farmers and buyers, fostering a more collaborative and transparent trading environment. This direct interaction not only strengthens relationships but also allows farmers to negotiate prices more effectively, ensuring a fair valuation for their produce.
- Transparency in Transactions: The transparency inherent in blockchain transactions brings clarity to the entire supply chain. Farmers can track the journey of their produce from farm to market, instilling confidence in both

buyers and consumers. This transparency reduces the likelihood of fraud and ensures that the quality and origin of the product are accurately represented.

- Reducing Intermediary Influence: By decentralizing marketplaces, blockchain reduces the influence of intermediaries in agricultural trade. This shift empowers farmers to negotiate directly with buyers, cutting through layers of bureaucracy and enabling fairer pricing structures. The resulting efficiency benefits the entire agricultural ecosystem by optimizing processes and minimizing delays.
- Strengthening Agricultural Livelihoods: The adoption of blockchain in decentralized marketplaces is not just a technological shift; it is a catalyst for strengthening the livelihoods of farmers. Fair compensation, reduced market uncertainties, and improved access to buyers contribute to the overall resilience and sustainability of agricultural practices.

### 2.1.6. The Complexity of Certification

For farmers committed to organic or sustainable farming practices, navigating the intricate web of regulatory and certification requirements is a formidable task. The traditional certification process involves exhaustive documentation, rigorous inspections, and compliance checks. This complexity not only consumes valuable time and resources but also introduces uncertainties and challenges in proving adherence to stringent standards.

Blockchain's Certification Revolution Includes:

- Transparent and Auditable Farming Records: Blockchain technology emerges as a catalyst for transforming the certification landscape. By providing a transparent and auditable record of farming practices, blockchain ensures that every step of the agricultural journey is securely documented and accessible. This immutable ledger becomes a reliable source of truth, reducing the complexities associated with meeting regulatory and certification requirements.
- Streamlining the Certification Process: Blockchain's decentralized nature allows for the creation of a shared platform where all stakeholders, including farmers, certification bodies, and consumers, can access and verify relevant information. This transparency streamlines the certification process, making it more efficient and less prone to errors. Farmers can upload real-time data on their farming practices, creating a living record that evolves with each cultivation cycle.
- Proving Compliance with Certifications: One of the significant advantages of leveraging blockchain for certification is the ability to prove compliance with regulations and certifications. The transparent and immutable nature

of blockchain ensures that the information recorded is trustworthy and tamper-resistant. Farmers can easily demonstrate their adherence to organic or sustainable practices, fostering trust among consumers who seek products with specific certifications.

- Consumer Empowerment Through Transparency: Blockchain's impact extends beyond regulatory compliance to consumer empowerment. With access to a transparent and auditable record of farming practices, consumers can make informed choices about the products they purchase. This increased transparency not only enhances consumer trust but also encourages a deeper understanding and appreciation for the sustainable and organic efforts of farmers.
- Collaboration for Success: Implementing blockchain solutions for agricultural certification requires collaboration among farmers, certification bodies, technology developers, and regulatory authorities. The successful integration of blockchain hinges on creating standardized protocols, ensuring data accuracy, and fostering a shared commitment to the adoption of innovative technologies in agriculture.
- Education and Adoption: To unlock the full potential of blockchain in certification, educating stakeholders about the benefits and functionalities of this technology is paramount. Farmers need to be equipped with the knowledge and tools to seamlessly integrate blockchain into their existing practices, fostering a culture of innovation and sustainability.

## 2.2. Importance of traceability and transparency in the food supply chain

### 2.2.1. Consumer demand for traceability and transparency

*a. The Rise of consumer consciousness*

Consumers today are increasingly concerned about the social and environmental impact of their purchasing decisions. They want to make informed choices that align with their values, whether it's supporting ethically sourced products, sustainable manufacturing processes, or fair labor practices. This heightened awareness has given rise to a new era of consumer consciousness, where individuals are no longer content with simply buying products off the shelf without knowing their origins.

*b. The call for traceability*

Traceability has become a key component of the consumer demand for transparency. Consumers want to be able to trace a product's journey from its source to the store shelf. They want to know if it was grown organically, if it contains any genetically modified ingredients, or if it was produced in a facility that follows fair trade practices. This information allows consumers to make informed decisions about the products they purchase and empowers them to support businesses that align with their values.

*c. Benefits of traceability and transparency*

The demand for traceability and transparency is not just a passing fad. Businesses that embrace these principles stand to benefit in several ways. First and foremost, transparency builds trust. When consumers have access to information about a product's origins, they feel more confident in their purchasing decisions and are more likely to develop loyalty towards a brand. Moreover, transparency can serve as a competitive advantage, setting a business apart from its competitors and attracting a growing segment of conscious consumers.

*d. Challenges and opportunities*

While meeting the demand for traceability and transparency presents challenges for businesses, it also creates opportunities for innovation and improvement. Implementing robust tracking systems, adopting sustainable and ethical practices, and engaging in open communication with consumers are some of the ways businesses can rise to the occasion. Embracing traceability not only helps businesses align with consumer expectations but also encourages continuous improvement and accountability within the supply chain.

*e. Technological solutions*

Advancements in technology have played a pivotal role in enabling businesses to meet the demand for traceability and transparency. From blockchain technology to mobile applications, various tools have emerged to facilitate the tracking and sharing of information throughout the supply chain. These technologies provide a means to collect and disseminate data about a product's journey, making it accessible to both businesses and consumers alike (Ahmed et al., 2022; Cottrill et al., 2018; Sander, Semeijn, & Mahr, 2018).

*f. Looking ahead*

The demand for traceability and transparency shows no signs of slowing down. As consumers become more educated and aware, businesses will need to continuously adapt to meet their expectations. The future will likely bring even greater emphasis on sustainability, responsible sourcing, and accountability. Companies that prioritize transparency and proactively address consumer concerns will not only thrive in this new consumer landscape but also contribute to a more sustainable and ethical global marketplace.

### 2.2.2. Food safety and quality assurance

*a. The significance of food safety*

Food safety is paramount to protect consumers from foodborne illnesses and hazards. Contaminated food can lead to severe health consequences, ranging from mild discomfort to life-threatening conditions. As a result, governments, regulatory agencies, and industry stakeholders have implemented stringent standards and regulations to prevent, detect, and respond to food safety risks. Ensuring the safety of food products not only protects public health but also safeguards the reputation and trust of businesses in the food industry.

*b. Understanding the challenges*

Maintaining food safety is a complex endeavor due to various challenges that businesses face. These challenges include the risk of microbial contamination, chemical hazards, allergens, improper handling, and inadequate sanitation practices. Additionally, globalization and the increasing complexity of supply chains have introduced new challenges, such as ensuring the safety of imported ingredients and managing risks across multiple locations. Addressing these challenges requires a comprehensive and proactive approach to food safety management.

*c. The role of quality assurance*

Quality assurance is closely intertwined with food safety and goes beyond ensuring that products are safe to consume. It encompasses all aspects of the production process, from sourcing ingredients to packaging and distribution. Quality assurance systems are designed to maintain consistency, meet regulatory requirements, and exceed customer expectations. These systems include robust quality control measures, adherence to good manufacturing practices, and regular inspections and audits to identify and rectify any deviations or potential risks.

*d. Implementing food safety management systems*

To achieve and maintain high standards of food safety, businesses often adopt food safety management systems. These systems, such as Hazard Analysis and Critical Control Points (HACCP) and ISO 22000, provide a systematic approach to identify, assess, and control food safety hazards at every stage of the food production process. By implementing these systems, businesses can establish preventive measures, set critical control points, monitor operations, and take corrective actions to ensure food safety and quality.

*e. Embracing technology and innovation*

Technology plays a pivotal role in enhancing food safety and quality assurance efforts. Advances in data collection, analytics, and automation have revolutionized the way businesses monitor and manage food safety risks. From real-time monitoring of temperature and humidity to blockchain-based traceability systems, technology enables businesses to identify and respond to potential risks more efficiently. Moreover, emerging technologies like artificial intelligence and machine learning offer promising opportunities to further improve food safety practices (Detwiler, 2020).

*f. Collaboration and continuous improvement*

Achieving the highest standards of food safety and quality assurance requires collaboration among all stakeholders involved in the food supply chain. This includes farmers, processors, distributors, retailers, regulatory agencies, and consumers. By working together, sharing knowledge, and implementing best practices, the industry can collectively address challenges, adapt to emerging risks, and continuously improve food safety systems.

*g. Looking ahead*

As the global food landscape continues to evolve, so do the challenges and opportunities related to food safety and quality assurance. The emergence of new pathogens, changing consumer preferences, and evolving regulatory frameworks necessitate a proactive and adaptive approach. Businesses must stay vigilant, embrace emerging technologies, and prioritize a culture of continuous improvement to meet the ever-increasing expectations for safe and high-quality food products.

### 2.2.3. Compliance with regulations and certifications

*a. Understanding compliance*

Compliance refers to the act of conforming to established guidelines, laws, regulations, and industry standards that govern specific areas of business operations. These requirements may be set by governmental bodies, industry regulators, or international organizations. Compliance encompasses a wide range of areas, such as data protection, environmental sustainability, labor practices, financial reporting, and product safety, among others.

*b. The importance of compliance*

Compliance serves as a safeguard against legal and reputational risks. Failure to comply with applicable regulations can result in severe consequences, including fines, lawsuits, damage to brand reputation, and even criminal charges. By prioritizing compliance, organizations demonstrate their commitment to upholding legal and ethical standards and fostering trust among customers, investors, and other stakeholders.

*c. Challenges in maintaining compliance*

Compliance can be a complex and challenging endeavor for businesses. The regulatory landscape is dynamic, with laws and regulations constantly evolving. Staying up to date with these changes requires continuous monitoring, interpretation, and implementation of new requirements. Additionally, compliance often involves significant documentation, reporting, and record-keeping obligations, which can be resource-intensive and time-consuming.

Moreover, organizations operating in multiple jurisdictions face the added complexity of navigating diverse regulatory frameworks and ensuring consistency across different regions. Compliance requirements may also vary across industries, making it necessary for businesses to understand and address sector-specific regulations.

*d. The role of certifications*

Certifications serve as formal recognition that an organization meets specific standards and requirements. They offer an additional layer of assurance to stakeholders, including customers, partners, and investors. Certifications are typically awarded by independent third-party organizations that assess an organization's compliance with predetermined criteria.

Certifications can cover various aspects, such as quality management systems (e.g., ISO 9001), information security (e.g., ISO 27001), environmental management (e.g., ISO 14001), and social responsibility (e.g., B Corp certification). Achieving certifications often involves rigorous assessments, audits, and ongoing compliance monitoring.

*e. Benefits of compliance*

While compliance may pose challenges, it also offers significant benefits to businesses. These include:

- Risk Mitigation: Compliance helps mitigate legal, financial, and reputational risks associated with non-compliance, ensuring a more secure business environment.

- Enhanced Reputation: Demonstrating commitment to compliance enhances an organization's reputation, building trust among stakeholders and differentiating it from competitors.
- Access to Markets: Compliance with certain regulations and certifications may be a prerequisite for entering specific markets or securing contracts with government entities or large corporations.
- Operational Efficiency: Compliance frameworks often require organizations to adopt best practices, resulting in improved operational efficiency, standardized processes, and reduced costs.
- Sustainability and Responsibility: Compliance with environmental, social, and governance (ESG) regulations and certifications reinforces an organization's commitment to sustainable practices and responsible business conduct.
- Customer Confidence: Compliance reassures customers that their data, products, and services are handled with care, fostering loyalty and long-term relationships.

*f. Building a compliance culture*

To ensure sustained compliance, organizations must foster a culture that prioritizes ethical conduct and regulatory adherence. This involves:

- Leadership Commitment: Leaders must lead by example, demonstrating a strong commitment to compliance and establishing it as a core value within the organization.
- Employee Education and Training: Regular training programs help employees understand their compliance obligations, empowering them to make informed decisions and contribute to a compliant environment.
- Robust Compliance Frameworks: Establishing clear policies, procedures, and internal controls enables effective compliance management, monitoring, and reporting.
- Continuous Improvement: Regular reviews and assessments of compliance programs help identify areas for improvement and ensure ongoing alignment with evolving regulations.

### 2.2.4. Sustainability and ethical sourcing

*a. Understanding sustainability*

Sustainability encompasses practices that meet present needs without compromising the ability of future generations to meet their own needs. It involves finding a balance between economic development, environmental protection, and social well-being. In the context of business operations, sustainability refers to integrating environmentally friendly practices, social responsibility, and long-term economic viability.

*b. Importance of sustainability*

Sustainability is crucial for several reasons. First, it addresses the urgent need to mitigate and adapt to climate change, reduce carbon emissions, and protect natural resources. Second, sustainable practices help minimize environmental impacts, such as pollution, deforestation, and habitat destruction. Third, sustainability contributes to social progress by ensuring fair labor practices, promoting community engagement, and supporting local economies. Moreover, embracing sustainability enhances a company's reputation, attracts socially conscious consumers, and can lead to cost savings through energy efficiency, waste reduction, and process optimization. In an increasingly competitive marketplace, sustainability has become a key differentiator and a driver of innovation and business growth (OECD, 2020; Watson & Achinelli, 2008).

*c. Ethical sourcing*

Ethical sourcing refers to the procurement of goods and services in a manner that respects human rights, ensures fair treatment of workers, and upholds social and ethical standards. It involves evaluating suppliers based on criteria such as labor conditions, worker rights, child labor, health and safety, and adherence to international labor standards. By adopting ethical sourcing practices, organizations aim to eliminate or minimize exploitative labor practices, promote fair trade, and support suppliers who share their commitment to responsible business conduct. Ethical sourcing extends beyond direct suppliers and includes tracing the entire supply chain to identify and address any potential violations or risks.

*d. Challenges in sustainability and ethical sourcing*

Implementing sustainability and ethical sourcing practices presents various challenges to organizations (Guggisberg, 2022; Huq & Stevenson, 2020; Leire & Mont, 2010; Schneider & Wallenburg, 2012). These challenges can be:

- Supply Chain Complexity: Global supply chains can be complex, involving multiple tiers of suppliers and subcontractors. Ensuring sustainability and ethical practices throughout this web of relationships requires transparency, traceability, and collaboration.
- Compliance Verification: Verifying compliance with sustainability and ethical standards across supply chains often requires extensive auditing,

monitoring, and verification processes. Organizations must develop robust mechanisms to assess supplier compliance and address any identified issues effectively.

- Cost Considerations: While sustainability initiatives can lead to long-term cost savings, the initial investments and transition costs may pose financial challenges for some organizations, particularly smaller ones. Balancing sustainability goals with profitability is a critical consideration.
- Stakeholder Engagement: Engaging with various stakeholders, including suppliers, employees, customers, and communities, is vital for successful sustainability and ethical sourcing efforts. It requires clear communication, collaboration, and aligning interests to drive meaningful change.

*e. Benefits of sustainability and ethical sourcing*

Despite the challenges, embracing sustainability and ethical sourcing offers numerous benefits:

- Reputation and Brand Value: Organizations that prioritize sustainability and ethical sourcing cultivate a positive brand image, build trust with customers, and differentiate themselves from competitors.
- Risk Mitigation: By proactively addressing environmental and social risks in their supply chains, companies can mitigate reputational, legal, and operational risks associated with non-compliance and negative social impacts.
- Innovation and Efficiency: Sustainability often drives innovation, leading to the development of new products, services, and business models. It also promotes efficiency through resource optimization, waste reduction, and improved supply chain management.
- Access to Markets and Customers: Increasingly, customers and businesses are seeking products and services that align with their values. Embracing sustainability and ethical sourcing opens doors to new markets, partnerships, and customer segments.
- Regulatory Compliance: Sustainable practices align with evolving regulations focused on environmental protection, labor rights, and supply chain transparency. By being proactive, organizations can stay ahead of regulatory requirements and avoid legal penalties.
- Employee Engagement: A commitment to sustainability and ethical sourcing can boost employee morale, attract and retain top talent, and foster a sense of purpose and pride in the workplace.

*f. Implementing sustainability and ethical sourcing*

To effectively integrate sustainability and ethical sourcing into their operations, organizations can take the following steps:

- Develop a Sustainability Strategy: Set clear sustainability goals, create metrics for tracking progress, and integrate sustainability considerations into business strategies and decision-making processes.
- Engage Suppliers: Collaborating with suppliers to ensure shared values and adherence to sustainability and ethical sourcing standards. Develop supplier codes of conduct, conduct audits, and establish mechanisms for ongoing communication and improvement.
- Enhance Transparency and Traceability: Implement systems to trace and monitor supply chains, enabling organizations to identify potential risks, assess compliance, and address issues promptly.
- Educate and Involve Employees: Raise awareness among employees about sustainability and ethical sourcing, providing training and empowering them to contribute to sustainable practices within their roles.
- Communicate and Engage Stakeholders: Share sustainability initiatives and progress transparently with stakeholders, including customers, employees, investors, and communities. Actively seek feedback and incorporate stakeholder perspectives into decision-making processes.

### 2.2.5. Building consumer trust and confidence

*a. The significance of consumer trust and confidence*

Consumer trust is the belief that an organization will act in the best interests of its customers, deliver on its promises, and uphold ethical standards. Confidence, on the other hand, relates to customers' certainty that a business is competent, reliable, and capable of meeting their expectations. Both trust and confidence are crucial for attracting and retaining customers, fostering loyalty, and differentiating a brand from competitors.

*b. Factors influencing consumer trust and confidence*

Several key factors contribute to building consumer trust and confidence:

- Consistency and Reliability: Consistently delivering quality products, services, and customer experiences builds trust and confidence. Reliability in meeting commitments and resolving issues in a timely manner is essential.

- Transparency and Honesty: Open and transparent communication, honest marketing practices, and clear terms and conditions foster trust. Consumers appreciate businesses that are forthcoming about their practices, pricing, and policies.
- Data Privacy and Security: Safeguarding customer data and ensuring robust security measures demonstrate a commitment to protecting consumers' privacy, earning their trust and confidence in sharing personal information.
- Excellent Customer Service: Providing prompt, attentive, and personalized customer service builds confidence and reassures customers that their needs will be addressed promptly and effectively.
- Social Responsibility: Demonstrating a commitment to social and environmental causes can enhance trust, as consumers increasingly expect businesses to operate ethically and contribute to positive societal impact.

*c. Strategies for building consumer trust and confidence*

- Delivering on Promises: Consistently meeting or exceeding customer expectations builds trust and confidence. This includes delivering high-quality products, fulfilling orders promptly, and providing accurate and reliable information.
- Transparent Communication: Openly sharing information about products, services, pricing, and policies builds transparency and establishes credibility. Clear and honest communication also extends to addressing any issues or mistakes proactively.
- Prioritising Data Privacy and Security: Implementing robust data protection measures, complying with applicable privacy regulations, and being transparent about data handling practices instill confidence in consumers that their information is secure.
- Exceptional Customer Service: Investing in well-trained and responsive customer service teams that are empowered to resolve issues quickly and effectively fosters trust and confidence. Listening to customer feedback and addressing concerns demonstrates a commitment to customer satisfaction.
- Social and Environmental Responsibility: Engaging in sustainable practices, supporting social causes, and communicating these efforts to consumers can enhance trust and confidence, particularly among socially conscious customers.

- Building Relationships: Cultivating long-term relationships with customers through loyalty programs, personalized experiences, and ongoing communication creates a sense of trust, loyalty, and confidence.

*d. Benefits of building consumer trust and confidence*

- Customer Loyalty and Repeat Business: When customers trust and have confidence in a brand, they are more likely to become loyal advocates and repeat purchasers.
- Positive Word-of-mouth and Reputation: Satisfied and confident customers are more likely to recommend a business to others, contributing to positive word-of-mouth and enhancing the brand's reputation.
- Increased Sales and Market Share: Building trust and confidence attracts new customers, expands the customer base, and helps capture a larger market share.
- Competitive Differentiation: Trust and confidence differentiate a brand from competitors by emphasizing reliability, transparency, and customer-centric values.
- Resilience During Crisis: Businesses that have built strong trust and confidence with customers are better equipped to navigate and recover from crises or reputational challenges.
- Collaboration and Partnerships: Trusted brands are more likely to establish successful collaborations and partnerships with other businesses, leading to mutual growth and success.

## 2.3. Potential applications of blockchain technology in agriculture

The potential applications of blockchain technology in agriculture, as explored by Lin et al. (2020), span across various critical domains. These include supply chain traceability and provenance, enhanced product authentication, anti-counterfeiting measures, improved inventory management and logistics, smart contracts facilitating transparent and efficient transactions, decentralized marketplaces for agricultural trade, and fostering data sharing and collaboration among stakeholders. By leveraging blockchain, the agricultural sector can enhance transparency, traceability, and efficiency throughout the entire value chain, ultimately benefiting producers, distributors, and consumers alike.

### 2.3.1. Supply chain traceability and provenance

*a. Understanding supply chain traceability and provenance*

Supply chain traceability refers to the ability to track and trace the movement of products and components throughout the supply chain. It involves capturing and recording information about the origin, processes, intermediaries, and final destinations of goods. Provenance, on the other hand, focuses on establishing the authenticity and history of a product, including its origin, production methods, and any certifications or verifications it may have.

*b. Importance of supply chain traceability and provenance*

Supply chain traceability and provenance provide transparency and accountability in the supply chain, benefiting both businesses and consumers (Feng, Wang, Duan, Zhang, & Zhang, 2020):

- Consumer Demand: Increasingly, consumers want to make informed purchasing decisions based on factors such as ethical sourcing, sustainability, and product safety. Supply chain traceability and provenance enable them to verify claims made by businesses and align their choices with their values.
- Risk Mitigation: Traceability helps identify and address potential risks within the supply chain, such as counterfeit products, unauthorized intermediaries, or non-compliant materials. It enhances quality control, reduces the risk of product recalls, and mitigates reputational and legal risks.
- Compliance and Regulation: Traceability supports compliance with regulations related to product safety, environmental impact, labor practices, and ethical sourcing. It ensures that businesses can demonstrate adherence to relevant standards and regulations.
- Supply Chain Resilience: By mapping and understanding the entire supply chain, businesses can identify vulnerabilities, anticipate disruptions, and develop contingency plans. This enhances supply chain resilience and enables prompt response to unforeseen events.
- Market Access and Competitive Advantage: Supply chain traceability and provenance can be a competitive differentiator, enabling businesses to access new markets, secure partnerships, and differentiate their products from competitors.

*c. Challenges in achieving supply chain traceability and provenance*

Implementing effective supply chain traceability and provenance systems can present several challenges (Alshehri, 2023; Baralla, Ibba, Marchesi, Tonelli, & Missineo, 2019; Mirabelli & Solina, 2020):

- Complex Supply Chains: Global supply chains often involve multiple intermediaries, diverse geographies, and different regulatory frameworks. Tracing products throughout such complex networks can be challenging, requiring collaboration and information sharing across stakeholders.
- Data Collection and Management: Gathering and managing data related to product origins, processes, and certifications necessitates robust information systems and standardized data formats. Interoperability among different systems and organizations is crucial.
- Supply Chain Visibility: Achieving end-to-end visibility within the supply chain requires cooperation and transparency from suppliers, subcontractors, and other stakeholders. Resistance to sharing information or lack of traceability infrastructure can hinder progress.
- Verification and Auditing: Ensuring the accuracy and authenticity of information within the supply chain often involves verification processes, audits, and certifications. These can be resource-intensive and require coordination with external parties.
- Technology Adoption: Implementing traceability technologies, such as blockchain, RFID (Radio-frequency Identification), or IoT (Internet of Things), may require upfront investments, training, and overcoming technological barriers. Integration with existing systems and collaboration with suppliers may be necessary.

*d. Benefits of supply chain traceability and provenance*

The advantages of supply chain traceability and provenance are manifold, encompassing various critical aspects. These include fostering consumer trust and loyalty, reducing and mitigating risks associated with product quality and safety, and ensuring compliance with regulatory standards while promoting sustainability. Additionally, enhancing supply chain traceability and provenance contributes to operational efficiency and resilience, fosters collaboration and partnerships among stakeholders, and opens up opportunities for innovation and expansion into new markets, ultimately driving growth and competitiveness within the industry (Borah et al., 2020; Caro, Ali, Vecchio, & Giaffreda, 2018; Chun-Ting, Meng-Ju, Nen-Fu, Jhong-Ting, & Jia-Jung, 2020; Sunny, Undralla, & Pillai, 2020).

- Consumer Trust and Loyalty: Supply chain traceability and provenance enhance transparency, enabling businesses to earn consumer trust, build brand loyalty, and meet the demands of socially conscious consumers.
- Risk Reduction and Mitigation: Traceability helps identify and address supply chain risks promptly, reducing the likelihood of product recalls, reputational damage, and legal issues. It improves quality control and promotes responsible sourcing practices.
- Compliance and Sustainability: Achieving supply chain traceability and provenance supports compliance with regulations and sustainability initiatives related to product safety, environmental impact, labor practices, and ethical sourcing.
- Operational Efficiency and Resilience: Enhanced supply chain visibility improves operational efficiency, enables informed decision-making, and facilitates quicker response to disruptions. It fosters supply chain resilience and agility.
- Collaboration and Partnerships: Supply chain traceability and provenance can facilitate collaboration and partnerships with suppliers, industry peers, and other stakeholders. It strengthens relationships, promotes best practices, and drives industry-wide improvements.
- Innovation and Market Opportunities: Embracing traceability technologies and establishing robust traceability systems can foster innovation, create new market opportunities, and unlock potential for value-added services.

### 2.3.2. Enhanced product authentication and anti-counterfeiting measures

*a. Understanding product authentication*

Product authentication refers to the process of verifying the genuineness and integrity of a product. It involves implementing measures to ensure that a product is authentic, originating from a legitimate source, and has not been tampered with during the supply chain journey. Product authentication provides consumers with confidence in their purchases and helps businesses protect their brand reputation.

*b. The challenges posed by counterfeiting*

Counterfeiting has emerged as a pervasive problem affecting various industries, including fashion, electronics, pharmaceuticals, and luxury goods. The challenges posed by counterfeiting include:

- Revenue Loss: Counterfeiting leads to significant revenue losses for businesses as counterfeit products often compete directly with genuine ones, eroding market share and brand value.
- Consumer Health and Safety Risks: Counterfeit products, especially in sectors such as pharmaceuticals and electronics, can pose severe health and safety risks to consumers due to substandard materials, lack of quality control, and non-compliance with safety regulations.
- Brand Dilution: Counterfeit products damage brand reputation by diluting brand value, confusing consumers, and diminishing trust in genuine products.
- Supply Chain Integrity: Counterfeiting undermines the integrity of the supply chain by introducing unauthorized or substandard products, leading to reputational and legal risks for businesses.

*c. Anti-counterfeiting measures*

To combat counterfeiting and enhance product authentication, businesses employ various anti-counterfeiting measures:

- Holograms and Security Labels: Holograms and security labels with unique designs, tamper-evident features, or serialized numbers are commonly used to authenticate products. They are difficult to replicate and provide visible proof of authenticity.
- Track and Trace Technologies: Implementing track and trace technologies, such as RFID (Radio-frequency Identification), QR codes, or unique identifiers, enables real-time monitoring of product movements and helps verify authenticity at each stage of the supply chain.
- Authentication Technologies: Advanced authentication technologies, including invisible inks, microtaggants, or embedded security features, are integrated into products or packaging to provide covert authentication measures that are difficult to replicate.
- Digital Solutions: Leveraging digital solutions, such as mobile apps, blockchain, or online verification systems, allows consumers to verify product authenticity using digital platforms and provides a transparent record of the product's journey.
- Collaboration and Enforcement: Collaborating with law enforcement agencies, industry associations, and intellectual property rights organizations helps strengthen enforcement efforts against counterfeiting, leading to legal action against counterfeiters and disrupting their operations.

*d. Benefits of enhanced product authentication and anti-counterfeiting measures*

Implementing robust product authentication and anti-counterfeiting measures brings several benefits to businesses and consumers:

- Consumer Trust and Satisfaction: Enhanced product authentication assures consumers of the authenticity and quality of the products they purchase, fostering trust and satisfaction.
- Brand Protection and Reputation: Effective anti-counterfeiting measures safeguard brand reputation by preventing the circulation of counterfeit products, preserving brand value, and maintaining consumer confidence.
- Revenue Protection: By reducing the prevalence of counterfeit products, businesses can protect their revenues, market share, and intellectual property rights.
- Consumer Safety and Confidence: Authenticating products helps ensure consumer safety by mitigating the risks associated with counterfeit goods, such as faulty electronics or counterfeit medications.
- Supply Chain Integrity: Anti-counterfeiting measures enhance supply chain integrity by minimizing the infiltration of unauthorized or counterfeit products, protecting brand reputation, and reducing legal and regulatory risks.
- Market Competitiveness: Strong product authentication measures can differentiate a brand from counterfeit competitors, positioning it as a trusted and reliable choice for consumers.

### 2.3.3. Improved inventory management and logistics

*a. The significance of improved inventory management and logistics*

Inventory management involves overseeing the procurement, storage, and utilization of materials, components, and finished goods. Logistics, on the other hand, encompasses the planning, execution, and control of the flow of goods, information, and resources across the supply chain. Improved inventory management and logistics offer several advantages:

- Cost Optimization: Efficient inventory management minimizes inventory-holding costs, reduces the risk of stockouts and obsolescence, and improves cash flow. Effective logistics practices optimize transportation, warehousing, and distribution costs, resulting in cost savings.
- Customer Satisfaction: Timely availability of products, accurate order fulfillment, and efficient delivery processes enhance customer satisfaction

and loyalty. Meeting customer demands promptly and accurately improves overall customer experience.

- Operational Efficiency: Streamlined inventory management and logistics processes increase operational efficiency by reducing bottlenecks, improving order accuracy, and enhancing overall productivity. This leads to improved resource utilization and reduced wastage.
- Demand Responsiveness: Improved inventory management enables businesses to respond effectively to changes in customer demand, market trends, and supply chain disruptions. Agile logistics practices ensure the timely movement of goods and adaptability to evolving customer requirements.
- Collaboration and Integration: Effective inventory management and logistics require collaboration and integration among various stakeholders, such as suppliers, manufacturers, distributors, and transportation providers. Strengthening these relationships fosters better coordination, visibility, and performance across the supply chain.

*b. Strategies for improved inventory management and logistics*

To achieve improved inventory management and logistics, businesses can employ the following strategies (Babbar & Prasad, 1998; Frank, 2005; Lambert & Stock, 1993):

- Demand Forecasting and Planning: Accurate demand forecasting helps determine optimal inventory levels, anticipate customer requirements, and align procurement and production processes accordingly. Planning tools and techniques, such as just-in-time (JIT) or lean methodologies, assist in minimizing excess inventory and reducing lead times.
- Inventory Optimization: Employing inventory optimization techniques, such as ABC analysis, economic order quantity (EOQ), or safety stock optimization, helps maintain optimal stock levels, reduce carrying costs, and prevent stockouts.
- Real-time Visibility and Tracking: Implementing advanced technologies, such as RFID (Radio-frequency Identification), barcode scanning, or IoT (Internet of Things), provides real-time visibility into inventory levels, location tracking, and product movement. This enhances inventory accuracy, reduces manual errors, and facilitates proactive decision-making.
- Warehouse Management Systems (WMS): Utilizing WMS software streamlines warehouse operations by optimizing space utilization, automating inventory tracking, improving order picking accuracy, and facilitating efficient inventory replenishment.

- Efficient Transportation and Routing:

  Implementing effective transportation management systems (TMS) allows for optimized route planning, load consolidation, carrier selection, and real-time tracking. This minimizes transportation costs, improves delivery speed, and enhances overall logistics efficiency.

- Collaboration and Information Sharing: Enhancing communication and collaboration among supply chain partners through shared data, integrated systems, and collaborative forecasting leads to improved coordination, visibility, and efficiency throughout the supply chain.

*c. Benefits of improved inventory management and logistics*

- Cost Reduction: Optimized inventory levels, efficient transportation, and streamlined warehouse operations result in reduced inventory carrying costs, transportation expenses, and operational inefficiencies.
- Improved Customer Service: Accurate order fulfillment, faster delivery times, and effective response to customer demands enhance customer satisfaction, loyalty, and retention.
- Increased Operational Efficiency: Streamlined processes, reduced lead times, and effective resource utilization improve operational efficiency, productivity, and profitability.
- Supply Chain Resilience: Effective inventory management and logistics practices enhance supply chain resilience by enabling businesses to respond to disruptions, minimize stockouts, and maintain continuity in operations.
- Competitive Advantage: Improved inventory management and logistics give businesses a competitive edge by ensuring product availability, delivering faster than competitors, and meeting customer expectations consistently.
- Enhanced Sustainability: Optimized inventory management and logistics practices contribute to sustainability by reducing waste, minimizing carbon footprint through efficient transportation, and supporting responsible procurement and sourcing.

### 2.3.4. Smart contracts for transparent and efficient transactions

*a. Understanding smart contracts*

Smart contracts are computer programs that automatically execute predefined actions once specific conditions, agreed upon by the parties involved, are met. They are built on blockchain technology, which provides security, immutability, and transparency. Smart contracts eliminate the need for intermediaries, as the code itself ensures compliance and enforces the agreed-upon terms.

*b. Benefits of smart contracts*

- Transparency: Smart contracts enhance transparency by providing a tamper-proof and immutable record of transactions on the blockchain. All parties involved can access and verify the details of the contract, ensuring trust and eliminating the need for reliance on a centralized authority.
- Efficiency: Smart contracts automate the execution of contractual obligations, reducing the need for manual intervention and paperwork. This streamlines processes, minimizes errors, and accelerates transaction times.
- Security: Smart contracts utilize cryptography and distributed ledger technology, making them highly secure. The decentralized nature of blockchain ensures that transactions are resistant to hacking, fraud, and unauthorized modifications.
- Cost Savings: By eliminating intermediaries and reducing manual processes, smart contracts lead to cost savings in terms of fees, administrative overheads, and potential disputes.
- Accuracy and Reliability: Smart contracts are programmed to execute actions precisely as defined, leaving no room for misinterpretation or human error. This ensures accuracy and reliability in executing transactions.

*c. Applications of smart contracts*

Smart contracts find applications in various industries and sectors (Torky & Hassanein, 2020), including:

- Supply Chain Management: Smart contracts enable transparent and efficient management of the supply chain by automating processes such as inventory tracking, order fulfillment, and payment settlements. They provide real-time visibility and traceability, reducing delays and improving efficiency.
- Financial Services: Smart contracts streamline financial transactions, such as payments, loans, and insurance claims. They automate processes, verify compliance with regulations, and facilitate immediate and accurate settlements.
- Real Estate: Smart contracts simplify real estate transactions by automating tasks such as property transfers, escrow services, and rental agreements. They reduce the need for intermediaries and ensure transparent and secure transactions.

- Intellectual Property Rights: Smart contracts help protect intellectual property rights by establishing clear ownership, licensing terms, and royalty distribution mechanisms. They enable automated verification and enforcement of these rights.
- Healthcare: Smart contracts facilitate the secure sharing of patient data among healthcare providers, ensuring privacy and accuracy. They streamline billing and insurance processes, reducing administrative burdens and enhancing patient care.

*d. Promoting transparency and efficiency*

- Immutable Record-keeping: Smart contracts leverage the immutability of blockchain to create an unalterable record of transactions. This transparency enables all parties to access and verify the details, promoting trust and accountability.
- Automation of Processes: Smart contracts automate the execution of tasks, reducing the need for manual intervention and potential errors. This improves efficiency, speeds up transaction times, and minimizes the need for intermediaries.
- Real-time Settlements: Smart contracts facilitate immediate settlements by automatically executing payment terms once conditions are met. This eliminates delays associated with traditional settlement processes, such as clearinghouses or manual verifications.
- Trust and Security: The use of cryptography and decentralized blockchain technology ensures that transactions conducted through smart contracts are secure, transparent, and resistant to tampering or fraud.
- Auditable and Compliant Transactions: Smart contracts allow for easy auditing of transactions, as all transaction details are recorded on the blockchain. This promotes compliance with regulatory requirements and reduces the risk of fraudulent activities.

### 2.3.5. Decentralized marketplaces for agricultural trade

*a. Understanding decentralized marketplaces*

Decentralized marketplaces for agricultural trade leverage blockchain technology to create transparent, secure, and efficient platforms for buying and selling agricultural products. These marketplaces enable direct peer-to-peer interactions, eliminating intermediaries and fostering trust among participants. Smart contracts facilitate automated and verifiable transactions, ensuring compliance with agreed-upon terms.

*b. Benefits of decentralized marketplaces*

- Transparency: Decentralized marketplaces provide transparent access to market data, including pricing, product quality, and transaction history. All participants have visibility into the entire supply chain, promoting trust and fair trade practices.
- Efficiency: By removing intermediaries and automating processes, decentralized marketplaces streamline agricultural trade. They reduce paperwork, minimize transaction costs, and enable faster settlements, resulting in improved efficiency and liquidity.
- Market Access: Decentralized marketplaces offer expanded market access for farmers and producers, allowing them to reach a broader network of buyers and consumers. This promotes fair competition, eliminates geographical barriers, and improves market opportunities.
- Traceability and Quality Assurance: Blockchain technology enables the recording of every transaction and movement of agricultural products. This ensures traceability throughout the supply chain, enhances food safety measures, and provides assurance of product quality and authenticity.
- Financial Inclusion: Decentralized marketplaces enable direct peer-to-peer transactions, potentially empowering small-scale farmers and producers by bypassing traditional financial intermediaries. This enhances financial inclusion and supports economic growth in agricultural communities.

*c. Key features of decentralized marketplaces*

- Smart Contracts: Smart contracts automate and enforce the terms of trade agreements, ensuring trust and transparency. They facilitate secure and efficient transactions by automatically executing actions when predetermined conditions are met.
- Supply Chain Visibility: Blockchain-based decentralized marketplaces provide participants with real-time visibility into the entire supply chain. This transparency allows stakeholders to track the origin, quality, and movement of agricultural products, mitigating risks and enhancing trust.
- Reputation Systems: Decentralized marketplaces often incorporate reputation systems where participants can rate and provide feedback on their trading partners. These reputation systems help establish trust and enable informed decision-making when engaging in transactions.
- Tokenization and Digital Payments: Some decentralized marketplaces utilize digital tokens or cryptocurrencies for trade settlements. Tokenization simplifies cross-border transactions, reduces transaction costs, and increases financial inclusivity.

- Data Analytics and Insights: Decentralized marketplaces can leverage data analytics to generate valuable insights for participants. These insights may include market trends, price fluctuations, demand patterns, and supply chain efficiency, empowering stakeholders to make informed business decisions.

*d. Potential impact on the agricultural industry*

Decentralized marketplaces have the potential to revolutionize the agricultural industry in several ways (Rana, Tricase, & De Cesare, 2021; Xiong, Dalhaus, Wang, & Huang, 2020):

- Market Efficiency: By facilitating direct transactions between producers and buyers, decentralized marketplaces eliminate intermediaries, reduce costs, and promote price transparency. This leads to fairer market conditions and increased efficiency.
- Trust and Quality Assurance: Transparent supply chains and traceability foster trust among participants. Consumers gain confidence in the origin and quality of agricultural products, while producers can differentiate their offerings based on reliable information.
- Market Access for Small-scale Producers: Decentralized marketplaces offer opportunities for small-scale farmers and producers to access larger markets and connect with a broader range of buyers. This reduces reliance on local middlemen and enables fairer market participation.
- Sustainable Agriculture: Decentralized marketplaces can support sustainable agriculture practices by promoting traceability and responsible sourcing. Participants can prioritize products that meet specific sustainability criteria, leading to positive environmental and social impacts.
- Resilience and Risk Mitigation: Blockchain-based decentralized marketplaces provide resiliency against supply chain disruptions and external shocks. The transparency and traceability of transactions enable efficient risk mitigation strategies, such as identifying alternative suppliers or adjusting production plans.

### 2.3.6. Data sharing and collaboration among stakeholders

*a. The significance of data sharing and collaboration*

- Holistic Insights: Data sharing enables stakeholders to gain a comprehensive and holistic understanding of complex issues. By pooling together diverse datasets from multiple sources, organizations can derive valuable insights and uncover hidden patterns or correlations that would otherwise be difficult to identify.

- Informed Decision-making: Access to shared data empowers stakeholders to make more informed decisions. By leveraging a broader range of information, decision-makers can assess risks, identify opportunities, and develop strategies that are based on a more accurate and comprehensive understanding of the situation at hand.
- Collaboration and Synergy: Data sharing encourages collaboration among stakeholders. By sharing data, organizations can work together to address common challenges, share resources, and leverage collective expertise. This collaboration promotes synergy, fosters innovation, and enables the development of solutions that are more robust and effective.
- Efficiency and Resource Optimization: Data sharing minimizes duplication of efforts and resources. By pooling together data, organizations can avoid redundant data collection and analysis processes, leading to increased efficiency and optimized resource utilization.
- Societal Impact: Data sharing and collaboration have the potential to generate significant societal impact. By sharing data and collaborating on key challenges, stakeholders can address pressing issues such as healthcare, climate change, and poverty eradication more effectively, ultimately leading to positive social change.

*b. Challenges and considerations*

- Data Privacy and Security: Data sharing necessitates careful consideration of privacy and security concerns. Stakeholders must ensure that appropriate data protection measures are in place to safeguard sensitive information and comply with relevant regulations.
- Interoperability and Standardization: Achieving effective data sharing requires interoperability and standardization. Stakeholders need to agree on common data formats, protocols, and metadata standards to enable seamless integration and exchange of data.
- Trust and Governance: Establishing trust among stakeholders is crucial for successful data sharing. Clear governance frameworks, data-sharing agreements, and mechanisms for data access control and consent management are necessary to build trust and ensure responsible data-sharing practices.
- Data Quality and Reliability: Data sharing relies on the availability of high-quality and reliable data. Stakeholders should address data accuracy,

completeness, and timeliness to ensure that shared data is fit for its intended purpose.

- Cultural and Organizational Barriers: Data sharing often requires overcoming cultural and organizational barriers. Stakeholders may have different data-sharing practices, incentives, or concerns. Building a culture of collaboration and trust is essential to encourage active participation and foster a data-sharing mindset.

*c. Benefits of data sharing and collaboration*

The advantages of data sharing and collaboration encompass multiple facets, such as improved decision-making, accelerated innovation, and enhanced efficiency and productivity. Additionally, leveraging shared data enables organizations to harness predictive analytics and insights, address complex challenges more effectively, and empower data-driven policy and regulation. By fostering collaboration and sharing data resources, stakeholders can unlock the full potential of information exchange, driving positive outcomes across various sectors and promoting sustainable growth and development (Kshetri, 2017; Mukherjee, Singh, Mishra, & Bag, 2021).

- Improved Decision-making: Data sharing enables stakeholders to make more informed decisions, leading to better outcomes and reduced risks.
- Accelerated Innovation: Collaboration and data sharing foster innovation by combining diverse perspectives, knowledge, and resources. Shared data can serve as a catalyst for developing new solutions and advancing research and development efforts.
- Enhanced Efficiency and Productivity: Data sharing reduces redundancy and promotes efficient use of resources. By leveraging shared data, organizations can streamline processes, avoid duplication, and optimize resource allocation.
- Predictive Analytics and Insights: Shared data allows for the application of advanced analytics techniques, such as predictive modeling and machine learning. This enables stakeholders to derive valuable insights, make accurate forecasts, and anticipate trends or challenges.
- Addressing Complex Challenges: Data sharing and collaboration enable stakeholders to tackle complex challenges that require collective action. By combining expertise and resources, organizations can develop comprehensive strategies and interventions to address societal issues effectively.
- Empowering Data-driven Policy and Regulation: Shared data can inform evidence-based policymaking and regulation. Stakeholders can leverage

shared data to understand the impact of policies, evaluate outcomes, and make data-driven decisions that benefit society.

*d. Best practices for data sharing and collaboration*

The best practices for data sharing and collaboration include: establishing clear objectives and governance frameworks, ensuring data privacy and security, and promoting interoperability and standardization. Additionally, fostering a culture of collaboration, promoting data sharing incentives, and fostering data literacy and skills are crucial for successful collaboration initiatives. Embracing technology and tools that facilitate efficient data sharing and collaboration further enhances the effectiveness of these practices, enabling stakeholders to unlock the full potential of shared data resources (Zhao et al., 2019).

- Establish Clear Objectives and Governance Frameworks: Define the purpose and scope of data-sharing initiatives, establish clear governance structures, and develop data-sharing agreements that outline roles, responsibilities, and data access policies.
- Ensure Data Privacy and Security: Implement robust data protection measures, including encryption, access controls, and anonymization techniques, to safeguard privacy and mitigate security risks.
- Promote Interoperability and Standardization: Foster the use of common data formats, metadata standards, and interoperable systems to facilitate seamless data sharing and integration.
- Encourage a Culture of Collaboration: Foster a collaborative mindset among stakeholders through open communication, trust-building, and the recognition of shared goals and benefits.
- Promote Data Sharing Incentives: Provide incentives for data sharing, such as financial rewards, recognition, or access to additional resources or expertise, to encourage active participation.
- Foster Data Literacy and Skills: Enhance data literacy and skills among stakeholders to promote effective data sharing, interpretation, and utilization.
- Embrace Technology and Tools: Leverage emerging technologies, such as data sharing platforms, blockchain, and secure data exchange protocols, to facilitate efficient and secure data sharing.

Chapter 3

# Blockchain-based supply chain management

*From Seed to Plate: Unlocking Transparency and Efficiency in Agricultural Supply Chains* - Unknown


**Abstract**

This chapter focuses on the integration of blockchain technology in supply chain management within the agriculture industry. It provides a detailed examination of how blockchain can enhance supply chain operations, emphasizing the benefits of traceability, transparency, and accountability. The chapter begins by delving into the potential of blockchain technology to revolutionize supply chain management in agriculture. It explores the key features of blockchain, such as immutability and decentralization, which enable enhanced traceability and transparency throughout the supply chain. Furthermore, the chapter highlights the benefits of implementing blockchain-based supply chain management in the agriculture industry. It examines how traceability enabled by blockchain can improve food safety, quality control, and sustainability practices. The transparency provided by blockchain fosters trust among consumers, facilitates compliance with regulations, and enables fair trade practices. Additionally, blockchain's accountability mechanisms can help mitigate issues related to fraud, ethical sourcing, and supply chain disruptions.



**Keywords**: Blockchain technology, Supply chain management, Agriculture industry, Traceability, Transparency, Accountability, Food safety, Quality control, Sustainability practices, Fair trade practices


**Chapter contents**

- Supply chain challenges in agriculture and food production
- Traceability and provenance of agricultural products using blockchain
- Implementing smart contracts for transparent and efficient transactions

***

## 3.1. Supply chain challenges in agriculture and food production

The agricultural and food production supply chain is currently confronted with a host of critical challenges, including issues related to transparency, trust, complexity, efficiency, and authenticity. These challenges encompass various aspects of the supply chain, ranging from visibility and accountability to process efficiency and product authenticity. Addressing these challenges requires comprehensive strategies aimed at enhancing transparency, fostering trust, streamlining processes, and combating fraudulent activities throughout the supply chain (Blokdyk, 2021).

### 3.1.1. Lack of transparency and visibility

Lack of transparency and visibility emerge as formidable challenges plaguing the agricultural sector, hindering efficient operations and impeding stakeholders' ability to make informed decisions. Without clear visibility into various aspects of the supply chain, such as production processes, inventory levels, and market demand, agricultural businesses struggle to optimize resource allocation and respond effectively to fluctuations in demand. Addressing these challenges requires implementing robust systems for data collection, sharing, and analysis to enhance transparency and visibility throughout the agricultural value chain, ultimately improving efficiency and decision-making capabilities (Tomu, 2020).

*a. Consequences of limited transparency and visibility*

- Trust Deficit: When stakeholders lack access to relevant information and cannot verify the accuracy of data or actions, trust diminishes. This trust deficit can affect relationships between organizations, customers, employees, and other stakeholders, hindering collaboration and hindering business growth.
- Inefficiency and Ineffective Decision-making: Limited visibility into operations and processes can lead to inefficiencies and suboptimal decision-making. Without a comprehensive understanding of the entire value chain, organizations may struggle to identify bottlenecks, streamline processes, or make informed choices based on accurate and real-time information.
- Increased Risk and Compliance Challenges: Insufficient transparency increases the risk of non-compliance with regulations and internal policies. Organizations may face difficulties in identifying and addressing risks promptly, potentially resulting in legal and reputational consequences.
- Lack of Accountability: Without transparency, it becomes challenging to hold individuals or entities accountable for their actions. Accountability is

essential for maintaining ethical standards, ensuring responsible behavior, and building a culture of integrity.

- Missed Opportunities for Improvement: Limited visibility inhibits the identification of areas for improvement and innovation. Organizations may miss valuable insights, customer preferences, or emerging trends that could lead to competitive advantages or operational efficiencies.

*b. Causes of limited transparency and visibility*

- Fragmented Systems and Data Silos: Data stored in separate systems or departmental silos can impede transparency. The lack of integration between different systems and data sources makes it difficult to have a holistic view of operations and processes.
- Legacy Technologies and Manual Processes: Outdated technologies and reliance on manual processes hinder visibility. Manual data entry, paper-based documentation, or legacy systems lacking real-time data integration create information gaps and slow down decision-making.
- Lack of Data Standardization: Inconsistent data formats, naming conventions, and data quality issues make it challenging to integrate and analyze data effectively. Discrepancies in data interpretation can lead to misunderstandings and misaligned decisions.
- Information Asymmetry: Unequal access to information between stakeholders can contribute to a lack of transparency. Imbalances in information sharing can create power dynamics that hinder trust and collaboration.
- Cultural and Organizational Barriers: Organizational culture and resistance to change can impede transparency efforts. A lack of openness, fear of accountability, or hierarchical structures that restrict information flow can hinder transparency and visibility.

*c. Solutions to improve transparency and visibility*

- Integrated Systems and Data Sharing: Implementing integrated systems and breaking down data silos enable a unified view of operations. Centralized data repositories and data-sharing protocols facilitate transparency and collaboration across departments and stakeholders.
- Real-time Reporting and Analytics: Leveraging real-time reporting and analytics tools provides up-to-date insights into operations and processes. Real-time data integration and visualization dashboards enable quick and informed decision-making.

- Standardized Data and Documentation: Establishing data standardization protocols and documentation frameworks enhances visibility. Consistent data formats, naming conventions, and documentation practices ensure clarity and reduce interpretation errors.
- Stakeholder Engagement and Communication: Actively involving stakeholders in transparent and open communication fosters trust and promotes transparency. Clear and consistent communication channels allow for the exchange of information, feedback, and collaboration.
- Technology Adoption and Automation: Embracing emerging technologies such as blockchain, artificial intelligence, and automation can enhance transparency and visibility. These technologies offer secure and efficient ways to track and trace transactions, automate processes, and improve data accuracy.
- Regulatory Compliance and Ethical Practices: Adhering to regulations and ethical practices promotes transparency and accountability. Organizations should establish robust compliance programs, adhere to industry standards, and embrace responsible data management practices.
- Data Governance and Security: Implementing data governance frameworks ensures data integrity, privacy, and security. Data access controls, encryption, and regular audits protect sensitive information while promoting transparency and trust.
- Cultural Shift towards Transparency: Foster a culture that values transparency and open communication. Encourage leaders to lead by example, promote a learning environment, and reward transparency and ethical behavior.

### 3.1.2. Limited trust and accountability

*a. Implications of limited trust and accountability*

- Decreased Collaboration: When trust is lacking, collaboration among individuals or groups becomes challenging. Without a foundation of trust, people may be reluctant to share information, cooperate, or work towards common goals, hindering productivity and stifling innovation.
- Weakened Relationships: Limited trust erodes relationships between stakeholders, such as employees, customers, partners, and communities. Without trust, relationships become strained, and the willingness to engage in mutually beneficial interactions diminishes.
- Reduced Transparency: Limited trust often leads to a lack of transparency. When people are hesitant to share information or be open about their

actions, it becomes difficult to gain a comprehensive understanding of processes, decision-making, and outcomes.

- Impaired Decision-making: Without trust and accountability, decision-making processes may suffer. People may hesitate to take ownership of decisions or rely on inaccurate or incomplete information, resulting in suboptimal choices and missed opportunities.
- Reputational Damage: Limited trust and accountability can tarnish an organization's reputation. When stakeholders perceive a lack of trustworthiness, it can lead to negative public perception, reduced customer loyalty, and potential financial repercussions.

*b. Causes of limited trust and accountability*

- Lack of Communication: Inadequate or ineffective communication contributes to limited trust and accountability. When information is withheld, miscommunicated, or poorly disseminated, trust erodes, and accountability is compromised.
- Unethical Behavior: Unethical practices undermine trust and accountability. When individuals or organizations act in a manner that goes against moral or ethical principles, it erodes trust and creates skepticism about accountability.
- Inconsistent Actions and Promises: Inconsistencies between words and actions erode trust. When promises are broken, commitments are not fulfilled, or there is a lack of follow-through, stakeholders lose faith in the accountability of others.
- Lack of Transparency: Insufficient transparency in decision-making processes and operations can create skepticism and hinder trust-building efforts. When stakeholders perceive that information is being withheld or manipulated, trust diminishes.
- Past Negative Experiences: Previous instances of betrayal, dishonesty, or unethical behavior can create long-lasting effects on trust. Negative experiences linger in people's minds, making it difficult to rebuild trust and establish accountability.

*c. Strategies to enhance trust and foster accountability*

- Open and Transparent Communication: Foster a culture of open and transparent communication, where information flows freely and stakeholders are encouraged to voice their opinions and concerns. Establish effective channels for communication and ensure that information is shared promptly and accurately.

- Lead by Example: Leaders play a crucial role in establishing trust and accountability. Demonstrate integrity, ethical behavior, and accountability in all actions. When leaders set a positive example, it cascades throughout the organization and influences the behavior of others.
- Establish Clear Expectations and Responsibilities: Define clear expectations, roles, and responsibilities for individuals and teams. Ensure that everyone understands their obligations and is aware of the consequences of their actions or inactions.
- Promote Collaboration and Teamwork: Foster a collaborative environment that encourages teamwork and cooperation. Establish cross-functional teams, promote knowledge-sharing, and reward collaborative efforts to build trust and accountability among team members.
- Encourage Feedback and Learning: Create mechanisms for feedback and continuous learning. Encourage individuals to provide constructive feedback, engage in self-reflection, and support ongoing professional development to foster a culture of accountability and growth.
- Hold People Accountable: Establish mechanisms for holding individuals accountable for their actions. Implement performance evaluation systems, establish clear metrics and targets, and ensure consequences for unethical or irresponsible behavior.
- Build Trust in Relationships: Invest time and effort in building trust with stakeholders. Foster relationships based on transparency, reliability, and mutual respect. Consistently deliver on commitments, address concerns promptly, and engage in open dialogue to nurture trust.
- Embrace Ethical Practices: Emphasise the importance of ethics in decision-making and operations. Establish a code of conduct and ethics policies, provide training on ethical practices, and hold individuals accountable for upholding these standards.

### 3.1.3. Complexity and fragmentation

*a. Implications of complexity and fragmentation*

- Operational Inefficiencies: Complexity and fragmentation can lead to operational inefficiencies. As organizations grow and diversify, their processes, systems, and structures often become more intricate, making it challenging to streamline operations and achieve optimal productivity.
- Communication Breakdowns: Complexity and fragmentation can hinder effective communication. When information flows across multiple channels,

departments, or geographies, there is a higher risk of miscommunication, information gaps, and misunderstandings, leading to errors and delays.

- Decision-making Challenges: Complex and fragmented environments make decision-making more difficult. With numerous interdependencies, diverse stakeholders, and intricate systems, decision-makers may struggle to access timely and accurate information resulting in delayed or suboptimal decisions.
- Reduced Agility and Adaptability: Complexity and fragmentation can impede an organization's agility and adaptability. In rapidly changing markets, organizations need to respond quickly to emerging trends and opportunities. However, navigating complex and fragmented structures can slow down decision-making and hinder agility.
- Customer Dissatisfaction: Complexity and fragmentation can negatively impact the customer experience. Customers expect seamless interactions and consistent service across channels, but when complexity and fragmentation exist internally, it often translates into a disjointed and inconsistent experience for customers.

*b. Causes of complexity and fragmentation*

- Organizational Growth and Expansion: As organizations grow, they often face increased complexity due to expanding operations, diversifying product lines, and entering new markets. Growth brings new challenges related to managing larger teams, coordinating processes, and integrating disparate systems.
- Technological Advancements: While technology has the potential to streamline operations, it can also contribute to complexity and fragmentation. Organizations may adopt multiple systems, platforms, and tools that do not integrate seamlessly, resulting in data silos and disjointed processes.
- Globalisation and Supply Chain Complexity: Globalisation has brought numerous benefits, but it has also introduced complexities in supply chain management. Organizations must navigate diverse regulatory frameworks, cultural differences, and complex logistics networks, leading to increased fragmentation.
- Mergers and Acquisitions: Mergers and acquisitions often result in complex organizational structures and systems. Integrating different cultures, processes, and technologies can be challenging, leading to fragmentation and inefficiencies if not managed effectively.
- Lack of Standardization: Inconsistent processes, data formats, and documentation across departments or business units can contribute to

complexity and fragmentation. Without standardized practices and clear guidelines, coordination and collaboration become difficult.

*c. Strategies to navigate complexity and fragmentation*

- Streamline Processes: Evaluate existing processes and identify opportunities for streamlining and simplification. Eliminate unnecessary steps, automate manual tasks, and promote cross-functional collaboration to reduce complexity.
- Enhance Communication and Collaboration: Foster a culture of open communication and collaboration. Encourage knowledge-sharing, provide platforms for cross-departmental collaboration, and leverage technology to facilitate communication and information sharing.
- Invest in Integrated Systems: Implement integrated systems and technologies that promote data transparency and streamline operations. Centralized data repositories, enterprise resource planning (ERP) systems, and collaborative tools can help consolidate information and facilitate seamless communication.
- Develop Agile Decision-making Processes: Establish agile decision-making frameworks that enable quick, data-driven decisions. Empower decision-makers with timely and accurate information, foster collaboration in decision-making processes, and encourage experimentation and learning from failures.
- Standardize Processes and Documentation: Implement standardized processes, data formats, and documentation practices across the organization. Develop clear guidelines and training programs to ensure consistency and reduce fragmentation.
- Simplify the Customer Experience: Focus on simplifying the customer journey and ensuring a consistent experience across touchpoints. Remove unnecessary steps, integrate systems to enable seamless interactions, and proactively address customer pain points.
- Embrace Change Management: Effectively managing organizational change is crucial in navigating complexity and fragmentation. Provide training and support to employees, involve stakeholders in decision-making, and communicate the benefits of change to foster buy-in and minimize resistance.
- Continuous Improvement and Learning: Cultivate a culture of continuous improvement and learning. Encourage employees to identify and address inefficiencies, promote innovation, and provide opportunities for professional development to enhance skills in navigating complexity.

### 3.1.4. Inefficient processes and paperwork

*a. Implications of inefficient processes and paperwork*

- Reduced Productivity: Inefficient processes and excessive paperwork consume valuable time and resources, leading to decreased productivity. Employees spend significant time on manual, repetitive tasks or navigating convoluted processes, diverting their focus from more value-added activities.
- Increased Errors and Delays: Inefficient processes and paperwork contribute to a higher risk of errors and delays. When tasks are not streamlined or automated, human error becomes more prevalent, resulting in mistakes, rework, and potential bottlenecks that can disrupt workflow and customer satisfaction.
- Higher Operational Costs: Inefficiency in processes and paperwork can drive up operational costs. Inefficient processes often require more time, effort, and resources to complete tasks, leading to increased labor costs, wastage of materials, and higher administrative expenses.
- Employee Frustration and Burnout: Cumbersome processes and excessive paperwork can negatively impact employee morale and well-being. When employees feel overwhelmed by manual tasks, redundant paperwork, or unnecessary bureaucracy, their job satisfaction and motivation may decline, leading to increased stress and burnout.
- Customer Dissatisfaction: Inefficient processes and paperwork can impact the customer experience. Delays, errors, and lack of responsiveness due to inefficient processes or excessive paperwork can frustrate customers, erode trust, and damage the organization's reputation.

*b. Causes of inefficient processes and paperwork*

- Lack of Automation: Manual, paper-based processes that rely on manual data entry and approvals contribute to inefficiency. Without automation, tasks take longer to complete, and the risk of errors and delays increases.
- Redundant and Non-value-added Steps: Processes that include redundant or non-value-added steps can lead to inefficiency. When steps are not thoroughly examined and optimized, unnecessary tasks and handoffs can accumulate, slowing down the overall process.
- Lack of Standardization: Inconsistent processes across departments or business units can contribute to inefficiency. When different teams follow their own procedures or use disparate systems, it becomes challenging to streamline operations and achieve consistency.

- Complex Approval Processes: Complex approval processes, excessive layers of bureaucracy, and lengthy decision-making cycles can impede efficiency. Excessive reviews, unnecessary sign-offs, and multiple handoffs create bottlenecks and hinder timely progress.
- Paper-intensive Workflows: Organizations that heavily rely on paper-based workflows face inefficiencies related to physical storage, retrieval, and transportation of documents. Manual handling of paperwork can slow down processes and increase the risk of errors or misplacement.

*c. Strategies to improve efficiency and reduce paperwork*

- Process Optimization: Conduct a thorough analysis of existing processes to identify inefficiencies and areas for improvement. Streamline workflows, eliminate redundant steps, and leverage technology to automate manual tasks and approvals.
- Standardization and Simplification: Establish standardized processes across the organization. Develop clear guidelines, document procedures, and provide training to ensure consistency and reduce variations that can contribute to inefficiencies.
- Automation and Digitisation: Embrace automation and digitization to streamline processes and reduce paperwork. Implement digital document management systems, electronic signatures, and workflow automation tools to reduce manual effort and enhance efficiency.
- Simplify Approval Processes: Simplify approval processes by reducing unnecessary steps, clarifying decision-making authority, and leveraging technology for streamlined approvals. Empower employees to make decisions at appropriate levels, reducing the need for excessive reviews.
- Training and Skill Development: Provide training and skill development opportunities for employees to enhance their efficiency and adapt to new technologies and processes. Encourage a culture of continuous improvement and empower employees to identify and suggest process improvements.
- Collaborative Platforms and Tools: Implement collaborative platforms and project management tools that facilitate communication, coordination, and task tracking. Centralized platforms enable efficient collaboration, reducing the need for excessive paperwork and email exchanges.
- Regular Process Reviews: Conduct regular reviews of processes to identify bottlenecks, inefficiencies, and emerging issues. Engage employees in the review process, seek their feedback, and implement improvements based on their insights and suggestions.

- Lean and Six Sigma Principles: Apply Lean and Six Sigma principles to identify and eliminate process waste, reduce errors, and improve efficiency. Adopting these methodologies can help organizations identify root causes of inefficiency and implement-targeted improvements.

### 3.1.5. Counterfeit and fraudulent products

*a. Implications of counterfeit and fraudulent products*

- Economic Impact: Counterfeit and fraudulent products undermine legitimate businesses by diverting sales and revenue from genuine products. This illicit market deprives companies of their rightful profits, reduces tax revenues, and disrupts supply chains, leading to job losses and hindering economic growth.
- Consumer Health and Safety: Counterfeit products, particularly in sectors such as pharmaceuticals, electronics, and automotive parts, pose serious health and safety risks to consumers. These products often fail to meet quality standards, contain harmful substances, or lack proper labeling, putting consumers at risk of injury or illness.
- Brand Reputation Damage: Counterfeit products can tarnish the reputation of established brands. Poor-quality replicas or fraudulent products bearing a reputable brand's name erode consumer trust and loyalty. The negative association with counterfeit goods can have long-lasting effects on a brand's image and customer perception.
- Intellectual Property Infringement: Counterfeit products infringe upon intellectual property rights, including trademarks, patents, and copyrights. This undermines innovation and discourages companies from investing in research and development, leading to a decline in competitiveness and creativity.
- Criminal Activity and Organised Crime: Counterfeit trade often involves organized criminal networks engaged in illicit activities such as money laundering, human trafficking, and drug smuggling. The profits generated from counterfeit operations fund these criminal enterprises, perpetuating a cycle of illicit activities.

*b. Causes of counterfeit and fraudulent products*

- Global Supply Chain Complexity: The complexity of global supply chains presents opportunities for counterfeiters to infiltrate legitimate distribution networks. With numerous intermediaries involved, it becomes challenging to ensure the authenticity and integrity of products throughout the supply chain.

- Lack of Regulatory Oversight: Inadequate regulatory oversight and enforcement create loopholes that counterfeiters exploit. Insufficient resources, weak intellectual property protection laws, and corruption contribute to a favorable environment for counterfeit operations to thrive.
- Advancements in Technology: Technological advancements have made it easier for counterfeiters to replicate products and deceive consumers. Sophisticated printing, packaging, and manufacturing techniques, coupled with online marketplaces, enable the production and distribution of counterfeit goods on a global scale.
- Consumer Demand and Perception: The demand for cheap or discounted products, coupled with consumers' lack of awareness or indifference towards counterfeit goods, drives the market for fraudulent products. Consumers may unknowingly purchase counterfeit goods or choose them based on perceived value, contributing to the counterfeit industry's growth.
- Limited Collaboration and Information Sharing: Insufficient collaboration and information sharing among stakeholders, including governments, law enforcement agencies, and businesses, impede the effective detection and prevention of counterfeit activities. The lack of a coordinated approach allows counterfeiters to exploit gaps and continue their illicit operations.

*c. Strategies to combat counterfeit and fraudulent products*

- Strengthen Intellectual Property Rights: Governments should enact and enforce robust intellectual property laws and regulations to protect businesses' intellectual property rights. This includes implementing effective mechanisms for trademarks, patents, and copyrights, and imposing severe penalties for infringers.
- Enhanced Supply Chain Traceability: Implement supply chain traceability systems to track and authenticate products from their origin to the end consumer. Technologies such as blockchain, RFID, and unique product identifiers can provide visibility and accountability, making it easier to identify and eliminate counterfeit products.
- Collaboration and Information Sharing: Foster collaboration among government agencies, law enforcement, industry associations, and technology providers to share information, intelligence, and best practices. Establish platforms for reporting and sharing counterfeit incidents and coordinate efforts to disrupt counterfeit networks.
- Consumer Education and Awareness: Educate consumers about the risks and consequences of purchasing counterfeit products. Promote awareness

campaigns highlighting the importance of buying from authorized retailers and recognizing the signs of counterfeit goods.

- Proactive Monitoring and Enforcement: Increase the resources allocated to monitoring and enforcement activities targeting counterfeit operations. Invest in advanced technologies, train law enforcement personnel, and conduct regular inspections to identify and seize counterfeit products.
- Stronger E-commerce Regulations: Strengthen regulations and oversight for e-commerce platforms to prevent the sale of counterfeit goods. Implement stringent seller verification processes, counterfeit reporting mechanisms, and proactive monitoring of online marketplaces to detect and remove fraudulent listings.
- International Cooperation: Foster international cooperation and collaboration among governments, organizations, and law enforcement agencies to combat cross-border counterfeit activities. Share intelligence, harmonize legal frameworks, and collaborate on joint enforcement operations to disrupt counterfeit supply chains.

## 3.2. Traceability and provenance of agricultural products using blockchain

Blockchain technology enhances the transparency of agricultural products and the food supply chain, bolstering consumer trust and safety, as highlighted by FAO (2019) and Katsikouli, Wilde, Dragoni, and Høgh-Jensen (2021). By providing an immutable ledger of transactions and data, blockchain ensures that every step in the production and distribution process is recorded and accessible. This not only fosters greater accountability but also enables consumers to make more informed choices about the products they purchase and consume.

### 3.2.1. The role of blockchain in traceability

Blockchain technology serves as a cornerstone in enhancing traceability within various industries, including agriculture, as emphasized by Antonucci et al. (2019), Cottrill et al. (2018), Kamilaris et al. (2019) and O'Leary (2017). By providing an immutable and transparent ledger, blockchain facilitates the tracking of products throughout the supply chain, ensuring greater accountability and authenticity. However, despite its potential benefits, blockchain encounters several challenges that warrant careful consideration.

*a. Understanding blockchain technology*

- Decentralized Ledger: Blockchain is a distributed ledger that records transactions across multiple nodes or computers. Each transaction, or "block," is linked to the previous one, forming a chronological chain of information.

- Transparency and Immutability: Once a transaction is recorded on the blockchain, it becomes permanent and cannot be altered. This immutability ensures data integrity and enables transparency, as all participants can access the same information.
- Consensus Mechanism: Blockchain relies on a consensus mechanism, where network participants agree on the validity of transactions. This consensus ensures that all participants have a synchronized and consistent view of the blockchain's state.

*b. Benefits of blockchain in traceability*

- Enhanced Transparency: Blockchain provides a transparent and tamper-proof record of transactions. Each step in the supply chain can be recorded on the blockchain, enabling stakeholders to trace the origin, movement, and transformation of products in real time.
- Improved Trust and Accountability: The immutability and transparency of blockchain build trust among stakeholders. They can verify the authenticity and integrity of data without relying on a centralized authority, fostering accountability and reducing the risk of fraud or manipulation.
- Efficient and Accurate Data Management: Blockchain eliminates the need for intermediaries and manual record-keeping, streamlining data management. Smart contracts can automate processes, trigger actions, and ensure compliance, reducing errors, delays, and paperwork.
- Supply Chain Visibility: Blockchain enables end-to-end visibility across complex supply chains. Participants can track and verify the movement of goods, validate certifications and compliance, and gain insights into inventory levels and product conditions in real time.
- Rapid Traceability and Recall Management: In the event of a product recall or quality issue, blockchain enables swift traceability. Stakeholders can quickly identify affected products, trace their origin, and take appropriate action, minimizing the impact on consumer safety and brand reputation.

*c. Applications of blockchain in traceability*

- Food and Agriculture: Blockchain can track the journey of food products from farm to fork, ensuring transparency, verifying certifications (e.g., organic or fair trade), and facilitating food safety recalls.
- Pharmaceuticals and Healthcare: Blockchain can track the provenance of pharmaceuticals, ensure the authenticity of drugs, prevent counterfeit

products, and enable secure sharing of patient records while preserving privacy.

- Luxury Goods and Counterfeit Prevention: Blockchain can verify the authenticity of luxury goods by recording their unique characteristics and ownership history, reducing the prevalence of counterfeit products.
- Ethical Sourcing and Sustainability: Blockchain can certify the ethical sourcing of materials, such as conflict-free minerals or sustainably harvested timber, providing consumers with verifiable information and promoting responsible practices.
- Supply Chain Finance: Blockchain can streamline supply chain finance by automating and verifying transactions, improving trust between trading partners, and reducing the risk of fraud.

*d. Challenges and considerations*

- Integration with Existing Systems: Integrating blockchain with existing systems and legacy infrastructure can be complex and require collaboration among stakeholders.
- Scalability and Speed: Blockchain technology still faces scalability and speed limitations. As more transactions are added to the blockchain, it can affect performance and transaction processing times.
- Data Privacy and Security: While blockchain ensures data integrity, privacy concerns arise when sensitive information is stored on a public blockchain. Hybrid or private blockchain solutions may be required for specific use cases.
- Governance and Standards: Establishing governance frameworks and industry-wide standards for blockchain implementation are essential to ensure interoperability, consistency, and trust among participants.

### 3.2.2. Transparent and immutable record-keeping

Transparent and immutable record-keeping is fundamental to enhancing trust and accountability throughout the agriculture and food supply chain, as emphasized by Chang, Chen, and Lu (2019) and Saurabh and Dey (2021). By providing a clear and unalterable trail of information, these record-keeping systems enable stakeholders to trace the journey of products from farm to fork, ensuring safety and quality assurance. Moreover, the utilization of such technology facilitates data-driven decision-making and process optimization, ultimately contributing to a more efficient and resilient supply chain ecosystem.

*a. Importance of transparent and immutable record-keeping*

- Trust and Accountability: Transparent and immutable record-keeping builds trust among stakeholders by providing a verifiable and auditable trail of information. It holds parties accountable for their actions and enables effective governance and compliance.
- Data Integrity and Accuracy: Transparent and immutable records ensure the integrity and accuracy of data. It prevents unauthorized changes, tampering, or manipulation of information, maintaining its reliability over time.
- Legal and Regulatory Compliance: Many industries have strict legal and regulatory requirements for record-keeping. Transparent and immutable records help organizations demonstrate compliance, mitigate risks, and facilitate audits or investigations.
- Decision-making and Analytics: Reliable and transparent records serve as a foundation for informed decision-making and data analytics. Accurate historical data enables organizations to identify trends, make predictions, and derive valuable insights.

*b. Benefits of transparent and immutable record-keeping*

- Data Integrity: Immutable records ensure that once data is recorded, it cannot be altered or deleted without leaving a trace. This integrity guarantees the authenticity and reliability of information, reducing the risk of fraud or disputes.
- Transparency and Auditability: Transparent records provide visibility into the entire lifecycle of data, allowing stakeholders to trace its origin, changes, and access history. This transparency fosters trust and enables effective auditing or compliance monitoring.
- Efficiency and Time Savings: Immutable record-keeping eliminates the need for manual reconciliation, verification, and validation processes. It reduces administrative overhead, improves operational efficiency, and saves time in data management and reporting.
- Security and Resilience: Immutable records, protected by cryptographic algorithms, enhance data security. The decentralized nature of some technologies ensures redundancy and resilience against single points of failure or cyber-attacks.

*c. Technologies enabling transparent and immutable record-keeping*

- Blockchain: Blockchain technology, with its decentralized and tamper-resistant nature, enables transparent and immutable record-keeping.

Each transaction is recorded as a block, linked to the previous one, forming a chain of information. Once recorded, data cannot be altered without consensus from network participants.

- Distributed Ledger Technology (DLT): DLT, similar to blockchain, provides transparent and immutable record-keeping. It distributes and replicates data across multiple nodes, ensuring consensus and data consistency.
- Cryptography: Cryptographic techniques, such as hash functions and digital signatures, play a crucial role in ensuring data integrity and authenticity. These techniques secure records and provide mathematical proof of their integrity.
- Digital Timestamping: Digital timestamping adds an additional layer of trust and accountability to records. It enables the verification of when specific information was created or modified, preventing retroactive changes.

*d. Considerations for implementation*

- Data Governance: Establishing clear data governance policies and procedures is essential for transparent and immutable record-keeping. It ensures data quality, access controls, retention periods, and compliance with applicable regulations.
- Collaboration and Interoperability: Collaboration among stakeholders is crucial to establish standardized formats, protocols, and interoperable systems. This collaboration enables seamless data exchange and ensures transparency across organizational boundaries.
- Scalability and Performance: Implementing transparent and immutable record-keeping technologies at scale requires careful consideration of performance and scalability. As the volume of data increases, organizations must ensure that the chosen technology can handle the load efficiently.
- Privacy and Data Protection: Balancing transparency with privacy is important when implementing transparent and immutable record-keeping. Organizations must consider the appropriate level of data anonymization or encryption to protect sensitive information.

### 3.2.3. Tracking the journey of agricultural products

Tracking the journey of agricultural products is indispensable within the agriculture and food supply chain. Nevertheless, this endeavor encounters numerous challenges that necessitate the adoption of suitable technologies and approaches, as underscored by Antonucci et al. (2019) Bumblauskas,

Mann, Dugan, and Rittmer (2020), Motta, Tekinerdogan, and Athanasiadis (2020), Pandey, Pant, and Snasel (2022) and van Wassenaer et al. (2023).

*a. Significance of tracking agricultural products*

- Food Safety: Tracking the journey of agricultural products is crucial for ensuring food safety. By tracing the origin and movement of products, stakeholders can identify potential contamination points, respond swiftly to foodborne illness outbreaks, and implement targeted recalls if necessary.
- Quality Assurance: Tracking allows for quality assurance throughout the supply chain. It enables stakeholders to monitor and verify adherence to quality standards, certifications, and production practices, ensuring that consumers receive safe and high-quality agricultural products.
- Regulatory Compliance: Many countries have stringent regulations regarding the traceability of agricultural products. Accurate tracking facilitates compliance with these regulations, including labeling requirements, country of origin documentation, and the verification of organic or sustainable certifications.
- Sustainability and Ethical Sourcing: Tracking agricultural products supports sustainability and ethical sourcing initiatives. It enables the identification of products produced under sustainable practices, fair trade principles, or specific geographical indications, providing consumers with transparent information about the products they purchase.

*b. Challenges in tracking agricultural products*

- Complex Supply Chains: Agricultural supply chains often involve numerous intermediaries, making it challenging to trace products accurately. Products may change hands multiple times, undergo processing or mixing, and be subject to diverse transportation routes, increasing the complexity of tracking.
- Data Standardization: The lack of standardized data formats and interoperable systems across stakeholders hinders effective tracking. Incompatible information systems and varying data collection practices create data silos, limiting the visibility and accuracy of product information.
- Limited Technology Adoption: Some agricultural sectors, particularly small-scale producers or regions with limited resources, may face barriers to adopting advanced tracking technologies. Cost, technical expertise, and infrastructure limitations can impede the implementation of traceability systems.

- Counterfeit and Fraudulent Products: The prevalence of counterfeit and fraudulent agricultural products poses challenges to tracking efforts. Illicit actors may exploit gaps in traceability systems, falsify documentation, or engage in mislabeling, making it difficult to accurately trace the journey of products.

*c. Technologies and approaches for effective traceability*

- Blockchain Technology: Blockchain offers decentralized and transparent record-keeping, enabling secure and immutable tracking of agricultural products. Each step in the supply chain can be recorded on the blockchain, providing stakeholders with a shared, tamper-proof ledger.
- Internet of Things (IoT): IoT devices, such as sensors and RFID tags, can be used to collect and transmit data in real time. This data, combined with cloud computing and analytics, enables accurate tracking of products, temperature monitoring, and quality control.
- Barcode and QR Code Systems: Barcode and QR code systems provide a cost-effective solution for tracking agricultural products. Each product can be assigned a unique code that contains relevant information, allowing stakeholders to scan and retrieve data throughout the supply chain.
- Traceability Platforms and Software: Traceability platforms and software provide end-to-end visibility and data management capabilities. These platforms enable the collection, integration, and analysis of data from various sources, facilitating efficient tracking and traceability reporting.

*d. Collaborative approaches and industry initiatives*

- Public-private Partnerships: Collaborative efforts between governments, industry associations, and private sector entities can drive the adoption of traceability solutions. These partnerships can foster standardization, knowledge sharing, and financial support to overcome barriers to implementation.
- Certification and Verification Programs: Certification schemes and verification programs help ensure the integrity of traceability systems. They provide independent validation of product claims, such as organic or fair trade certifications, adding credibility and trust to the tracked journey of agricultural products.
- Supply Chain Collaboration: Collaboration among supply chain stakeholders, including producers, processors, distributors, and retailers, is crucial for effective traceability. Sharing information, adopting common standards,

and integrating systems promote end-to-end visibility and traceability across the entire supply chain.

### 3.2.4. Verification of certifications and quality standards

*a. Significance of certification verification*

- Ensuring Product Integrity: Verification of certifications and quality standards ensures that products meet specific criteria and adhere to established benchmarks. It provides assurance to consumers that the products they purchase meet the claimed standards, such as organic, fair trade, or sustainable certifications.
- Trust and Credibility: Verification enhances trust and credibility in the marketplace. It allows consumers to make informed decisions based on reliable information, reinforcing their confidence in the products and the companies that produce them.
- Regulatory Compliance: In various industries, compliance with regulatory requirements is essential. Verification ensures that businesses meet the necessary standards, regulations, and legal obligations, avoiding penalties or reputational damage.
- Supply Chain Transparency: Verification processes contribute to the transparency of supply chains. By verifying certifications and quality standards, stakeholders can trace the origin and production practices of products, promoting ethical sourcing and sustainability.

*b. Challenges in certification verification*

- Complex Supply Chains: Global supply chains often involve multiple stakeholders, making it challenging to verify certifications consistently. Products may undergo various transformations and pass through several intermediaries, requiring effective coordination and documentation.
- Verification Resources and Expertise: Verification processes require sufficient resources and expertise to ensure accuracy and reliability. Some certifications demand specialized knowledge, training, and equipment, which can be costly or difficult to access, particularly for small-scale businesses.
- Counterfeit Certifications: The proliferation of counterfeit certifications poses a significant challenge to verification efforts. Illicit actors may falsify or misuse certifications, making it crucial to implement robust authentication measures to distinguish genuine certifications from fraudulent ones.
- Standardization and Harmonisation: Certification systems and quality standards vary across industries, regions, and countries. Achieving

standardization and harmonization is essential to streamline verification processes, minimize complexity, and facilitate cross-border trade.

*c. Strategies and technologies for effective verification*

- Accreditation and Auditing: Accredited certification bodies and independent auditors play a vital role in the verification processes. They assess compliance with standards, conduct on-site inspections, and issue certificates, ensuring impartiality and credibility.
- Documentation and Record-keeping: Effective documentation and record-keeping are fundamental to certification verification. Clear and organized records enable traceability and allow auditors to review and validate the necessary documentation for compliance.
- Technology-enabled Verification: Technology plays a crucial role in streamlining and enhancing verification processes. Digital platforms, blockchain, and distributed ledger technology (DLT) can provide transparent and tamper-proof records of certifications, enabling efficient verification across supply chains.
- Data Sharing and Collaboration: Collaboration among stakeholders, such as manufacturers, suppliers, certification bodies, and regulatory authorities, facilitates verification processes. Sharing relevant data and information helps ensure the accuracy and validity of certifications.

*d. Best Practices for certification verification*

- Clear Communication: Transparent communication about certifications and quality standards is essential. Clearly conveying the scope, criteria, and implications of certifications to consumers and stakeholders fosters understanding and trust.
- Independent Third-party Verification: Independent third-party verification adds credibility to certifications. Involving trusted and accredited certification bodies or auditors helps establish confidence in the verification process.
- Periodic Reviews and Audits: Regular reviews and audits of certifications contribute to ongoing compliance and quality assurance. Scheduled and surprise inspections, along with data analysis, help maintain the integrity of certifications over time.
- Continuous Improvement: Certification verification should be a continuous improvement process. Regular evaluation, feedback loops, and updates to verification methodologies ensure the relevance and effectiveness of verification procedures.

### 3.2.5. Enhancing consumer trust and safety

Ensuring consumer trust and safety is paramount within the agriculture and food supply chain. However, this endeavor encounters various obstacles that demand the implementation of appropriate technologies and approaches, including fostering collaboration and partnerships, as advocated by Stranieri, Riccardi, Meuwissen, and Soregaroli (2021). Overcoming these hurdles requires a concerted effort from stakeholders to integrate innovative technologies, enhance transparency, and prioritize consumer-centric practices, thereby bolstering trust and safety throughout the supply chain.

*a. Importance of consumer trust and safety*

- Brand Reputation: Consumer trust is closely tied to brand reputation. Establishing trust in a brand builds customer loyalty, positive word-of-mouth, and repeat business. Conversely, a lack of trust can lead to reputational damage and customer attrition.
- Consumer Confidence: Trust and safety contribute to consumer confidence in the products or services they choose. When consumers feel safe and trust a brand, they are more likely to engage in transactions, make purchases, and recommend the brand to others.
- Regulatory Compliance: Compliance with regulatory standards and safety requirements is essential to ensure consumer well-being. Adhering to safety regulations demonstrates a commitment to consumer protection and helps avoid legal consequences.
- Market Differentiation: Trust and safety can be key differentiators in a competitive market. Brands that prioritize consumer trust and safety have a unique selling proposition that sets them apart from competitors and attracts discerning consumers.

*b. Challenges in enhancing consumer trust and safety*

- Product Quality and Safety: Ensuring the quality and safety of products presents challenges for businesses, particularly when dealing with complex supply chains, diverse product portfolios, and evolving safety standards.
- Fraud and Counterfeit Products: The proliferation of counterfeit products and fraudulent practices undermines consumer trust and compromises safety. Businesses need to implement robust authentication measures and anti-counterfeiting strategies to protect consumers.
- Data Privacy and Security: The increasing digitization of consumer interactions raises concerns about data privacy and security. Businesses

must prioritize safeguarding consumer information and communicate transparently about data usage and protection.

- Crisis and Risk Management: Unexpected crises or safety incidents can quickly erode consumer trust. Organizations must be prepared to effectively manage and communicate during crises, demonstrating their commitment to consumer safety.

*c. Strategies and approaches to enhance consumer trust and safety*

- Product Testing and Certification: Independent product testing and certification by reputable organizations build trust by verifying product quality, safety, and compliance with standards. Displaying certifications and labels reassures consumers about the safety and reliability of products.
- Transparent Communication: Open and transparent communication with consumers is essential. Providing clear information about product ingredients, manufacturing processes, safety measures, and potential risks fosters trust and empowers consumers to make informed choices.
- Customer Feedback and Reviews: Actively listening to customer feedback and responding to concerns and inquiries shows that a brand values consumer opinions and prioritizes their safety. Encouraging and addressing customer reviews and inquiries builds trust and helps identify areas for improvement.
- Supply Chain Visibility and Traceability: Ensuring transparency and traceability throughout the supply chain enhances consumer trust. Implementing technologies like blockchain, IoT, and traceability systems allows consumers to track the journey of products, verify authenticity, and ensure safety.

*d. Collaboration and partnerships*

- Industry Collaboration: Collaboration among industry stakeholders, including businesses, regulatory bodies, and consumer advocacy groups, fosters shared responsibility in enhancing consumer trust and safety. Collaborative initiatives can develop industry-wide standards, best practices, and promote self-regulation.
- Government and Regulatory Support: Governments play a crucial role in setting safety regulations, enforcing compliance, and protecting consumer rights. Collaboration with regulatory bodies and active engagement in shaping policies contribute to consumer trust and safety.
- Consumer Education and Empowerment: Educating and empowering consumers about their rights, safety precautions, and how to identify

trustworthy products and services enhances their trust and safety awareness. Consumer education initiatives promote responsible consumption and informed decision-making.

- Continuous Improvement and Adaptation: Consumer trust and safety are ongoing commitments. Regularly reviewing and improving safety protocols, staying updated on industry best practices, and adapting to changing consumer expectations contribute to maintaining and strengthening trust.

## 3.3. Implementing smart contracts for transparent and efficient transactions

Smart contracts are indispensable in the realm of agriculture and the food supply chain, serving as a pivotal tool for ensuring transparency and accountability. These contracts not only facilitate efficient payment and settlement processes but also contribute to streamlining operations. Moreover, their implementation significantly mitigates the risks associated with fraud and disputes, as noted by Dutta, Choi, Somani, and Butala (2020).

### 3.3.1. Introduction to smart contracts

A smart contract is a self-executing agreement with the terms of the contract written directly into code. These contracts are built on blockchain technology, a distributed ledger that ensures transparency, security, and immutability. Unlike traditional contracts, which rely on intermediaries such as lawyers or brokers to enforce the terms, smart contracts automatically execute predefined actions once the specified conditions are met. This automation eliminates the need for intermediaries and enables trustless interactions between parties (Bhardwaj et al., 2021; Giancaspro, 2017; Triana Casallas, Cueva Lovelle, & Rodríguez Molano, 2020).

One of the key features of smart contracts is their ability to facilitate peer-to-peer transactions without the need for a centralized authority. By leveraging blockchain's decentralized nature, smart contracts provide a level playing field for individuals and businesses, enabling direct interactions and reducing costs associated with intermediaries. This decentralized approach has the potential to democratize access to various services, such as financial transactions, supply chain management, intellectual property protection, and more.

The underlying code of a smart contract is typically written in a programming language specifically designed for this purpose. Ethereum, the most well-known blockchain platform for smart contracts, introduced the Solidity programming language, which allows developers to create complex agreements with ease. Other blockchain platforms, such as Cardano and Polkadot, have also emerged, offering alternative programming languages and innovative features to enhance the capabilities of smart contracts.

Smart contracts operate based on a set of rules defined by the involved parties. These rules determine the conditions under which the contract should execute specific actions. For example, in a simple escrow agreement, the smart contract could release funds to the seller once the buyer confirms receipt of the goods. By automating this process, smart contracts provide efficiency, transparency, and reduce the risk of fraud.

However, it's important to note that while smart contracts offer numerous advantages, they are not without limitations. The immutability of blockchain means that once a smart contract is deployed, it cannot be easily modified or revoked. This feature, although essential for security, requires careful consideration during the development phase to ensure that the contract's logic is flawless. Furthermore, smart contracts can only interact with data and systems that are part of the blockchain network on which they reside, limiting their scope to the digital realm.

### 3.3.2. Automating supply chain transactions

Traditionally, supply chain transactions have relied heavily on manual processes, involving substantial paperwork, human intervention, and time-consuming communication between stakeholders. However, with the rapid advancement of technology and the rise of innovative digital solutions, the automation of supply chain transactions has become a transformative force, revolutionizing the way businesses operate and interact within the supply chain ecosystem. Through automation, tasks that once required significant time and effort can now be streamlined and executed with greater efficiency, leading to reduced costs, improved accuracy, and enhanced collaboration among supply chain partners (Bhardwaj et al., 2021).

Automating supply chain transactions offers numerous benefits, including increased speed, accuracy, transparency, and cost efficiency. By leveraging cutting-edge technologies such as artificial intelligence (AI), machine learning, robotic process automation (RPA), and blockchain, organizations can streamline their supply chain operations and drive significant improvements in productivity and customer satisfaction.

One of the key areas where automation is making a profound impact is in procurement. Automated procurement systems enable businesses to electronically manage their purchasing activities, from supplier selection and contract negotiation to purchase order generation and invoice processing. These systems leverage AI algorithms to analyze historical data, market trends, and supplier performance, enabling organizations to make informed decisions, optimize their procurement processes, and ensure the timely delivery of goods and services.

Another critical aspect of supply chain automation is order fulfillment. Automated order management systems integrate seamlessly with e-commerce platforms, enabling businesses to receive, process, and fulfil customer orders in real time. These systems can automatically update inventory levels, allocate stock, generate picking lists, and coordinate shipping and delivery, eliminating manual errors and reducing order processing time. Furthermore, intelligent order fulfillment algorithms can optimize shipping routes, warehouse layouts, and delivery schedules, reducing transportation costs and improving overall supply chain efficiency.

Inventory management is yet another area where automation has proven to be immensely beneficial. By deploying automated inventory tracking systems that utilize sensors, RFID technology, and IoT devices, organizations can gain real-time visibility into their stock levels, location, and condition. This enables businesses to optimize inventory replenishment, reduce stockouts and excess inventory, and improve demand forecasting accuracy. Additionally, AI-powered inventory management systems can identify patterns and trends, enabling organizations to proactively manage demand fluctuations, enhance supply chain resilience, and mitigate potential disruptions.

Transportation is a vital component of the supply chain, and automation has revolutionized this area as well. Advanced transportation management systems leverage real-time data, GPS tracking, and route optimization algorithms to streamline the movement of goods, reduce transportation costs, and enhance delivery speed. Automated freight booking platforms connect shippers with carriers, facilitating seamless collaboration and ensuring efficient utilization of transportation resources. Furthermore, emerging technologies such as autonomous vehicles and drones hold the potential to further transform transportation by enabling autonomous deliveries, improving last-mile logistics, and enhancing supply chain visibility.

While the benefits of automating supply chain transactions are evident, organizations must also address the challenges and considerations associated with this transformation. These include data security and privacy concerns, integration complexities with existing systems, change management, and the need for skilled talent to manage and optimize automated processes. Organizations must carefully plan their automation strategies, assess the suitability of technologies, and consider the long-term implications to fully capitalize on the potential of automation in supply chain transactions.

### 3.3.3. Ensuring transparency and accountability

In recent years, there has been a growing demand from consumers, regulatory bodies, and stakeholders for increased visibility into supply chain practices. This demand is driven by various factors, including concerns about environmental sustainability, labor rights, ethical sourcing, product safety, and social responsibility. As a result, organizations are now expected to demonstrate a commitment to transparency and accountability by providing clear and comprehensive information about their supply chain operations.

One of the key drivers of transparency in supply chains is technology. Digital solutions such as blockchain, the Internet of Things (IoT), and data analytics have emerged as powerful tools to enable greater transparency and traceability. Blockchain technology, in particular, has gained significant attention for its ability to create an immutable and transparent record of transactions and activities across the supply chain. By leveraging blockchain, organizations can provide a secure and decentralized platform for recording and verifying information related to product origin, certifications, manufacturing processes, and logistics, among other critical details.

IoT devices, such as sensors and RFID tags, play a crucial role in capturing and transmitting real-time data about the movement, condition, and handling of goods throughout the supply chain. This data can be integrated with supply chain management systems, enabling organizations to monitor and track product flows, identify bottlenecks, ensure compliance with regulatory requirements, and promptly respond to any deviations or issues. The use of IoT technologies enhances transparency by providing stakeholders with accurate and up-to-date information about the status and location of goods at any given time.

Data analytics also contributes significantly to transparency and accountability in supply chains. By leveraging advanced analytics techniques, organizations can extract valuable insights from the vast amount of data generated within the supply chain. These insights can help identify patterns, trends, and potential risks, enabling proactive decision-making and risk mitigation strategies. Furthermore, data analytics can facilitate the monitoring and evaluation of supplier performance, compliance with sustainability standards, and adherence to ethical business practices, fostering a culture of accountability across the supply chain network.

Collaboration and partnerships are essential for ensuring transparency and accountability in supply chains. Organizations must actively engage and collaborate with their suppliers, customers, industry associations, and regulatory bodies to establish common standards, share information, and enforce best practices. Collaborative initiatives, such as supplier audits, certifications, and industry-wide sustainability programs, promote transparency by establishing clear expectations and benchmarks for responsible supply chain practices. By fostering strong partnerships and open lines of communication, organizations can collectively work towards improving transparency and accountability throughout the supply chain ecosystem.

Furthermore, it is essential for organizations to establish robust governance frameworks and internal controls to ensure accountability within their supply chain operations. This involves implementing clear policies and procedures, conducting regular audits and risk assessments, and holding individuals and entities accountable for non-compliance or unethical behavior. By establishing a culture of accountability, organizations can foster trust, mitigate reputational risks, and strengthen their relationships with stakeholders.

### 3.3.4. Streamlining payments and settlements

One of the key drivers of streamlined payments and settlements is the adoption of electronic payment systems. Traditional methods, such as checks and cash, are increasingly being replaced by digital payment solutions that offer greater speed, security, and convenience. Electronic funds transfer (EFT) systems allow businesses to transfer funds between bank accounts electronically, eliminating the need for physical checks and reducing processing time. Moreover, electronic payment platforms enable organizations to initiate payments online, track transaction status in real time, and automate payment reconciliations, reducing manual effort and improving accuracy (Shahab & Allam, 2020; Uddin, Salah, Jayaraman, Pesic, & Ellahham, 2021).

Another transformative technology in the realm of payments and settlements is blockchain. Blockchain, with its decentralized and immutable ledger system, holds immense potential to revolutionize financial transactions in supply chains. Smart contracts, powered by blockchain technology, can automate and streamline the execution of payment and settlement processes. These self-executing contracts enable the automatic release of funds or triggering of payments when predefined conditions, such as delivery confirmation or quality assurance, are met. By leveraging blockchain-based payment systems, organizations can reduce administrative overhead, eliminate intermediaries, and enhance the security and transparency of financial transactions.

Additionally, digital invoicing and electronic data interchange (EDI) systems play a significant role in streamlining payments and settlements. Digital invoicing eliminates the need for paper-based invoices, enabling businesses to generate, send, and process invoices electronically. This not only reduces manual errors and administrative costs but also expedites the invoicing cycle, enabling faster payment processing. EDI systems, on the other hand, facilitate the exchange of structured business documents, including invoices, purchase orders, and payment instructions, in a standardized electronic format. This standardized communication streamlines the reconciliation process, improves accuracy, and expedites payment settlements between trading partners.

Integration between supply chain management systems and financial systems is another critical aspect of streamlining payments and settlements. By integrating these systems, organizations can automate the flow of financial data, such as purchase orders, invoices, and payment information, between the two domains. This integration enables seamless communication, real-time visibility into financial transactions, and automatic synchronisation of payment details, reducing manual intervention and errors. Moreover, it facilitates the consolidation of financial data for reporting and analysis, providing organizations with valuable insights into cash flow, working capital, and financial performance.

Furthermore, the emergence of innovative fintech solutions, such as mobile payment platforms and digital wallets, has transformed the landscape of payments and settlements. These solutions allow for quick and secure payments through mobile devices, offering convenience and flexibility for both buyers and suppliers. Mobile payment platforms also support features like instant payment notifications, real-time fund transfers, and transaction histories, empowering organizations to manage their cash flow effectively and make informed financial decisions. Digital wallets, on the other hand, store payment credentials and enable seamless transactions across different channels, further simplifying the payment process.

While streamlining payments and settlements offers numerous benefits, organizations must also address certain challenges and considerations. These include data security and privacy concerns, regulatory compliance, interoperability between different payment systems, and the need for standardized formats and protocols. It is essential for organizations to carefully evaluate and select payment solutions that align with their specific requirements and integrate seamlessly with existing systems and processes.

### 3.3.5. Reducing fraud and disputes

One of the key approaches to reducing fraud and disputes is the implementation of robust authentication and verification systems. With the advent of digital technologies, organizations can adopt multi-factor authentication methods to ensure that only authorized individuals have access to sensitive supply chain information and systems. This can include measures such as biometric authentication, secure login credentials, and encryption protocols to protect data integrity. By implementing strong authentication practices, organizations can significantly reduce the risk of unauthorized access, data breaches, and fraudulent activities within their supply chain operations.

Transparency plays a crucial role in combating fraud and disputes. By fostering a culture of transparency and open communication, organizations can create an environment where all stakeholders are encouraged to report any suspicious activities or concerns. This can be achieved through regular audits, whistleblower programs, and clearly defined channels for reporting fraud or disputes. Furthermore, the use of emerging technologies such as blockchain can enhance transparency by providing an immutable and auditable record of supply chain activities. Blockchain can enable real-time tracking and verification of transactions, ensuring that information is trustworthy and tamper-proof, which in turn reduces the risk of fraudulent activities.

Implementing data analytics and artificial intelligence (AI) can also be instrumental in fraud detection and prevention. By analyzing vast amounts of supply chain data, organizations can identify patterns, anomalies, and potentially fraudulent activities. AI algorithms can continuously monitor transactions, flag suspicious behavior, and provide real-time alerts. Predictive analytics can also help organizations identify potential risks and proactively implement controls to mitigate fraud. By leveraging data analytics and AI, organizations can strengthen their fraud detection capabilities and take proactive measures to prevent fraudulent activities before they cause significant damage.

Establishing strong internal controls is another critical aspect of reducing fraud and disputes. Organizations should implement segregation of duties, requiring multiple individuals to be involved in critical supply chain processes to ensure checks and balances. Regular internal audits should be conducted to evaluate the effectiveness of control measures and identify any vulnerabilities. Additionally, robust supplier vetting and due diligence processes can help organizations minimize the risk of fraudulent suppliers entering their supply chain network. By implementing stringent controls and continuously monitoring compliance, organizations can significantly reduce the occurrence of fraud and disputes.

In the event of a dispute, organizations should have a well-defined and transparent dispute resolution process in place. This can include clear escalation procedures, mediation services, and arbitration clauses in contracts to facilitate timely and fair resolution of conflicts. Collaborative platforms and electronic dispute resolution systems can also be leveraged to streamline the dispute resolution process, ensuring efficient communication and documentation exchange between parties. By proactively addressing disputes in a transparent and equitable manner, organizations can maintain healthy relationships with trading partners and minimize disruptions to the supply chain.

Educating and training employees on fraud awareness and prevention is crucial. Organizations should provide comprehensive training programs that raise awareness about common fraud schemes, red flags to watch out for, and reporting mechanisms. By fostering a culture of vigilance and accountability, employees can play an active role in detecting and preventing fraud within the supply chain.

## 3.4. Blockchain and food safety standards

Blockchain technology offers a multifaceted approach to maintaining and enhancing food safety and quality. Through real-time monitoring of temperature and storage conditions, it ensures that products remain within optimal parameters. Additionally, it enables swift responses to food recalls and crises, thereby safeguarding public health. Moreover, blockchain enhances compliance with regulations and standards, streamlining certification and auditing processes (Dutta, Choi, et al., 2020).

### 3.4.1. Improving food safety and quality assurance

One of the foundational elements of enhancing food safety and quality assurance is the implementation of stringent food safety management systems. These systems, such as hazard analysis and critical control points (HACCP) or the global food safety initiative (GFSI) recognised schemes, provide a framework for identifying, assessing, and controlling food safety hazards at every stage of the supply chain. By implementing these systems, organizations can ensure that food safety risks are effectively managed and preventive measures are in place to maintain product integrity (Bumblauskas et al., 2020; Detwiler, 2020; Tse, Zhang, Yang, Cheng, & Mu, 2017).

Leveraging technology is key to improving food safety and quality assurance. One of the transformative technologies in this domain is the Internet of Things (IoT). IoT devices, such as temperature sensors, humidity monitors, and RFID tags, can provide real-time data on food storage conditions, transportation, and handling. This data can be integrated with supply chain management systems to enable proactive monitoring and alert systems that flag any deviations from optimal conditions. By leveraging IoT, organizations can identify potential risks, mitigate quality issues, and take corrective actions in a timely manner, thereby ensuring the safety and freshness of food products.

Blockchain technology also holds tremendous potential for enhancing food safety and quality assurance. By leveraging the immutability and transparency of blockchain, organizations can create a tamper-proof record of every step in the supply chain, including production, transportation, storage, and distribution. This enables stakeholders to trace the origin of products, verify certifications, and monitor the adherence to quality standards. In the event of a food safety issue or product recall, blockchain can facilitate swift and accurate traceability, enabling organizations to identify the source of the problem and take immediate actions to contain it, thereby minimizing the impact on consumer safety and brand reputation.

Furthermore, organizations can embrace data analytics and artificial intelligence (AI) to improve food safety and quality assurance. By analyzing vast amounts of data, including sensor data, laboratory test results, and consumer feedback, organizations can gain valuable insights into potential quality issues, trends, and patterns. AI algorithms can detect anomalies and deviations from expected norms, enabling early intervention and preventive measures. Predictive analytics can also help organizations forecast demand, optimize inventory levels, and enhance supply chain planning, minimizing the risk of quality issues due to stockouts or excess inventory.

Collaboration is another crucial aspect of improving food safety and quality assurance. Organizations should foster strong partnerships with suppliers, distributors, and regulatory bodies to establish common quality standards, share best practices, and facilitate information exchange. Collaboration can include regular audits, supplier performance evaluations, and joint initiatives to address quality challenges collectively. By working together, organizations can enhance transparency, identify areas for improvement, and ensure that quality standards are consistently met across the supply chain.

In addition, organizations should invest in training and education programs to enhance the knowledge and skills of employees involved in food production, handling, and quality assurance. Training should cover topics such as hygiene practices, proper handling and storage techniques, and the importance of following established quality procedures. By fostering a culture of quality and providing employees with the necessary knowledge and resources, organizations can empower them to play an active role in maintaining food safety and quality standards.

### 3.4.2. Real-time monitoring of temperature and storage conditions

The foundation of real-time monitoring lies in the deployment of IoT-enabled sensors. These sensors are strategically placed throughout the supply chain, including warehouses, distribution centers, refrigerated containers, and transportation vehicles. IoT sensors can measure and record temperature, humidity, light exposure, vibration, and other relevant parameters. The collected data is transmitted wirelessly to a centralized platform, where it is processed and analyzed in real time.

Data analytics plays a vital role in extracting actionable insights from the collected sensor data. By applying advanced analytics techniques, organizations can identify patterns, trends, and anomalies in temperature and storage conditions. Historical data can be used to establish benchmarks and set acceptable ranges for various products. Real-time analytics can then compare the current sensor readings against these benchmarks and generate alerts when deviations occur. This enables organizations to take immediate corrective actions before any significant damage or quality issues arise.

Automated alerts are a key component of real-time monitoring systems. When predefined thresholds are breached or unusual conditions are detected, the system triggers automated alerts to relevant stakeholders. These alerts can be sent via email, text message, or mobile app notifications, ensuring that responsible personnel are promptly informed. Alerts can also include detailed information about the nature of the deviation, its location, and potential implications. By receiving timely alerts, organizations can swiftly respond to temperature or storage deviations, minimize product spoilage, and ensure quality standards are met.

Integration with supply chain management systems is essential for effective real-time monitoring. By integrating sensor data with inventory management, order fulfillment, and logistics systems, organizations can gain comprehensive visibility into the impact of temperature and storage conditions on their supply chain operations. This integration enables real-time decision-making, such as rerouting shipments, adjusting storage parameters, or prioritising the handling of sensitive goods. It also facilitates traceability and accountability, as stakeholders can track the entire journey of a product and access historical data for auditing and compliance purposes.

Cloud-based platforms provide the scalability, storage capacity, and computing power necessary for real-time monitoring of temperature and storage conditions. By leveraging cloud infrastructure, organizations can securely store and process large volumes of sensor data. Cloud platforms also enable data visualisation, dashboards, and reporting capabilities, allowing stakeholders to access real-time insights and historical trends. Furthermore, cloud-based solutions facilitate collaboration by enabling authorized stakeholders to access and analyze the data from anywhere, at any time, using connected devices.

Real-time monitoring of temperature and storage conditions offers several significant benefits to organizations. First and foremost, it helps ensure compliance with regulatory requirements and industry standards. Organizations can demonstrate adherence to strict temperature control regulations, such as those set by the FDA for pharmaceutical products or food safety agencies for perishable goods. Compliance not only avoids penalties but also reinforces trust among customers and trading partners.

Additionally, real-time monitoring minimizes product losses and waste. By promptly addressing temperature deviations, organizations can prevent spoilage, extend product shelf life, and optimize inventory management. This translates into cost savings, reduced write-offs, and improved operational efficiency. Moreover, organizations can enhance product quality and customer satisfaction by consistently delivering products that meet specified storage conditions and maintain their integrity throughout the supply chain.

### 3.4.3. Rapid response to food recalls and crises

A fundamental aspect of rapid response is the development of a comprehensive recall and crisis management plan. This plan should outline the roles and responsibilities of key personnel, establish communication protocols, and provide step-by-step procedures to guide the response process. The plan should be regularly reviewed, tested through simulated scenarios, and updated as necessary to ensure its effectiveness in addressing various types of crises (Bumblauskas et al., 2020; Yiannas, 2018).

Effective communication is paramount during a food recall or crisis. Organizations should establish clear lines of communication both internally and externally. Internally, designated crisis management teams should be formed, comprising representatives from relevant departments such as quality assurance, legal, public relations, and senior management. These teams should meet regularly to review the recall plan, discuss potential scenarios, and ensure a coordinated response. Externally, organizations should establish communication channels with regulatory agencies, customers, suppliers, and the public. Timely and transparent communication is essential to mitigate public concerns, inform stakeholders about the situation, and provide guidance on actions to be taken.

Leveraging technology can greatly enhance the speed and efficiency of the recall and crisis management process. Digital systems and software solutions can streamline data collection, traceability, and communication. By maintaining accurate and up-to-date records of products, suppliers, and customers, organizations can quickly identify affected batches, trace their distribution, and communicate targeted recall instructions. Furthermore, digital platforms can facilitate real-time collaboration among stakeholders, enabling swift information sharing and coordination.

Traceability systems play a critical role in rapid response during a food recall or crisis. By implementing robust traceability mechanisms, organizations can effectively track the movement of products throughout the supply chain. This includes capturing data on product origins, manufacturing processes, distribution channels, and customer destinations. In the event of a recall, traceability systems enable organizations to quickly identify the affected products, their locations, and the parties involved. This enables targeted recall actions, minimizes the scope of the recall, and reduces the impact on unaffected products.

Collaboration with supply chain partners is essential during a recall or crisis. Organizations should establish strong relationships with suppliers, distributors, and retailers, and ensure open lines of communication. This collaboration allows for the swift exchange of information, coordination of recall efforts, and the ability to trace the affected products across the supply chain. By working together, organizations can expedite the recall process, mitigate the risk of contaminated products reaching consumers, and minimize financial losses.

In addition to the immediate response, organizations should conduct thorough investigations and root cause analyses following a recall or crisis. This helps identify the source of the problem and implement corrective actions to prevent similar incidents in the future. It is essential to evaluate existing processes, review quality control measures, and update protocols to enhance product safety and prevent recurrences. Transparency and accountability are crucial during this phase, as stakeholders expect organizations to take responsibility and demonstrate their commitment to continuous improvement.

Training and preparedness are critical elements of a rapid response strategy. Employees should be trained on the recall and crisis management plan, their roles and responsibilities, and effective communication protocols. Regular drills and simulations should be conducted to test the organization's readiness and identify areas for improvement. By investing in training and preparedness, organizations can minimize response time, enhance decision-making capabilities, and instill a culture of preparedness throughout the organization.

### 3.4.4. Enhancing compliance with regulations and standards

A crucial step in enhancing compliance is the development of a comprehensive compliance program. This program should encompass policies, procedures, and controls that align with relevant regulations and industry standards. It should clearly articulate the organization's commitment to compliance, define roles and responsibilities, and provide guidance on ethical conduct and best practices. The compliance program should be communicated effectively to all employees, and regular training sessions should be conducted to ensure awareness and understanding.

Staying up to date with regulations and standards is essential for maintaining compliance. Organizations should establish a systematic process for monitoring and assessing changes in laws, regulations, and industry guidelines that impact their operations. This may involve subscribing to regulatory news alerts, engaging legal experts, and participating in industry associations or forums. By proactively identifying changes, organizations can adapt their compliance programs, update policies and procedures, and ensure ongoing alignment with the latest requirements.

Implementing robust governance and risk management processes is crucial for compliance. Organizations should establish clear lines of accountability, oversight, and reporting. This includes designating compliance officers or committees responsible for monitoring and managing compliance-related activities. Regular risk assessments should be conducted to identify areas of vulnerability and implement appropriate mitigation measures. By embedding compliance into the organization's governance framework, organizations can foster a culture of responsibility and ensure that compliance considerations are integrated into decision-making processes.

Leveraging technology can significantly enhance compliance efforts. Compliance management systems, document control software, and data analytics tools can streamline compliance processes, facilitate reporting, and automate compliance-related tasks. These technologies can help organizations monitor adherence to policies, track compliance activities, and generate accurate and timely reports. By leveraging technology, organizations can reduce manual efforts, improve data accuracy, and gain real-time visibility into compliance performance.

Regular internal audits are essential to assess the effectiveness of compliance programs and identify areas for improvement. Internal audit teams, or external auditors when necessary, should conduct independent assessments of the organization's compliance controls, processes, and procedures. These audits provide valuable insights into gaps, weaknesses, or non-compliance issues, allowing organizations to take corrective actions and strengthen their compliance programs. Audits also demonstrate a commitment to continuous improvement and provide assurance to stakeholders, including regulatory bodies and customers.

Building strong relationships with regulatory authorities is crucial for enhancing compliance. Organizations should establish open lines of communication, proactively engage with regulators, and seek guidance when needed. By maintaining transparent and cooperative relationships, organizations can gain valuable insights into regulatory expectations, demonstrate a commitment to compliance, and address any compliance-related concerns promptly and effectively. Collaboration with regulators also allows organizations to stay informed about upcoming regulatory changes and actively participates in the development of new regulations or standards.

Supplier management is another critical aspect of enhancing compliance throughout the supply chain. Organizations should establish robust supplier qualification processes, conduct due diligence, and assess the compliance performance of their suppliers. This includes evaluating suppliers' adherence to regulatory requirements, quality standards, and ethical practices. By selecting and partnering with compliant suppliers, organizations can minimize supply chain risks, ensure the integrity of their products and services, and meet customer expectations.

### 3.4.5. Certification and auditing processes

Certification is a formal recognition that an organization meets specific standards set by regulatory bodies, industry associations, or international organizations. Certifications can cover various aspects, including quality management, environmental sustainability, information security, occupational health and safety, and social responsibility. The certification process typically involves an external certification body conducting a thorough assessment of the organization's systems, processes, and performance against the established standards. This assessment may include document reviews, on-site inspections, interviews, and performance evaluations. Upon successful completion, the organization is awarded the certification, which is valid for a specified period and subject to regular surveillance audits.

Certifications offer several benefits to organizations. First and foremost, they provide assurance to customers, stakeholders, and regulatory authorities that the organization has implemented effective systems and processes to meet industry standards. Certifications enhance the organization's reputation, instill confidence, and can be a prerequisite for doing business with certain partners or entering specific markets. Certifications also drive continuous improvement, as organizations must maintain compliance and strive for excellence to retain their certifications. Additionally, certifications can lead to operational efficiencies, cost savings, and improved risk management by implementing best practices and standardized processes.

Auditing is the systematic examination and evaluation of an organization's operations, processes, and systems to ensure compliance with regulations, internal policies, and industry standards. Audits can be conducted internally by an organization's own audit team or externally by independent auditing firms. Audits may cover a wide range of areas, including financial records, operational procedures, data security, ethical practices, and regulatory compliance. The audit process involves collecting evidence, performing tests, interviewing personnel, and analyzing findings. Audits can be scheduled or surprise inspections, depending on the nature of the audit and its objectives.

Regular auditing provides several benefits to organizations. It helps identify areas of non-compliance, potential risks, and opportunities for improvement.

Audits provide an objective assessment of the organization's performance, internal controls, and adherence to policies and regulations. They offer insights into operational inefficiencies, process gaps, and areas where corrective actions are needed. By addressing these findings, organizations can enhance their operations, mitigate risks, and ensure ongoing compliance. Audits also contribute to a culture of accountability, transparency, and continuous improvement within the organization.

Certification and auditing processes go hand in hand. Certifications often require organizations to undergo regular audits to maintain compliance with the standards. Audits serve as a validation mechanism for the certification body, ensuring that the organization continues to meet the requirements over time. Audits can also identify areas where additional improvements are necessary to meet certification criteria. The combination of certifications and audits provides a robust framework for organizations to demonstrate their commitment to quality, compliance, and continual improvement.

To prepare for certifications and audits, organizations should establish strong documentation systems, maintain accurate records, and implement effective internal controls. They should conduct self-assessments to identify potential gaps and areas for improvement before undergoing external assessments. Regular internal audits can help organizations monitor their compliance status and address any deficiencies proactively. Collaboration with external auditors or certification bodies is essential to understand the requirements, prepare the necessary documentation, and ensure a smooth certification or audit process.

## 3.5. Collaborative supply chain management with blockchain

Collaborative supply chain management empowered by blockchain technology presents a multitude of advantages. It facilitates seamless data sharing among stakeholders, fostering trust and collaboration throughout the network. Moreover, it efficiently coordinates inventory management and logistics, ensuring optimal resource allocation. Additionally, by streamlining order and delivery coordination, blockchain enhances supply chain resilience, as noted by Chang et al. (2019) and Dutta, Choi, et al. (2020).

### 3.5.1. Enabling data sharing among stakeholders

Data sharing among stakeholders involves the exchange of relevant information and insights across organizational boundaries. These stakeholders can include suppliers, customers, partners, regulatory agencies, and other entities involved in the supply chain or industry ecosystem. Data shared may encompass a wide range of types, such as product information, inventory levels, quality metrics, demand forecasts, production schedules, and compliance data. By sharing this data, stakeholders can gain a holistic view of the supply chain, identify areas for improvement, and collaborate on joint initiatives (Krzyzanowski Guerra & Boys, 2022).

One of the key challenges in data sharing is ensuring the security and privacy of the shared information. Organizations must implement robust data protection measures, such as encryption, access controls, and data anonymization, to safeguard sensitive data from unauthorized access or misuse. Compliance with applicable data protection regulations, such as GDPR (General Data Protection Regulation) or HIPAA (Health Insurance Portability and Accountability Act), is essential. Clear data-sharing agreements, outlining the rights, responsibilities, and limitations, should be established between stakeholders to ensure transparency and mutual understanding.

Technological advancements have enabled the development of sophisticated data-sharing platforms and frameworks. Cloud-based solutions, secure APIs (Application Programming Interfaces), and blockchain technology offer secure and scalable infrastructure for data sharing among stakeholders. These platforms facilitate real-time data exchange, enable seamless integration of disparate systems, and provide a centralized repository for shared information. Data analytics and machine learning capabilities can be leveraged to extract valuable insights from shared data, improving decision-making and optimizing supply chain performance.

Establishing trust and building collaborative relationships among stakeholders is fundamental for successful data sharing. Organizations should foster an environment of transparency, open communication, and mutual benefit. Clearly defined roles and responsibilities regarding data ownership, access rights, and data governance should be established. Trust can be further enhanced through the implementation of audit trails, data validation mechanisms, and periodic reviews of data-sharing practices. Building trust among stakeholders encourages a culture of data sharing and fosters innovation and collaboration.

Standardization plays a significant role in facilitating data sharing among stakeholders. Adopting industry-wide standards for data formats, terminology, and communication protocols ensures compatibility and seamless integration across different systems and organizations. Standards such as GS1 (Global Standard One) for product identification or EDI (Electronic Data Interchange) for structured data exchange enable efficient and accurate data sharing, reducing errors, delays, and the need for manual interventions.

Data sharing among stakeholders can yield numerous benefits. Improved supply chain visibility allows for better demand forecasting, optimized inventory management, and enhanced production planning. Collaboration on product development and innovation leads to faster time-to-market and the creation of customer-centric solutions. Sharing data related to quality metrics and compliance enables proactive risk management and ensures adherence to industry standards. Furthermore, shared data can facilitate traceability and accountability, especially in industries where regulatory compliance and consumer safety are paramount.

To enable effective data sharing, organizations should develop data governance frameworks that define data ownership, data quality standards, and data sharing policies. Collaboration platforms and digital ecosystems should be implemented to facilitate secure and seamless data exchange. Cross-functional teams should be established to oversee data-sharing initiatives, monitor performance, and address any challenges or concerns that arise.

### 3.5.2. Facilitating trust and collaboration

Building trust among stakeholders is a foundational element of successful collaboration. Trust is established through consistent and transparent communication, reliable and ethical behavior, and the fulfillment of commitments. Organizations should prioritize integrity, honesty, and ethical conduct in their interactions with stakeholders. By delivering on promises, being transparent about goals, strategies, and challenges, and demonstrating a commitment to mutual benefit, organizations can foster trust and create an environment conducive to collaboration (Litke, Anagnostopoulos, & Varvarigou, 2019).

Open and effective communication is key to building trust and facilitating collaboration. Clear and timely communication helps to align expectations, share information, and address concerns. Organizations should encourage open dialogue and active listening, ensuring that all stakeholders have a voice and are actively engaged. Regular meetings, feedback sessions, and collaborative platforms can facilitate communication and foster a sense of inclusiveness. By promoting transparency and fostering a culture of communication, organizations can establish trust and strengthen collaborative relationships.

Collaboration requires a mindset of shared goals and mutual benefits. Organizations should seek to create a collaborative culture that encourages cross-functional and cross-organizational collaboration. Breaking down silos and promoting teamwork helps to leverage diverse perspectives, knowledge, and expertise. Collaborative environments should be supported by structures, processes, and technologies that facilitate the exchange of ideas, knowledge sharing, and joint decision-making. By fostering a collaborative culture, organizations can tap into the collective intelligence and creativity of stakeholders, leading to innovative solutions and improved outcomes.

Trust and collaboration can be further strengthened through the establishment of strategic partnerships and alliances. Collaborative partnerships allow organizations to pool resources, expertise, and capabilities, creating synergies and enhancing competitiveness. When selecting partners, organizations should prioritize shared values, complementary strengths, and a common vision. Clear agreements and well-defined roles and responsibilities should be established to manage expectations and ensure a mutually beneficial relationship. Regular communication, joint planning, and ongoing evaluation of partnership performance are crucial for nurturing trust and maintaining successful collaborations.

Technology plays a vital role in facilitating trust and collaboration among stakeholders. Digital collaboration tools, project management software, and shared platforms enable real-time communication, document sharing, and collaborative workspaces. These technologies promote efficient collaboration, streamline processes, and enhance productivity. Additionally, emerging technologies like blockchain can provide secure and transparent systems for tracking and verifying collaborative activities, ensuring trust and accountability among stakeholders.

To facilitate trust and collaboration, organizations should invest in training and development programs that focus on building trust, fostering effective communication, and developing collaboration skills. These programs can enhance interpersonal skills, promote understanding of cultural differences, and provide tools for conflict resolution and problem-solving. By equipping employees with the necessary skills and knowledge, organizations can foster a collaborative mindset throughout the workforce.

Recognising and celebrating collaborative achievements is essential for sustaining trust and motivating stakeholders. Organizations should acknowledge and reward collaborative efforts, both at the individual and team levels. Celebrating successes and sharing best practices across the organization encourages a culture of collaboration, inspires others to embrace collaboration, and reinforces the value placed on collaborative behavior.

### 3.5.3. Coordinating inventory management and logistics

Inventory management involves overseeing the flow of goods within an organization, including procurement, storage, and distribution. Effective inventory management aims to strike a balance between having enough stock to meet customer demand while minimizing excess inventory and associated costs. It requires careful planning, forecasting, and monitoring of inventory levels across the supply chain.

Coordinating inventory management with logistics activities begins with accurate demand forecasting. By analyzing historical data, market trends, and customer insights, organizations can estimate future demand and plan their inventory levels accordingly. This enables them to align procurement and production processes with anticipated demand, reducing the risk of stockouts or excess inventory.

Inventory visibility is a key aspect of coordination. Organizations should implement robust inventory tracking systems that provide real-time information on stock levels, location, and movement. This visibility enables better decision-making regarding inventory allocation, replenishment, and order fulfillment. By sharing inventory data with logistics partners, organizations can optimize transportation routes, plan for efficient warehouse operations, and improve order fulfillment processes.

Collaboration between inventory management and logistics teams is crucial for effective coordination. Cross-functional teams should be established to facilitate communication, information sharing, and joint decision-making. Regular meetings, performance reviews, and shared KPIs (Key Performance Indicators) help align goals, address bottlenecks, and identify opportunities for improvement. By fostering a culture of collaboration and teamwork, organizations can enhance coordination between inventory management and logistics functions.

Technological advancements play a significant role in coordinating inventory management and logistics. Inventory management systems, warehouse management systems, and transportation management systems provide integrated platforms to track and manage inventory across the supply chain. These systems automate processes, improve accuracy, and enable real-time visibility, facilitating seamless coordination between inventory and logistics activities. Additionally, IoT (Internet of Things) devices, such as RFID (Radio Frequency Identification) tags, can be used to track inventory movement and improve inventory accuracy.

Effective coordination of inventory management and logistics also involves optimizing transportation and distribution networks. Organizations should analyze transportation routes, modes of transportation, and carrier performance to ensure efficient and cost-effective movement of goods. Collaboration with logistics partners, such as carriers and third-party logistics providers, can help streamline logistics operations, reduce transportation costs, and improve delivery timelines.

Furthermore, implementing lean inventory management practices, such as Just-in-time (JIT) or Vendor-managed Inventory (VMI), can enhance coordination between inventory management and logistics. JIT aims to minimize inventory holding costs by receiving and producing goods as close as possible to the time they are needed. VMI involves suppliers managing inventory levels based on demand information shared by the organization, reducing the need for manual inventory management and ensuring timely replenishment.

Data analytics and supply chain visibility tools play a crucial role in coordinating inventory management and logistics. By analyzing data on demand patterns, lead times, and transportation performance, organizations can optimize inventory levels, reduce stockouts, and improve order fulfillment. Predictive analytics can also help identify potential supply chain disruptions, enabling proactive actions to mitigate risks and maintain smooth operations.

Continuous improvement is essential for ongoing coordination between inventory management and logistics. Organizations should regularly review and evaluate inventory and logistics processes, seeking opportunities to optimize efficiency, reduce costs, and enhance customer service. Performance metrics and key performance indicators should be monitored to assess the effectiveness of coordination efforts and identify areas for improvement.

### 3.5.4. Efficient coordination of orders and deliveries

Order coordination begins with effective order management systems and processes. Organizations should implement robust order management systems that centralize and automate order processing activities. These systems should capture order details, track order status, and facilitate seamless communication between different stakeholders involved in the order fulfillment process. By integrating order management systems with other supply chain components, such as inventory management and logistics systems, organizations can ensure real-time visibility and enable proactive decision-making.

Efficient order coordination requires accurate order capture and processing. Organizations should establish clear and standardized order entry processes to minimize errors and discrepancies. This involves capturing complete and accurate customer information, product specifications, quantities, and delivery requirements. Automated order validation mechanisms can help identify any inconsistencies or missing information, reducing the risk of order errors and subsequent delays.

Real-time order visibility is crucial for effective coordination. Organizations should strive to provide customers and internal stakeholders with access to real-time order status updates. This visibility enables customers to track their orders, anticipate delivery dates, and make informed decisions. Internally, real-time order visibility allows teams to proactively address any potential issues, allocate resources efficiently, and provide accurate delivery commitments to customers.

Effective communication and collaboration among stakeholders are key to successful order coordination. Clear and timely communication between sales teams, customer service representatives, logistics providers, and other relevant parties is essential. Regular meetings, shared platforms, and standardized communication channels can facilitate seamless information exchange, address queries, and resolve potential issues. By fostering collaboration, organizations can ensure that everyone involved in the order fulfillment process is aligned and working towards a common goal.

Optimizing delivery logistics is another vital aspect of efficient order coordination. Organizations should establish well-defined delivery processes, considering factors such as transportation modes, routes, carrier selection, and scheduling. Collaborating with logistics partners, such as carriers and third-party logistics providers, can help optimize delivery operations, improve transit times, and reduce costs. Additionally, leveraging technology, such as route optimization software and real-time tracking systems, enables organizations to make data-driven decisions and proactively manage delivery schedules.

Efficient order coordination also requires effective inventory management. Organizations should maintain accurate and up-to-date inventory records to ensure product availability for order fulfillment. Integration between order management systems and inventory management systems enables real-time inventory visibility and helps prevent stockouts or overstocks. Just-in-time (JIT) or Vendor-managed Inventory (VMI) practices can further enhance order coordination by aligning inventory levels with anticipated demand.

Automation and digitisation play a significant role in efficient order coordination. Leveraging technology, such as electronic data interchange (EDI), automated order processing, and electronic invoicing, streamlines the order-to-cash process and reduces manual errors and delays. Advanced technologies, including artificial intelligence and machine learning, can also be employed to predict demand patterns, optimize order fulfillment, and improve delivery accuracy.

Continuous monitoring and performance measurement are crucial for ongoing improvement in order coordination. Organizations should establish key performance indicators (KPIs) to track order cycle time, on-time delivery rates, order accuracy, and customer satisfaction. Regular performance reviews, root cause analysis of any issues, and the implementation of corrective actions contribute to the continuous enhancement of order coordination processes.

### 3.5.5. Enhancing supply chain resilience

Supply chain resilience refers to an organization's ability to anticipate, absorb, and recover from disruptions while maintaining essential operations and meeting customer demands. It involves identifying vulnerabilities, implementing strategies to manage risks, and building adaptive capabilities. By enhancing supply chain resilience, organizations can minimize the impact of disruptions, quickly recover from setbacks, and continue serving customers effectively.

The first step in enhancing supply chain resilience is to conduct a comprehensive risk assessment. Organizations need to identify potential risks and vulnerabilities throughout their supply chain network. This includes evaluating risks associated with suppliers, transportation routes, geopolitical factors, natural disasters, demand volatility, and cybersecurity threats. By understanding these risks, organizations can develop effective risk mitigation strategies and contingency plans.

Diversification and redundancy are key principles in building supply chain resilience. Organizations should assess their supplier base and distribution network to avoid overreliance on a single source or location. By diversifying suppliers and establishing backup options, organizations can reduce the vulnerability to disruptions in one specific area. Redundancy can be created through multiple distribution centers, alternative transportation modes, and duplicate inventory storage. This ensures that if one link in the supply chain is disrupted, alternative channels can be activated to maintain the flow of goods and services.

Establishing strong relationships with suppliers and other stakeholders is vital for supply chain resilience. Collaboration and open communication foster trust and facilitate effective responses during disruptions. Organizations should work closely with suppliers to understand their capabilities, assess their risk management practices, and establish clear communication channels. Maintaining strong relationships enables better coordination, information sharing, and joint problem-solving during challenging times.

Technological advancements play a crucial role in enhancing supply chain resilience. The adoption of digital technologies such as the Internet of Things (IoT), cloud computing, and data analytics provides real-time visibility and enables proactive risk management. IoT devices can monitor supply chain assets, detect anomalies, and provide early warning signs of potential disruptions. Cloud-based platforms allow for remote access to critical data and enable collaboration among stakeholders, even in times of physical disruptions. Data analytics can help identify patterns, anticipate risks, and support decision-making for effective risk mitigation strategies.

Continuous monitoring and real-time information are essential for managing supply chain disruptions. Organizations should implement robust monitoring systems to track key performance indicators (KPIs), such as delivery times, inventory levels, and supplier performance. This helps identify potential disruptions and enables proactive measures to address them promptly. Early detection of disruptions allows organizations to activate contingency plans, reallocate resources, and minimize the impact on operations.

Resilience testing and simulation exercises are valuable tools for enhancing supply chain resilience. By conducting scenario-based simulations and stress tests, organizations can assess their preparedness and identify areas for improvement. These exercises help validate the effectiveness of contingency plans, uncover potential vulnerabilities, and enable organizations to refine their response strategies. Regular reviews and updates of contingency plans based on lessons learned from simulations contribute to continuous improvement in supply chain resilience.

Supply chain visibility and transparency are critical elements for resilience. Organizations should strive for end-to-end visibility across their supply chain, including suppliers, manufacturing facilities, transportation, and distribution networks. Real-time information about inventory levels, production schedules, and transportation status enables proactive decision-making and facilitates rapid response to disruptions. Transparency among supply chain partners enhances collaboration, trust, and joint problem-solving.

Investing in employee training and development is crucial for building supply chain resilience. Employees need to be equipped with the knowledge and skills to manage disruptions effectively. Training programs should focus on risk awareness, crisis management, decision-making under uncertainty, and communication skills. Empowering employees to act swiftly and decisively during disruptions enhances the overall resilience of the supply chain.

# Chapter 4

# Farming and crop management

*Harvesting the Power of Connectivity: Efficiency, Accountability, and Fairness in Modern Agriculture* - Unknown

**Abstract**

This chapter explores the applications of blockchain technology in optimizing farming processes and crop management. It discusses how blockchain can enhance transparency, efficiency, and trust in the agriculture sector. The chapter focuses on utilizing smart contracts for agreements between farmers and stakeholders. The chapter begins by examining the potential of blockchain technology to optimize farming processes and crop management. It highlights how blockchain's decentralized nature and immutability can enable secure and transparent data sharing among farmers, suppliers, distributors, and consumers. The chapter then delves into the utilization of smart contracts, a key feature of blockchain technology, for efficient and transparent agreements in the agricultural domain. It discusses how smart contracts can automate and enforce agreements between farmers and stakeholders, eliminating intermediaries and reducing administrative overhead. By leveraging smart contracts, farmers can establish trust, ensure fair compensation, and enhance collaboration with suppliers, buyers, and service providers.

**Keywords**: Blockchain technology, Farming processes, Crop management, Transparency, Efficiency, Trust, Smart contracts, Data sharing, Agreements, Stakeholders

**Chapter contents**

- Internet of Things (IoT) integration with blockchain technology in precision farming
- Traceability and authentication of agricultural inputs and machinery
- Decentralized marketplaces for buying and selling agricultural products

***

## 4.1. Internet of Things integration with blockchain technology in precision farming

The integration of Internet of Things (IoT) and Blockchain technologies presents a myriad of benefits for precision farming, including enhanced precision and efficiency, improved crop management, and increased yields. Real-time environmental monitoring further aids in optimizing farming practices. Moreover, this integration facilitates data collection and monitoring, leveraging blockchain for secure and decentralized data storage, ultimately enhancing efficiency and decision-making in agriculture. These advancements are underscored by research from Ferrández-Pastor, Mora-Pascual, and Díaz-Lajara (2022), Kim, Hilton, Burks, and Reyes (2018), Kumar et al. (2022), Tomu (2020), Torky and Hassanein (2020) and (Xiong et al., 2020).

### 4.1.1. Introduction to precision farming and IoT

*a. Precision farming*

Precision farming, also known as precision agriculture or smart farming, is an approach that leverages technology to optimize agricultural practices. It involves the use of various tools such as sensors, drones, satellite imagery, and data analytics to gather detailed information about the farming environment. This data-driven approach enables farmers to make informed decisions and take precise actions to enhance productivity, reduce costs, and minimize environmental impact (Blokdyk, 2021; Detwiler, 2020).

*b. Internet of Things*

The Internet of Things (IoT) serves as a critical enabler for precision farming. It connects physical devices and sensors, enabling them to communicate and share data seamlessly. IoT devices can be deployed across farmlands to monitor and collect real-time information about soil moisture levels, weather conditions, crop health, livestock behavior, and much more. This data is then processed and analyzed to provide valuable insights, allowing farmers to make data-driven decisions in near real time (Alshehri, 2023; Blokdyk, 2021; Detwiler, 2020; Pincheira, Vecchio, Giaffreda, & Kanhere, 2021; Rejeb, Simske, Rejeb, Treiblmaier, & Zailani, 2020).

*c. Advantages of precision farming and the Internet of Things*

One of the key advantages of precision farming and IoT is the ability to apply inputs, such as water, fertilizers, and pesticides, with precision. Instead of uniformly applying these inputs across the entire field, farmers can use sensor data to identify specific areas that require attention. This targeted approach reduces waste, minimizes environmental impact, and optimizes resource utilization. By fine-tuning their farming practices, farmers can improve crop yields, quality, and overall efficiency (Alshehri, 2023; Blokdyk, 2021; Detwiler, 2020; Pincheira et al., 2021; Rejeb, Simske, et al., 2020).

Moreover, precision farming and IoT enable proactive monitoring and early detection of issues that could impact crop health or yield. For instance, by analyzing sensor data, farmers can identify signs of pest infestation, nutrient deficiencies, or disease outbreaks at an early stage. This early warning system allows farmers to take timely action, preventing potential crop losses and ensuring optimal plant health.

Another significant benefit of precision farming and IoT is the automation and remote-control capabilities they offer. By integrating IoT devices with farm machinery and equipment, farmers can automate tasks such as irrigation, fertilization, and harvesting. They can remotely monitor and control farm operations from anywhere, using their smartphones or computers. This level of automation not only saves time and labor but also enhances operational efficiency and reduces human errors.

*d. Challenges and considerations*

Despite the promising benefits of technological advancements like precision farming and IoT, several challenges and considerations must be addressed for successful implementation. Reliable connectivity infrastructure and robust cybersecurity measures are essential to safeguard sensitive data collected through these technologies. Moreover, providing farmers with adequate training and support is crucial to ensure their effective utilization of these tools and accurate interpretation of the generated data, maximizing the potential benefits for agricultural productivity and sustainability (Blokdyk, 2021; Detwiler, 2020).

### 4.1.2. Benefits of IoT integration in agriculture

*a. Enhanced precision and efficiency*

The IoT enables precise monitoring and control of various agricultural processes. Through the deployment of sensors, farmers can gather real-time data on crucial parameters such as soil moisture levels, temperature, humidity, and crop growth patterns. This granular data empowers farmers to make data-driven decisions with a high degree of accuracy, ensuring that resources such as water, fertilizers, and pesticides are applied precisely where and when they are needed. By eliminating guesswork and optimizing resource utilization, IoT integration enhances operational efficiency and minimizes waste (Pranto, Noman, Mahmud, & Haque, 2021).

*b. Improved crop management and yield*

IoT devices provide continuous and detailed insights into crop health and growth. With the help of sensors, farmers can monitor crucial factors like nutrient levels, pH balance, and disease presence. This information allows them to take timely actions such as adjusting irrigation schedules, applying appropriate fertilizers, or implementing pest control measures. By detecting potential issues early and addressing them promptly, farmers can mitigate crop losses, maximize yields, and improve overall crop management (Pranto et al., 2021).

*c. Real-time environmental monitoring*

In the realm of precision farming and IoT, real-time environmental monitoring plays a pivotal role in enabling farmers to make informed decisions and take proactive measures to optimize agricultural practices. By leveraging a network of sensors, data analytics, and connectivity, real-time environmental monitoring provides farmers with valuable insights into the conditions that impact crop growth, soil health, and overall farm sustainability. Armed with this information, farmers can implement targeted interventions, such as adjusting irrigation schedules or applying fertilizers, to maximize yields, minimize resource usage, and enhance long-term environmental stewardship (Rejeb, Simske, et al., 2020).

Environmental monitoring systems are designed to capture data from various sources in the farm's ecosystem. These sources can include weather stations, soil sensors, water quality sensors, air quality sensors, and more. Each sensor is strategically placed to collect data on specific environmental parameters, allowing farmers to gain a comprehensive understanding of the conditions influencing their crops.

Weather stations are a fundamental component of real-time environmental monitoring. They continuously measure variables such as temperature, humidity, rainfall, wind speed, and solar radiation. This data helps farmers track weather patterns, detect extreme events like frost or heatwaves, and anticipate changes that could affect crop growth and yield. With this information at their fingertips, farmers can adjust irrigation schedules, apply protective measures, and optimize planting and harvesting windows.

Soil sensors are another vital aspect of real-time environmental monitoring. These sensors are embedded in the soil and measure parameters such as moisture content, nutrient levels, pH balance, and temperature. By continuously monitoring these factors, farmers can ensure that crops receive the ideal conditions for growth. For instance, if soil moisture levels are too low, irrigation can be triggered automatically to provide the necessary hydration. Similarly, nutrient deficiencies or imbalances can be detected promptly, allowing farmers to adjust fertilizer application accordingly.

Water quality sensors play a crucial role in monitoring the health and safety of irrigation water sources. These sensors assess key indicators like pH, electrical conductivity, dissolved oxygen levels, and the presence of contaminants. Monitoring water quality ensures that crops are not exposed to harmful substances that could compromise their health or contaminate the food supply. Real-time data from water quality sensors enables farmers to take swift action if any anomalies or issues arise, preserving the integrity of their irrigation systems.

Air quality sensors contribute to a comprehensive understanding of the farm's environment by monitoring factors such as air temperature, humidity, carbon dioxide ($CO_2$) levels, and particulate matter. High $CO_2$ levels can inhibit photosynthesis and negatively impact plant growth, while excessive particulate matter can affect air quality and crop health. With real-time air quality data, farmers can implement measures to mitigate these risks, such as adjusting ventilation systems or deploying preventive measures to minimize dust or pollutant emissions.

The integration of these real-time environmental monitoring systems is made possible by IoT connectivity, which enables seamless communication between the sensors and a central data hub. This hub collects, analyzes, and presents the data in a user-friendly format, providing farmers with actionable insights. Advanced data analytics algorithms can identify patterns, correlations, and anomalies in the data, allowing farmers to make data-driven decisions and take timely action.

Real-time environmental monitoring has numerous benefits for farmers and the agricultural industry as a whole. By leveraging this technology, farmers can optimize resource usage, conserve water, reduce the application of fertilizers and pesticides, and minimize the environmental impact of their operations. Additionally, early detection of potential issues through real-time monitoring can help prevent crop diseases, mitigate pest infestations, and improve overall yield and quality.

### 4.1.3. Data collection and monitoring in precision farming

Precision farming relies on a diverse array of equipment and activities for effective data collection and monitoring. This includes the deployment of sensors and IoT devices, leveraging remote sensing and imaging technologies, ensuring seamless connectivity and data transmission, efficient data management and storage solutions, and the integration of systems for enhanced interoperability. These components collectively contribute to the advancement of precision farming practices (Rejeb, Simske, et al., 2020).

*a. Sensors and IoT devices*

Data collection in precision farming relies heavily on a wide range of sensors and IoT devices strategically placed throughout the farm. These devices are designed to capture real-time data on various parameters, including soil moisture, temperature, nutrient levels, weather conditions, crop health, and more. Soil sensors, weather stations, drones equipped with cameras, and satellite imagery are just a few examples of the devices used for data collection. These sensors interact with the environment, continuously collecting data and transmitting it to a central database or cloud platform.

*b. Remote sensing and imaging*

Remote sensing technologies, such as aerial imagery and satellite-based sensors, have revolutionized the way farmers gather information about their fields. Aerial drones equipped with high-resolution cameras and multispectral sensors can capture detailed images, allowing farmers to monitor crop health, detect stress, and identify areas of potential concern. Satellite imagery provides a broader perspective, offering insights into large-scale trends, vegetation indices, and overall field conditions. These remote sensing techniques enable farmers to detect early signs of disease, assess nutrient deficiencies, and optimize resource allocation.

*c. Connectivity and data transmission*

To enable real-time monitoring and analysis, data collected by sensors and devices need to be transmitted to a central system. IoT connectivity plays a crucial role in facilitating this data transmission. Wireless technologies such as Wi-Fi, cellular networks, and low-power wide-area networks (LPWANs) enable seamless communication between sensors and the central database. This connectivity ensures that data is captured and transmitted reliably, allowing farmers to access real-time information from any location.

*d. Data management and storage*

Managing and storing the vast amounts of data generated by precision farming systems require robust data management practices. Cloud-based platforms and databases provide scalable solutions for securely storing and organising agricultural data. These platforms allow farmers to store historical data, access real-time information, and integrate data from multiple sources. Data management systems often include data visualisation tools, dashboards, and analytics capabilities, enabling farmers to gain insights and make data-driven decisions.

*e. Data analytics and decision support*

Collecting data is only the first step; extracting meaningful insights from the data is equally important. Data analytics techniques such as machine learning algorithms and predictive modelling, help farmers uncover patterns, correlations, and trends in their data. By analyzing historical data and combining it with real-time information, farmers can gain valuable insights into crop performance, optimize resource allocation, predict yield, and make informed decisions. Decision support systems can be developed to provide recommendations based on the analyzed data, assisting farmers in taking actions that maximize productivity and profitability.

*f. Integration and interoperability*

To fully harness the benefits of data collection and monitoring in precision farming, integration and interoperability between different systems and data sources are essential. Integrating data from multiple sensors and devices allows for a holistic understanding of the farm's ecosystem. Additionally, interoperability enables seamless data sharing and collaboration among different stakeholders, such as farmers, agronomists, and researchers. Open data standards and application programming interfaces (APIs) facilitate this integration, ensuring compatibility and enabling the development of innovative farming solutions.

### 4.1.4. Leveraging blockchain for secure and decentralized data storage

Harnessing blockchain technology offers a myriad of benefits, including bolstered data security, decentralization, and preservation of data integrity. Additionally, it promotes transparency and auditability throughout processes, facilitated by the implementation of smart contracts and automation. Moreover, blockchain fosters seamless data sharing and collaboration, contributing to improved efficiency and trust within systems (FAO, 2019).

*a. Understanding blockchain technology*

Blockchain is a distributed ledger technology that enables the secure and transparent storage of data across a network of computers, known as nodes. Each transaction or data entry is recorded in a block, which is linked to the previous block, forming a chain of blocks. This decentralized nature of blockchain eliminates the need for a central authority, making it resistant to tampering and fraud.

*b. Enhanced data security*

One of the significant advantages of leveraging blockchain in precision farming is its ability to enhance data security. Traditional data storage methods often rely on centralized servers, making them vulnerable to data breaches and cyberattacks. In contrast, blockchain employs cryptographic algorithms and consensus mechanisms to secure data. Each block in the blockchain is encrypted and linked to the previous block, creating a tamper-evident record. Additionally, consensus mechanisms, such as proof of work or proof of stake, ensure that data stored on the blockchain is verified and validated by multiple nodes, increasing the overall security of the system.

*c. Decentralization and data integrity*

By leveraging blockchain technology, precision farming can move away from centralized data storage and embrace decentralization. In a decentralized blockchain network, data is distributed among multiple nodes, eliminating the single point of failure and reducing the risk of data loss. This decentralized architecture enhances data integrity as each node in the network independently verifies and stores a copy of the blockchain. Any attempt to tamper with data in one node would be immediately detected and rejected by the other nodes, ensuring the integrity of the information stored on the blockchain.

*d. Transparency and auditability*

Blockchain's transparency and immutability features provide a powerful tool for auditing and verifying data in precision farming. Each transaction or data entry recorded on the blockchain is visible to all participants in the network, ensuring transparency and accountability. This transparency fosters trust among stakeholders, as they can independently verify the authenticity and integrity of the data. Additionally, the audit trail provided by the blockchain enables traceability of data, allowing farmers to track the origin and history of information, such as the source of seeds, fertilizers, or pesticides used in their operations.

*e. Smart contracts and automation*

Blockchain technology also introduces the concept of smart contracts, which are self-executing contracts with predefined rules and conditions. Smart contracts can automate various processes in precision farming, such as verifying the authenticity of certifications, managing supply chain transactions, or automatically triggering actions based on predefined conditions. This automation streamlines operations, reduces administrative overhead, and ensures compliance with predefined rules and standards.

*f. Data sharing and collaboration*

Blockchain technology facilitates secure data sharing and collaboration among different stakeholders in the agricultural ecosystem. With blockchain, farmers, researchers, agronomists, and suppliers can securely share data, such as crop performance metrics, soil quality reports, or weather data. By leveraging smart contracts and predefined permissions, data sharing can be controlled and governed, ensuring that sensitive information is shared only with authorized parties. This increased collaboration and data sharing can foster innovation, enable better decision-making, and drive the development of new farming practices.

Despite its potential, it is essential to consider the challenges associated with blockchain implementation in precision farming. These challenges include scalability, energy consumption, interoperability, and the need for standardization. However, as blockchain technology continues to evolve, these challenges are being addressed through the development of scalable consensus mechanisms, energy-efficient protocols, and interoperability standards.

### 4.1.5. Improving efficiency and decision-making in farming practices

Modern farming practices, particularly precision farming, encompass a diverse range of inputs and activities essential for success. These include the utilization of data-driven decision-making processes, the integration of farm management software to optimize operations, the adoption of automation and robotics for enhanced efficiency, and the seamless integration of supply chains. Such comprehensive approaches to agriculture are vital for meeting the challenges of the contemporary agricultural landscape (Dey & Shekhawat, 2021).

*a. Precision farming*

Precision farming, also known as precision agriculture or smart farming, is a key approach to improving efficiency and decision-making in farming practices. It involves the integration of advanced technologies such as remote sensing, GPS, IoT devices, and data analytics to optimize resource allocation and farm management. By utilizing precision farming techniques, farmers can analyze and manage their fields at a granular level, considering variations in soil conditions, crop health, and environmental factors. This targeted approach enables the precise application of inputs such as fertilizers, water, and pesticides, reducing waste and maximizing crop productivity.

*b. Data-driven decision-making*

Data plays a crucial role in improving decision-making in farming practices. With the proliferation of sensors, drones, satellite imagery, and other data collection tools, farmers have access to an abundance of information about their fields. By leveraging data analytics techniques and advanced algorithms, farmers can derive valuable insights from this data. For example, historical weather data combined with real-time forecasts can inform irrigation scheduling, enabling farmers to optimize water usage and reduce costs. Similarly, analyzing crop growth patterns and pest infestation data can help farmers implement timely interventions and mitigate crop losses. Data-driven decision-making empowers farmers to make informed choices based on evidence and trends, increasing efficiency and productivity.

*c. Farm management software*

Farm management software systems are powerful tools that enable farmers to streamline their operations and make informed decisions. These software solutions integrate various aspects of farming, including crop planning, inventory management, equipment scheduling, financial tracking, and more. By centralizing data and providing real-time insights, farm management software enhances efficiency by automating routine tasks, reducing paperwork, and facilitating data-driven decision-making. Farmers can track expenses, monitor yields, analyze profitability, and optimize resource allocation using these software solutions, resulting in improved operational efficiency and financial management.

*d. Automation and robotics*

Automation and robotics technologies are transforming farming practices by reducing labor requirements and increasing operational efficiency. Automated systems can perform tasks such as planting, harvesting, and irrigation with precision and consistency. Robotics technologies, including drones and autonomous vehicles, enable farmers to collect data, monitor crops, and conduct aerial surveys efficiently. These technologies not only save time and labor but also provide valuable data for decision-making. For instance, drones equipped with multispectral sensors can capture detailed crop health data, enabling farmers to identify areas of concern and take proactive measures.

*e. Supply chain integration*

Efficient farming practices go beyond the boundaries of individual farms and encompass the entire supply chain. Integrating supply chain management systems enables farmers to optimize logistics, reduce waste, and respond to market demands effectively. By connecting with processors, distributors, retailers, and consumers, farmers can streamline the flow of products, manage inventory, and ensure timely delivery. Supply chain integration facilitates better coordination, reduces inefficiencies, and enables farmers to make data-driven decisions regarding production volumes, packaging, and transportation, resulting in improved profitability and customer satisfaction.

*f. Continuous learning and innovation*

To improve efficiency and decision-making in farming practices, it is crucial to embrace continuous learning and innovation. Farmers should stay updated with the latest agricultural research, industry trends, and technological advancements. Participating in training programs, attending workshops, and collaborating with experts and peers can provide valuable insights and foster innovation. By experimenting with new techniques, adopting emerging technologies, and implementing best practices, farmers can optimize their operations, minimize risks, and stay ahead in an ever-evolving agricultural landscape.

## 4.2. Traceability and authentication of agricultural inputs and machinery

Traceability and authentication mechanisms for agricultural inputs and machinery are instrumental in safeguarding the authenticity and quality of products. These systems enable supply chain traceability for critical components like seeds, fertilizers, and pesticides, ensuring transparency in their origin and production methods. Furthermore, they bolster trust and transparency in the procurement process while effectively combating the proliferation of counterfeit and substandard agricultural inputs. This holistic approach to traceability and authentication is supported by research from Ferrández-Pastor et al. (2022), Kumar et al. (2022), Tomu (2020), Torky and Hassanein (2020) and (Xiong et al., 2020).

### 4.2.1. Ensuring authenticity and quality of agricultural inputs

*a. The significance of authentic and quality agricultural inputs*

Authenticity and quality in agricultural inputs are essential for several reasons. Firstly, farmers heavily rely on inputs to optimize crop growth, yield, and resistance to pests and diseases. By using genuine and high-quality inputs, farmers can enhance their chances of obtaining healthier, more productive crops and livestock. Secondly, consumers expect the food they consume to be safe, nutritious, and free from harmful residues. Authentic and quality inputs minimize the risk of contamination, ensuring that agricultural produce meets the required standards and regulations. Additionally, sustainable agricultural practices are dependent on the proper use of inputs, as their judicious application prevents environmental degradation and promotes long-term agricultural viability.

*b. Challenges in maintaining authenticity and quality*

Maintaining authenticity and quality in agricultural inputs is not without challenges. One of the primary obstacles is the presence of counterfeit or adulterated inputs in the market. Counterfeit products, such as fake seeds or substandard fertilizers, not only deceive farmers but also lead to poor crop performance and financial losses. Adulterated inputs, containing impurities or subpar components, can be detrimental to the environment, human health, and livestock well-being. Moreover, the complex supply chains in the agricultural sector pose challenges in tracking and verifying the origin, composition, and quality of inputs, making it difficult to ensure their authenticity.

*c. Strategies and technologies for ensuring authenticity and quality*

To address the challenges of maintaining authenticity and quality in agricultural inputs, several strategies and technologies have been developed and implemented. These include:

- Quality Assurance Programs: Governments, agricultural institutions, and industry organizations have established rigorous quality assurance programs. These programs involve setting and enforcing standards, conducting inspections and audits, and certifying inputs that meet the defined criteria. Regular testing and analysis of inputs for purity, potency, and safety are integral parts of these programs.
- Traceability Systems: Implementing robust traceability systems allows for the tracking and monitoring of inputs throughout the supply chain. Technologies such as barcodes, QR codes, and RFID tags enable accurate identification, verification, and documentation of inputs, facilitating transparency and accountability. By tracing inputs back to their source, any inconsistencies or irregularities can be quickly identified and addressed.
- Certification and Labeling: Certification schemes and labeling initiatives play a crucial role in assuring the authenticity and quality of agricultural inputs. Certifications, such as organic or fair trade labels, provide consumers with confidence in the inputs used in agricultural production. These schemes often involve strict compliance with specific criteria related to environmental sustainability, social responsibility, and product safety.
- Testing and Quality Control Laboratories: The establishment of testing laboratories dedicated to analyzing agricultural inputs is vital in ensuring their authenticity and quality. These laboratories employ advanced techniques and equipment to assess the composition, potency, and potential contaminants in inputs. Regular testing not only helps in identifying substandard or adulterated inputs but also provides valuable data for research and development efforts.

- Education and Awareness: Promoting education and awareness among farmers, agribusinesses, and consumers about the importance of authentic and quality agricultural inputs is crucial. Training programs, workshops, and campaigns can equip stakeholders with the knowledge and skills necessary to make informed decisions regarding input selection, usage, and verification.

### 4.2.2. Supply chain traceability for seeds, fertilizers, and pesticides

*a. The importance of supply chain traceability*

Supply chain traceability, particularly concerning seeds, fertilizers, and pesticides, carries substantial implications for a diverse array of stakeholders. Farmers rely on traceability measures to ensure the quality and authenticity of inputs, safeguarding their crops and investments. Manufacturers benefit from traceability systems as they enhance transparency and accountability throughout production processes, fostering trust among partners and consumers alike. Regulators leverage traceability data to enforce compliance with safety and sustainability standards, promoting responsible practices across the agricultural sector. Additionally, consumers increasingly prioritize products with transparent supply chains, underscoring the pivotal role of traceability in meeting evolving market demands.

*b. Key reasons why traceability is crucial:*

- Product authenticity and quality: Traceability enables the verification of the authenticity and quality of agricultural inputs. By tracking the origin and composition of seeds, fertilizers, and pesticides, stakeholders can ensure that they meet the required standards, are not counterfeit or adulterated, and are safe for use in agricultural practices (FAO, 2019).
- Compliance with regulations: Many countries have regulations and standards in place to safeguard the environment, human health, and food safety. Supply chain traceability facilitates compliance with these regulations by providing a mechanism to monitor and control the movement of seeds, fertilizers, and pesticides, ensuring that they meet the necessary criteria and are used responsibly.
- Product recalls and safety: In the event of a product recall or safety issue, traceability allows for the rapid identification and removal of affected batches. Timely traceability can help mitigate potential risks, protect consumers, and minimize financial losses for manufacturers and farmers.
- Sustainability and environmental impact: Traceability plays a vital role in promoting sustainable agriculture and reducing the environmental impact of

inputs. By tracking the source and usage of seeds, fertilizers, and pesticides, it becomes possible to identify and address issues such as excessive use, improper disposal, or the presence of environmentally harmful substances.

*c. Challenges in supply chain traceability*

Implementing effective supply chain traceability for seeds, fertilizers, and pesticides comes with its own set of challenges. Some of the key challenges include:

- Complex Supply Chains: Agricultural supply chains are often complex, involving multiple intermediaries, transportation, and storage facilities. This complexity can make it difficult to track and trace the movement of inputs accurately. The involvement of various stakeholders, each with their own record-keeping systems and practices, adds to the challenge.
- Data Management and Standardization: The success of traceability relies on efficient data management and standardization. However, achieving uniform data collection and sharing practices across different entities can be challenging, as they may use different formats, systems, or levels of technological sophistication.
- Verification and Authentication: Ensuring the authenticity and accuracy of data throughout the supply chain is critical. Verification mechanisms need to be in place to confirm that the information provided regarding the origin, composition, and handling of seeds, fertilizers, and pesticides is reliable and not subject to manipulation or fraud.

*d. Strategies and technologies for supply chain traceability*

To overcome the challenges associated with supply chain traceability for seeds, fertilizers, and pesticides, several strategies and technologies have emerged (Ben-Daya, Hassini, Bahroun, & Banimfreg, 2020; Casino et al., 2021; Cottrill et al., 2018; FAO, 2019; Huang, Zhou, & Liu, 2019; Mirabelli & Solina, 2020; Nesarani, Ramar, & Pandian, 2020; Prashar, Jha, Jha, Lee, & Joshi, 2020; Tripathi, Akul Krishnan, & Pandey, 2023). These include:

- Digital Tracking Systems: Utilizing digital tracking systems, such as barcodes, QR codes, or RFID tags, enables the unique identification and tracking of individual products or batches. These technologies facilitate real-time data capture, storage, and sharing, enhancing transparency and efficiency in supply chain management.
- Blockchain Technology: Blockchain technology offers a decentralized and immutable ledger that records every transaction or movement of a product. This technology provides a secure and transparent platform for

tracking and tracing inputs throughout the supply chain. It can enhance trust, integrity, and data accuracy, as information is shared and verified by multiple participants.

- Integration of Information Systems: Integrating information systems among different stakeholders, including manufacturers, distributors, retailers, and regulatory authorities, promotes seamless data exchange and enhances traceability capabilities. This integration can be achieved through standardized data formats, electronic data interchange (EDI), or application programming interfaces (APIs).
- Third-party Certification and Auditing: Involving independent third-party certification bodies and auditors can provide additional assurance regarding the authenticity and quality of inputs. These entities conduct inspections, audits, and sampling to verify compliance with standards and regulations, bolstering confidence in the traceability process.

### 4.2.3. Verifying the origin and production methods of machinery and equipment

*a. The importance of verifying origin and production methods*

Verifying the origin and production methods of machinery and equipment carries significant importance for several reasons:

- Product Quality and Performance: Verifying the origin and production methods helps ensure the quality and performance of machinery and equipment. Authentic products that adhere to high-quality manufacturing standards are more likely to offer reliability, durability, and optimal performance, leading to improved productivity and reduced downtime for farmers.
- Safety and Compliance: Verification enables the assessment of whether machinery and equipment comply with relevant safety regulations and standards. This ensures that the products meet necessary safety requirements and minimize risks to operators and the environment.
- Ethical and Sustainable Practices: Verifying the production methods of machinery and equipment allows for the evaluation of ethical and sustainable practices employed in their manufacturing. This includes considerations such as the use of environmentally friendly materials, energy efficiency, fair labor practices, and adherence to social responsibility standards.
- Warranty and Support: Verifying the origin helps farmers and buyers ensure that they are purchasing genuine products and that they will have access to after-sales support, warranties, and spare parts. Authentic

products from reputable manufacturers are more likely to provide reliable customer service and long-term support.

*b. Challenges in verifying origin and production methods*

Verifying the origin and production methods of machinery and equipment comes with its own set of challenges. Here are some key obstacles:

- Complex Supply Chains: Machinery and equipment often have complex supply chains that involve multiple suppliers, distributors, and manufacturers. This complexity can make it challenging to trace the origin and production processes accurately. Lack of transparency and information gaps within the supply chain pose difficulties in verifying authenticity and production practices.
- Counterfeit and Substandard Products: The presence of counterfeit and substandard products in the market is a significant challenge. Counterfeit machinery and equipment can deceive buyers with inferior quality, leading to performance issues, safety risks, and financial losses.
- Limited Information and Documentation: Insufficient documentation and limited availability of detailed information about the origin and production methods of machinery and equipment hinder the verification process. Incomplete or unreliable records make it difficult to confirm the authenticity and assess the adherence to quality standards.

*c. Strategies and technologies for verification*

To address the challenges associated with verifying the origin and production methods of machinery and equipment, several strategies and technologies have emerged. These include:

- Supply Chain Transparency: Promoting transparency within the supply chain is crucial. Encouraging manufacturers to disclose information about their suppliers, production processes, and materials used enhances traceability and allows for better verification. Collaboration between manufacturers, suppliers, and regulatory bodies can foster transparency and shared responsibility.
- Certification and Standards: Certifications and standards specific to machinery and equipment manufacturing provide a framework for verification. Industry-specific certifications and adherence to internationally recognised quality standards, such as ISO certifications, help establish trust and assurance in the authenticity and quality of products.
- Independent Audits and Inspections: Independent audits and inspections conducted by third-party organizations or regulatory bodies play a crucial

role in verifying origin and production methods. These audits assess compliance with quality standards, ethical practices, and safety regulations, providing reliable verification for buyers.

- Blockchain Technology: Blockchain technology offers potential solutions for verifying the origin and production methods of machinery and equipment. By leveraging a decentralized and immutable ledger, blockchain can provide a secure and transparent platform for recording and sharing information across the supply chain. This technology enhances traceability, establishes trust, and facilitates reliable verification.

### 4.2.4. Enhancing transparency and trust in the procurement process

*a. The importance of transparency and trust in procurement*

Transparency and trust in the procurement process hold significant importance for various stakeholders, including farmers, suppliers, and regulatory bodies. Here are some key reasons why enhancing transparency and trust is essential:

- Fair Competition: Transparency in the procurement process ensures fair competition among suppliers, allowing them to compete on an equal footing. It helps prevent favoritism, nepotism, and undue influence, enabling a level playing field for all participants. This fosters a competitive marketplace that encourages innovation, quality, and cost-effectiveness.
- Accountability and Governance: Transparency is a cornerstone of good governance and accountability. It enables stakeholders to monitor the procurement process and ensure compliance with regulations, policies, and ethical standards. Transparent processes discourage corruption, unethical practices, and mismanagement, promoting integrity in the agricultural industry.
- Efficient Resource Allocation: Transparent procurement processes facilitate efficient allocation of resources. When all stakeholders have access to relevant information about the procurement needs, criteria, and decision-making process, resources can be allocated effectively, optimizing the utilization of financial and material resources.
- Sustainability and Social Responsibility: Transparency in procurement contributes to promoting sustainable and socially responsible practices. It allows for the consideration of factors such as environmental impact, social equity, and ethical sourcing when selecting suppliers and procuring goods and services. This leads to more sustainable agricultural operations and responsible supply chain management.

*b. Challenges in enhancing transparency and trust*

Enhancing transparency and trust in the procurement process comes with its own set of challenges. Here are some key obstacles:

- Information Asymmetry: Information asymmetry occurs when one party possesses more or better information than others, leading to an imbalance of power and reduced transparency. Suppliers and buyers may have varying levels of knowledge and access to information about procurement needs, evaluation criteria, and decision-making processes, hindering transparency.
- Complex Procurement Procedures: Procurement processes in the agricultural industry can be complex, involving multiple steps, stakeholders, and regulations. These complexities can create confusion, delays, and opportunities for non-transparent practices. Streamlining and simplifying procurement procedures can enhance transparency and reduce administrative burdens.
- Lack of Data Management Systems: Inadequate data management systems pose a challenge to transparency in procurement. Manual record-keeping, fragmented information, and lack of data integration hinder the ability to track and analyze procurement activities. Implementing robust data management systems and utilizing technology can improve transparency and enable better data-driven decision-making.

*c. Strategies and technologies for enhancing transparency and trust*

To address the challenges associated with enhancing transparency and trust in the procurement process, several strategies and technologies have emerged (FAO, 2019). These include:

- Clear Procurement Policies and Guidelines: Establishing clear and comprehensive procurement policies and guidelines is crucial. These documents outline the procurement process, criteria, evaluation methods, and ethical standards. They provide a framework for transparent and accountable procurement practices.
- E-procurement Systems: Implementing e-procurement systems digitises the procurement process, enhancing transparency and efficiency. These systems enable online bidding, electronic document management, and real-time tracking of procurement activities. E-procurement promotes transparency by providing a centralized platform for stakeholders to access information and participate in the procurement process.
- Supplier Qualification and Evaluation: Implementing robust supplier qualification and evaluation processes ensures transparency and trust in

procurement. Transparent criteria, evaluation methods, and selection processes enable suppliers to understand the requirements and compete fairly. Regular supplier performance evaluation ensures accountability and facilitates continuous improvement.

- Stakeholder Engagement and Feedback Mechanisms: Engaging stakeholders, including farmers, suppliers, and regulatory bodies, in the procurement process promotes transparency and trust. Seeking feedback, conducting consultations, and maintaining open communication channels facilitate transparency and build trust among stakeholders.

### 4.2.5. Combating counterfeit and substandard agricultural inputs

*a. The importance of combating counterfeit and substandard inputs*

Combating counterfeit and substandard agricultural inputs is of utmost importance for several reasons:

- Productivity and Crop Health: Genuine and high-quality agricultural inputs, such as seeds, fertilizers, and pesticides, are essential for achieving optimal crop productivity and health. Counterfeit or substandard inputs can result in poor germination, weak plant growth, reduced yield, and increased vulnerability to pests and diseases.
- Environmental Protection: Counterfeit inputs may contain toxic or banned substances that can harm the environment, contaminate soil, water, and air, and disrupt ecosystem balance. Proper regulation and control of inputs are necessary to safeguard environmental sustainability and preserve natural resources.
- Farmer Livelihoods and Economic Impact: Counterfeit and substandard inputs have detrimental effects on farmer livelihoods and the agricultural economy. Farmers investing in counterfeit inputs face financial losses due to reduced yields or failed crops. Additionally, the agricultural sector as a whole suffers from decreased productivity, lower profitability, and compromised market competitiveness.
- Food Safety and Consumer Health: Substandard inputs can have residual effects on harvested crops, posing risks to food safety and consumer health. Contaminated or improperly applied pesticides, for instance, can leave harmful residues on food products, endangering consumers and undermining trust in the agricultural industry.

*b. Challenges in combating counterfeit and substandard inputs*

Combating counterfeit and substandard agricultural inputs presents several challenges. Some key obstacles include:

- Illicit Trade Networks: Counterfeit and substandard inputs often originate from complex illicit trade networks, involving multiple actors across national and international borders. These networks thrive due to limited regulation, weak enforcement, and lack of international cooperation, making it challenging to trace and disrupt their operations.
- Lack of Regulatory Oversight: Inadequate regulatory oversight and weak enforcement mechanisms contribute to the proliferation of counterfeit and substandard inputs. Insufficient resources, corruption, and gaps in monitoring and inspection systems limit the ability to identify and penalise offenders effectively.
- Limited Awareness and Education: Farmers and other stakeholders may lack awareness and knowledge about the risks associated with counterfeit and substandard inputs. Insufficient education and information-sharing platforms impede efforts to recognise and avoid fraudulent products, perpetuating the demand for such inputs.

*c. Strategies and Technologies for combating counterfeit and substandard inputs*

To address the challenges posed by counterfeit and substandard agricultural inputs, several strategies and technologies have been employed. These include:

- Strengthening Regulatory Frameworks: Enhancing regulatory frameworks and enforcement mechanisms is crucial. Governments should establish stringent regulations, quality standards, and penalties for producing, distributing, and selling counterfeit and substandard inputs. Strengthened enforcement, including targeted inspections, product testing, and prosecution of offenders, helps deter illicit activities.
- Quality Assurance and Certification Programs: Implementing quality assurance programs and certification schemes can provide assurance of the authenticity and quality of agricultural inputs. Third-party certification bodies and accreditation systems play a vital role in assessing and certifying inputs that meet prescribed standards, promoting consumer confidence and trust.
- Advanced Product Packaging and Authentication Technologies: Innovative packaging solutions and authentication technologies, such as tamper-evident seals, barcodes, holograms, and unique identifiers, help verify the authenticity of agricultural inputs. These technologies assist in distinguishing

genuine products from counterfeits and enable consumers and regulators to authenticate inputs easily.

- Supply Chain Traceability: Implementing traceability systems throughout the supply chain enhances transparency and accountability. Technologies like blockchain and RFID (Radio-frequency Identification) enable the tracking and recording of inputs from production to distribution, reducing the possibility of counterfeit or substandard products entering the market undetected.
- Awareness and Capacity Building: Conducting awareness campaigns and providing education and training programs are crucial to combat counterfeit and substandard inputs. Farmers, retailers, and other stakeholders need to be informed about the risks associated with counterfeit products, educated on product authentication methods, and empowered to make informed decisions.

### 4.3. Decentralized marketplaces for buying and selling agricultural products

Decentralized marketplaces offer innovative solutions to longstanding challenges in traditional agricultural markets. Through the implementation of peer-to-peer transactions facilitated by smart contracts, they effectively eliminate intermediaries, thereby reducing transaction costs. Moreover, these platforms enhance market accessibility for small-scale farmers while ensuring fair pricing and transparent transactions. Such advancements are evidenced in the research of Iqbal and Butt (2020) and Leduc, Kubler, and Georges (2021).

#### 4.3.1. Challenges in traditional agricultural marketplaces

*a. Key challenges in traditional agricultural marketplaces*

- Limited Market Access and Reach: One significant challenge in traditional agricultural marketplaces is limited market access and reach. Geographical constraints, inadequate transportation infrastructure, and distance from major urban centers often isolate farmers in remote areas. This restricts their ability to reach a broader customer base, resulting in limited market opportunities and reduced profitability.
- Price Volatility and Information Asymmetry: Price volatility and information asymmetry pose significant challenges for farmers in traditional marketplaces. Lack of transparent price discovery mechanisms and access to real-time market information can lead to unfair pricing practices, exploitation by intermediaries, and income instability for farmers. The absence of reliable market information hampers farmers' ability to make informed decisions about pricing and product diversification.

- Dependence on Intermediaries: Many traditional agricultural marketplaces heavily rely on intermediaries such as commission agents, brokers, and middlemen. While these intermediaries provide crucial services like logistics and market access, their presence can result in unfair practices, price manipulation, and increased transaction costs for farmers. The dominance of intermediaries can limit direct farmer-buyer interactions, leading to reduced profitability for farmers.
- Quality Control and Standardization: Ensuring quality control and standardization of agricultural produce is a challenge in traditional marketplaces. Variations in grading standards, lack of quality testing facilities, and inconsistent enforcement of regulations make it difficult for buyers to assess the quality and value of the products they purchase. This hinders marketability, reduces consumer trust, and restricts opportunities for farmers to access premium markets.
- Inefficient Supply Chain Management: Traditional agricultural marketplaces often struggle with inefficient supply chain management. Inadequate post-harvest handling, storage facilities, and transportation infrastructure result in significant post-harvest losses. Delays, spoilage, and quality deterioration during transportation and storage negatively impact farmers' profitability and overall market efficiency.

*b. Strategies and innovations to address challenges*

To address the challenges faced by traditional agricultural marketplaces, several strategies and innovations can be implemented:

- Market Information Systems: Developing and implementing market information systems that provide real-time data on prices, demand, and market trends can empower farmers with valuable information for decision-making. Such systems can be delivered through mobile applications, SMS services, or online platforms, ensuring wider access and reach.
- Direct Marketing and Farmer Producer Organizations (FPOs): Encouraging direct marketing channels and the formation of Farmer Producer Organizations (FPOs) enables farmers to bypass intermediaries and establish direct connections with buyers. This approach enhances market access, reduces transaction costs, and promotes fair pricing practices.
- Strengthening Infrastructure: Investments in transportation infrastructure, storage facilities, and cold chain systems are crucial to improve supply chain efficiency. Upgrading roads, establishing collection centers, and promoting cold storage facilities help minimize post-harvest losses and ensure the timely and safe delivery of agricultural products.

- Quality Assurance and Certification: Implementing quality assurance programs and certifications, such as Good Agricultural Practices (GAP), Organic Certification, or Geographical Indications (GI), can enhance the marketability of agricultural products. Standardized grading, quality testing, and certification systems build consumer trust and enable farmers to access premium markets.
- Digital Platforms and E-commerce: Embracing digital platforms and e-commerce solutions allows farmers to reach a broader customer base and facilitates direct transactions with buyers. Online marketplaces and platforms for agricultural trade enable farmers to showcase their products, negotiate prices, and establish direct connections with consumers or retailers.

### 4.3.2. Introduction to decentralized marketplaces

*a. Understanding decentralized marketplaces*

Decentralized marketplaces, also known as peer-to-peer (P2P) marketplaces, operate on decentralized networks, typically powered by blockchain technology. Unlike traditional marketplaces that rely on intermediaries and centralized control, decentralized marketplaces facilitate direct interactions between buyers and sellers. They eliminate the need for intermediaries and provide participants with greater control over their transactions.

*b. Key Features of decentralized marketplaces*

- Peer-to-peer Interactions: Decentralized marketplaces enable direct transactions between buyers and sellers, eliminating the need for intermediaries. Participants can engage in secure, transparent, and tamper-resistant exchanges of goods, services, or digital assets.
- Blockchain Technology: Decentralized marketplaces leverage blockchain technology to record and validate transactions. Blockchain's distributed ledger ensures transparency, immutability, and verifiability of transaction data, fostering trust among participants.
- Smart Contracts: Smart contracts are self-executing contracts with predefined terms and conditions encoded on the blockchain. They automate the execution of transactions, ensuring that agreed-upon conditions are met before funds or assets are transferred. Smart contracts enhance transparency, efficiency, and enforceability in decentralized marketplaces.
- Trust and Security: Decentralized marketplaces enhance trust and security through cryptographic encryption, decentralized consensus mechanisms,

and transparent transaction histories. Participants can verify the authenticity and integrity of transactions, reducing the risk of fraud and counterfeiting.

*c. Benefits of decentralized marketplaces*

Decentralized marketplaces offer a plethora of advantages, ranging from the elimination of intermediaries to fostering transparency, trust, and global market access. Additionally, they contribute to enhanced traceability, empowering small-scale farmers to participate more effectively in the market. These benefits underscore the transformative potential of decentralized marketplaces, as highlighted by Bhat, Huang, Sofi, and Sultan (2022).

- Elimination of Intermediaries: By removing intermediaries, decentralized marketplaces reduce transaction costs, increase efficiency, and enable participants to directly engage in trade. Farmers and buyers can interact directly, leading to fairer pricing and improved profitability for producers.
- Transparency and Trust: Decentralized marketplaces leverage blockchain's transparency and immutability to provide participants with verifiable transaction records. This transparency builds trust, fosters accountability, and reduces information asymmetry between buyers and sellers.
- Global Market Access: Decentralized marketplaces have the potential to connect participants from around the world, expanding market access for farmers. With Internet connectivity, farmers can reach a broader customer base, explore new markets, and diversify their revenue streams.
- Enhanced Traceability: Blockchain technology enables end-to-end traceability of products in decentralized marketplaces. This feature is particularly relevant in the agricultural sector, as it allows consumers to verify the origin, production methods, and supply chain of agricultural products, promoting food safety, sustainability, and ethical sourcing.
- Empowering Small-scale Farmers: Decentralized marketplaces empower small-scale farmers by providing them with direct access to markets and reducing their dependence on intermediaries. This enables them to capture a fair share of the value chain, improve their income, and strengthen their economic resilience.

*d. Potential impact on the agricultural sector*

Decentralized marketplaces hold significant potential to transform the agricultural sector in several ways, such as direct farmer-buyer relationships, market efficiency and fair pricing, and supply chain transparency and sustainability. By eliminating intermediaries, these marketplaces enable farmers to connect directly with buyers, fostering fairer pricing and improving market access for smaller producers. Additionally, decentralized platforms promote financial inclusion by providing access to markets and financial services for farmers who may have been previously marginalized or underserved (Greenow & Muñiz, 1988):

- Direct Farmer-buyer Relationships: Decentralized marketplaces enable direct interactions between farmers and buyers, fostering closer relationships and promoting a deeper understanding of consumer preferences. This facilitates customised production, reduces wastage, and promotes sustainable farming practices.
- Market Efficiency and Fair Pricing: By eliminating intermediaries and enabling transparent transactions, decentralized marketplaces promote market efficiency and fair pricing. Farmers can receive fair value for their produce, while buyers can access high-quality agricultural products at competitive prices.
- Supply Chain Transparency and Sustainability: Blockchain technology in decentralized marketplaces enhances supply chain transparency by recording every transaction and movement of products. This transparency enables stakeholders to track the origin, quality, and sustainability credentials of agricultural inputs, promoting responsible sourcing and enhancing consumer trust.
- Financial Inclusion: Decentralized marketplaces can facilitate access to financial services for farmers who are traditionally underserved by formal banking systems. Smart contracts can enable secure lending, insurance, and microfinance solutions, empowering farmers to invest in their businesses and manage risks effectively.

### 4.3.3. Enabling peer-to-peer transactions using smart contracts

*a. Understanding smart contracts*

Smart contracts are computer programs that run on blockchain networks. They are designed to facilitate, verify, and enforce the terms of an agreement between parties. Built on a code-based system, smart contracts eliminate the need for traditional legal documents and intermediaries by automating the execution of transactions based on predetermined conditions.

*b. Key features of smart contracts*

- Autonomy and Self-execution: Smart contracts are autonomous and self-executing. Once the conditions specified in the contract are met, the contract automatically executes the agreed-upon actions, such as transferring assets or releasing funds. This reduces the reliance on intermediaries and streamlines the transaction process.

- Trust and Transparency: Smart contracts operate on a decentralized blockchain network, ensuring transparency and immutability of transaction data. All parties involved can view and verify the terms and execution of the contract, fostering trust among participants.
- Security and Tamper-resistance: Smart contracts leverage cryptographic encryption to ensure the security of transactions. The decentralized nature of blockchain makes it highly resistant to tampering and fraud, enhancing the integrity of smart contract-based transactions.
- Efficiency and Cost Savings: Smart contracts eliminate the need for intermediaries, reducing transaction costs and streamlining the overall process. The automated execution of transactions eliminates manual errors and delays, increasing efficiency and saving time for all parties involved.

*c. Enabling peer-to-peer transactions*

Smart contracts play a crucial role in enabling peer-to-peer transactions in decentralized marketplaces. By leveraging the features of smart contracts, participants can engage in direct transactions with greater efficiency, transparency, and security.

- Direct Interaction: Smart contracts eliminate the need for intermediaries, allowing buyers and sellers to interact directly. This direct peer-to-peer interaction reduces transaction costs, eliminates delays, and enables participants to negotiate and agree upon terms without the involvement of third parties.
- Transparent and Secure Transactions: Smart contracts operate on a transparent and immutable blockchain, ensuring that all parties have access to the same information and transaction history. The transparency and security provided by smart contracts instill confidence in participants, reducing the risk of fraud and disputes.
- Automated Execution: Once the predefined conditions encoded in the smart contract are met, the contract automatically executes the transaction. For example, when a buyer transfers payment, the smart contract triggers the transfer of ownership of the asset or the release of funds to the seller. This automation eliminates the need for manual intervention, reducing human error and increasing transaction speed.
- Dispute Resolution: Smart contracts can also include dispute resolution mechanisms. By encoding predefined rules for resolving disputes within

the contract, participants can rely on the contract itself to mediate and resolve any conflicts that may arise, reducing the need for expensive and time-consuming legal proceedings.

*d. Benefits of smart contract-enabled peer-to-peer transactions*

The use of smart contracts to facilitate peer-to-peer transactions in decentralized marketplaces offers several benefits:

- Cost Efficiency: Smart contracts eliminate the need for intermediaries, reducing transaction costs associated with traditional marketplaces. Participants can save on fees and commissions typically charged by intermediaries, enhancing profitability and enabling more competitive pricing.
- Enhanced Trust and Security: The transparency and immutability of smart contracts on the blockchain build trust among participants. The secure nature of blockchain technology reduces the risk of fraud, providing a more secure environment for transactions.
- Speed and Efficiency: Smart contracts automate the execution of transactions, eliminating manual processes and reducing the time required for settlements. This increases the efficiency of peer-to-peer transactions and enables faster business operations.
- Streamlined Record-keeping: Smart contracts maintain a comprehensive and auditable record of all transaction details on the blockchain. This streamlines record-keeping and simplifies compliance with regulatory requirements, making audits and reporting more efficient.

### 4.3.4. Eliminating intermediaries and reducing transaction costs

*a. The role of intermediaries in traditional marketplaces*

Intermediaries play a significant role in traditional marketplaces, acting as middlemen between buyers and sellers. They provide services such as market access, logistics, quality control, and financial transactions. While intermediaries offer value in facilitating transactions, their involvement often comes at a cost.

- Transaction Fees: Intermediaries charge transaction fees or commissions for their services, which can add up and reduce the profitability for both buyers and sellers. These fees are typically based on a percentage of the transaction value and can significantly impact the overall costs.
- Information Asymmetry: Intermediaries have access to market information, which can create an information asymmetry between buyers and sellers. This information asymmetry can lead to unfair pricing practices, as

intermediaries may exploit their knowledge to manipulate prices to their advantage.

- Dependency and Limited Market Access: Relying on intermediaries can restrict market access for sellers, particularly smaller-scale producers who may face challenges in reaching a broader customer base. Intermediaries control the flow of goods and may prioritize larger suppliers, limiting opportunities for smaller players.

*b. Benefits of eliminating intermediaries*

Decentralized marketplaces offer a promising solution by eliminating intermediaries and reducing transaction costs. Here are some key benefits:

- Cost Savings: By removing intermediaries, decentralized marketplaces reduce or even eliminate transaction fees. Buyers and sellers can save on commission charges and enjoy greater profitability. These cost savings can have a significant positive impact on the bottom line, particularly for small-scale farmers and businesses.
- Direct Peer-to-peer Interactions: Decentralized marketplaces facilitate direct interactions between buyers and sellers, allowing them to negotiate terms, discuss requirements, and establish trust without the need for intermediaries. This fosters transparency, builds stronger relationships, and enhances the efficiency of transactions.
- Market Access for Small-scale Producers: Decentralized marketplaces provide small-scale producers with equal opportunities to participate in the market. By eliminating barriers imposed by intermediaries, these marketplaces empower farmers and smaller businesses to showcase their products to a wider customer base, expanding their market reach.
- Transparent Pricing: Intermediaries can create information asymmetry, resulting in unfair pricing practices. Decentralized marketplaces promote transparency by leveraging technologies like blockchain, enabling all participants to view transaction details and ensuring fairer pricing.

*c. Decentralized marketplaces and transaction cost reduction*

Decentralized marketplaces leverage blockchain technology and smart contracts to enable peer-to-peer transactions, reducing transaction costs in the following ways:

- Elimination of Intermediary Fees: In decentralized marketplaces, transactions occur directly between buyers and sellers, bypassing intermediaries. This eliminates the need for intermediary fees or commissions, resulting in significant cost savings for participants.

- Efficiency through Automation: Smart contracts automate the execution of transactions based on predefined conditions, reducing the need for manual processes and paperwork. This automation streamlines the transaction process, increases efficiency, and saves time and administrative costs.
- Transparent and Immutable Record-keeping: Blockchain technology ensures transparent and immutable record-keeping of all transactions. This eliminates the need for costly and time-consuming reconciliation processes, audits, and disputes, further reducing transaction costs.
- Direct Negotiation and Trust-building: Decentralized marketplaces enable direct peer-to-peer negotiation, reducing the need for intermediaries to facilitate communication and build trust. Direct interactions promote efficiency and reduce the chances of misunderstandings or conflicts that can lead to additional costs.

### 4.3.5. Improving access to markets for small-scale farmers

*a. The importance of market access for small-scale farmers*

Improving market access for small-scale farmers is essential for several reasons:

- Economic Empowerment: Access to markets allows small-scale farmers to generate income and improve their economic well-being. It enables them to sell their produce at fair prices, leading to increased profitability and livelihood improvement.
- Sustainable Farming Practices: Market access provides small-scale farmers with the incentive to adopt sustainable farming practices. By connecting with consumers who prioritize environmentally friendly and ethically sourced products, farmers are encouraged to implement sustainable agricultural techniques, such as organic farming and conservation practices.
- Food Security and Nutrition: Enhancing market access for small-scale farmers contributes to food security and nutrition. By diversifying market channels, farmers can increase the availability and accessibility of locally produced nutritious food, particularly in rural areas.

*b. Challenges in market access for small-scale farmers*

Small-scale farmers face several challenges in accessing markets:

- Limited Market Information: Farmers often lack timely and accurate market information, including prices, demand trends, and quality requirements. This information asymmetry hampers their ability to make informed decisions and negotiate fair deals.

- Infrastructure and Logistics: Inadequate infrastructure, such as roads, storage facilities, and transportation, poses challenges for small-scale farmers in reaching markets. Limited access to cold storage and efficient transportation systems can lead to post-harvest losses and reduced product quality.
- Quality Standards and Certification: Meeting quality standards and obtaining certifications can be costly and complex for small-scale farmers. Compliance with food safety regulations, organic certifications, and fair trade standards adds an additional layer of challenge, limiting their market opportunities.
- Limited Bargaining Power: Small-scale farmers often have limited bargaining power when negotiating with intermediaries or large buyers. This power imbalance can result in unfair pricing and unfavorable contract terms, reducing the profitability of farmers.

*c. Strategies to enhance market access for small-scale farmers*

Several strategies can help improve market access for small-scale farmers:

- Market Information Systems: Establishing market information systems that provide real-time data on prices, demand, and market trends can empower farmers with valuable information for decision-making. This enables them to identify market opportunities, align production with demand, and negotiate better prices.
- Infrastructure Development: Investing in rural infrastructure, including transportation networks, storage facilities, and marketplaces, can improve farmers' ability to reach markets efficiently and maintain product quality. Improved infrastructure reduces post-harvest losses, enhances value chain efficiency, and expands market opportunities.
- Capacity Building and Training: Providing training and capacity-building programs for small-scale farmers can enhance their knowledge and skills in areas such as quality control, post-harvest management, packaging, and marketing. This equips farmers with the necessary tools to meet market requirements and improve the marketability of their products.
- Collaboration and Collective Action: Encouraging small-scale farmers to form cooperatives or producer groups can enhance their bargaining power and negotiation capabilities. By collectively pooling resources, sharing knowledge, and collectively marketing their products, farmers can access larger markets and negotiate better terms with buyers.

- Market Diversification and Direct Marketing: Small-scale farmers can explore alternative marketing channels, such as farmers' markets, community-supported agriculture, online platforms, and farm-to-table initiatives. Direct marketing allows farmers to connect directly with consumers, bypassing intermediaries, and receiving fairer prices for their products.

### 4.3.6. Ensuring fair pricing and transparent transactions

*a. The importance of fair pricing and transparent transactions*

Fair pricing and transparent transactions in agricultural markets yield several benefits:

- Economic Sustainability: Fair pricing ensures that farmers receive reasonable compensation for their products, supporting their economic sustainability. It provides incentives for farmers to invest in quality production, adopt sustainable practices, and reinvest in their operations.
- Market Efficiency: Transparent transactions foster market efficiency by eliminating information asymmetry and enabling buyers and sellers to make informed decisions. Clear and accessible pricing information helps match supply and demand efficiently, reducing transaction costs and improving market performance.
- Trust and Collaboration: Fair pricing and transparent transactions build trust among market participants, fostering long-term relationships and collaboration. Trust is vital for the development of sustainable agricultural supply chains and encourages cooperation in addressing shared challenges.

*b. Challenges in ensuring fair pricing and transparent transactions*

Several challenges exist in achieving fair pricing and transparent transactions in agricultural markets:

- Information Asymmetry: Imbalances in market information between buyers and sellers can lead to unfair pricing practices. Farmers often lack access to timely and accurate market information, limiting their ability to negotiate fair prices and make informed decisions.
- Power Imbalances: Power imbalances between farmers and buyers, as well as intermediaries, can result in unfair pricing and unfavorable contract terms. Farmers with limited market options may have reduced bargaining power, making them more susceptible to exploitation.
- Lack of Price Discovery Mechanisms: In some agricultural markets, the absence of price discovery mechanisms contributes to opacity and

uncertainty in pricing. Limited transparency in setting and reporting prices makes it difficult for farmers to assess market conditions and negotiate fair prices.

- Inefficient Market Structures: Outdated market structures and inadequate infrastructure hinder fair pricing and transparent transactions. Limited access to markets, inefficient logistics, and inadequate storage facilities can disrupt price signals and create inefficiencies in the supply chain.

*c. Strategies to ensure fair pricing and transparent transactions*

To promote fair pricing and transparent transactions in agricultural markets, the following strategies can be employed:

- Market Information Systems: Establishing market information systems that provide real-time data on prices, supply and demand, and market trends can empower farmers with valuable information. Accessible information equips farmers with the knowledge needed to negotiate fair prices and make informed decisions.
- Strengthening Farmer Organizations: Supporting the formation and strengthening of farmer organizations, cooperatives, or producer groups enhances the bargaining power of farmers. Collectively negotiating contracts, pooling resources, and accessing markets can help balance power dynamics and promote fair pricing.
- Contractual Arrangements and Standards: Promoting the use of fair contractual arrangements and standards can ensure transparency and protect farmers' interests. Clear agreements outlining pricing mechanisms, quality requirements, and dispute resolution mechanisms enable fair transactions and reduce potential conflicts.
- Technology Solutions: Leveraging technological advancements such as blockchain and smart contracts can enhance transparency in agricultural transactions. Blockchain technology provides a decentralized and immutable record of transactions, while smart contracts automate execution based on predetermined conditions, ensuring transparency and efficiency.
- Capacity Building: Providing capacity-building programs and training to farmers on market dynamics, negotiation skills, and pricing strategies empowers them to actively participate in market transactions. Equipping farmers with the knowledge and skills necessary to navigate agricultural markets enhances their ability to achieve fair pricing.

## 4.4. Blockchain-based crop insurance and risk management

### 4.4.1. Enhancing crop insurance processes using blockchain

*a. Benefits of blockchain in crop insurance*

- Transparency and Trust: Blockchain provides a transparent and immutable ledger that records all transactions and activities related to crop insurance. This transparency builds trust among stakeholders, including farmers, insurers, and regulators, by ensuring that the data is tamper-proof and can be independently verified.
- Streamlined Claims Processing: Blockchain-based smart contracts enable automated claims processing, reducing paperwork, and eliminating intermediaries. Smart contracts execute predefined conditions, such as weather-based triggers or yield thresholds, and automatically initiate claim settlements, ensuring faster and more accurate compensation for farmers in case of losses.
- Data Security and Privacy: Blockchain's distributed ledger technology enhances data security and privacy by encrypting and storing information across multiple nodes. This reduces the risk of data breaches, fraud, and unauthorized access, safeguarding sensitive farmer and insurance-related data.
- Improved Efficiency: Blockchain streamlines administrative tasks associated with crop insurance, such as policy creation, premium calculation, and verification of farmer details. The decentralized nature of blockchain eliminates the need for reconciliations between different parties, minimizing paperwork, and reducing administrative costs.
- Enhanced Risk Assessment: Blockchain can facilitate access to reliable and transparent data sources, including historical weather patterns, soil conditions, and farming practices. By leveraging this data, insurers can make more accurate risk assessments, resulting in fairer premiums and improved underwriting processes.

*b. Challenges and considerations*

- Integration and Interoperability: Implementing blockchain in crop insurance requires collaboration between various stakeholders, including insurance companies, farmers, technology providers, and regulatory bodies. Achieving interoperability and seamless integration with existing systems can be challenging due to differing standards and infrastructure.

- Scalability and Performance: Blockchain networks need to handle a large volume of transactions and data in real-time to support the scale of crop insurance operations. Ensuring the scalability and performance of blockchain solutions remains a technical challenge that must be addressed for widespread adoption.
- Data Accuracy and Reliability: While blockchain provides immutable records, the accuracy and reliability of data fed into the system remain crucial. Garbage-in-garbage-out still applies, and efforts must be made to ensure the quality and integrity of data sources to avoid erroneous risk assessments or fraudulent claims.
- Regulatory and Legal Frameworks: The implementation of blockchain in crop insurance requires appropriate regulatory and legal frameworks to govern its use. Clear guidelines regarding data privacy, compliance, and dispute resolution need to be established to address concerns related to liability and legal enforceability.

### 4.4.2. Automating claims verification and settlement

*a. Benefits of automating claims verification and settlement*

- Speed and Efficiency: Automation significantly reduces the time required for claims verification and settlement. By leveraging AI and machine learning algorithms, insurers can automate the analysis of claim documents, policy terms, and historical data, expediting the overall process. Automated workflows ensure that claims are processed promptly, reducing customer waiting time and enhancing operational efficiency.
- Accuracy and Consistency: Manual processes are susceptible to human errors and inconsistencies. Automating claims verification and settlement minimizes the chances of mistakes by eliminating manual data entry and relying on standardized algorithms. By consistently applying predefined rules and criteria, insurers can achieve higher accuracy and ensure fair and consistent claims settlements.
- Fraud Detection and Prevention: Automated systems can employ sophisticated fraud detection algorithms to flag suspicious claims for further investigation. By analyzing patterns, anomalies, and historical data, insurers can identify potentially fraudulent claims more efficiently. Automation enables the integration of external databases and data analytics tools to validate claim information, ensuring that only legitimate claims are processed.
- Cost Savings: Automation reduces the reliance on manual labor and repetitive administrative tasks, leading to cost savings for insurers. By

optimizing resource allocation and streamlining processes, insurers can redirect human resources to more value-added activities, such as customer service or complex claim investigations. Additionally, automation minimizes the need for physical paperwork and storage, further reducing operational expenses.

- Enhanced Customer Experience: Automating claims verification and settlement improves the overall customer experience. By expediting the process, providing prompt updates, and ensuring fair settlements, insurers can enhance customer satisfaction and loyalty. Automation enables self-service portals, allowing customers to track their claims in real time, reducing the need for manual inquiries.

*b. Challenges and considerations*

- Data Quality and Integration: Automation relies heavily on accurate and structured data. Insurers need to ensure the availability and quality of data from various sources, such as claim forms, policy documents, and external databases. Integration with legacy systems and disparate data sources can pose challenges that require careful planning and implementation.
- Regulatory Compliance: Automating claims verification and settlement must adhere to regulatory requirements and compliance standards. Insurers need to ensure that automated processes meet legal and ethical obligations, such as data privacy, security, and anti-fraud measures. Compliance frameworks must be designed and integrated into automated workflows.
- Technological Implementation: Implementing automation technologies requires a comprehensive understanding of the existing infrastructure, systems, and processes. Insurers must invest in the necessary technological infrastructure, including AI platforms, machine learning algorithms, and robotic process automation tools. Adequate training and change management initiatives are essential to ensure a smooth transition and successful implementation.
- Human Expertise and Oversight: While automation offers significant benefits, human expertise remains crucial in handling complex claims, exceptions, and decision-making. Insurers should strike a balance between automation and human intervention, ensuring that experienced professionals oversee and manage automated processes effectively.

### 4.4.3. Leveraging smart contracts for parametric insurance

*a. Benefits of smart contracts in parametric insurance*

- Automated and Transparent Triggers: Smart contracts are self-executing agreements that automatically trigger predefined actions when specific conditions are met. In parametric insurance, these conditions can be encoded into smart contracts, such as weather data reaching a certain threshold or a specified index value being exceeded. The automation of triggers ensures timely and accurate payouts without the need for manual intervention or claims adjustment.
- Seamless Claims Processing: Smart contracts simplify the claims process in parametric insurance. Once the predetermined trigger conditions are met, the smart contract automatically initiates the payout process. This eliminates the need for time-consuming claim assessments, reducing administrative costs and improving the speed of compensation for policyholders.
- Trust and Transparency: Smart contracts operate on blockchain technology, providing a transparent and immutable ledger of all transactions and activities. This transparency builds trust among stakeholders by ensuring that the terms and conditions of the insurance contract are executed as agreed upon. All parties involved, including insurers, policyholders, and reinsurers, can independently verify the transactions recorded on the blockchain, enhancing transparency and reducing disputes.
- Cost Efficiency: Smart contracts streamline administrative processes and eliminate intermediaries, resulting in cost savings for insurers. By automating claims verification and settlement, insurers can reduce operational expenses associated with manual paperwork, claims adjustment, and processing. The removal of intermediaries also minimizes the administrative fees and commissions traditionally associated with insurance transactions.
- Enhanced Accessibility: Parametric insurance, powered by smart contracts, can extend insurance coverage to underserved markets or regions with limited access to traditional insurance. The automation and efficiency of smart contracts enable insurers to offer parametric products at more affordable premiums, making insurance more accessible to a wider population.

*b. Challenges and considerations*

- Data Accuracy and Reliability: Parametric insurance heavily relies on accurate and reliable data sources, such as weather data or indices. The accuracy and quality of the data used to trigger smart contracts are crucial to ensure the fairness and effectiveness of parametric insurance. Data sources must be trustworthy and verifiable to avoid erroneous triggers or false claims.

- Integration and Standardization: Integrating smart contracts into existing insurance systems and infrastructure can be challenging. Insurers need to ensure compatibility and interoperability between different platforms and networks. Additionally, standardizing the parameters, triggers, and data sources used in parametric insurance contracts is essential to facilitate seamless integration and promote market adoption.
- Legal and Regulatory Frameworks: Smart contracts in parametric insurance must adhere to legal and regulatory frameworks specific to each jurisdiction. The legality, enforceability, and compliance of smart contracts must be carefully considered and aligned with local laws, contractual obligations, and consumer protection regulations.
- Technological Complexity: Developing and implementing smart contracts require expertise in blockchain technology, programming languages, and contract design. Insurers need to invest in the necessary technological infrastructure, skilled resources, and partnerships with technology providers to ensure successful adoption and integration.

### 4.4.4. Reducing fraud and improving trust in insurance coverage

*a. Detecting and preventing fraud*

- Advanced Data Analytics: Insurance companies are leveraging advanced data analytics techniques to detect patterns, anomalies, and red flags associated with fraudulent activities. By analyzing vast amounts of structured and unstructured data, insurers can identify suspicious claims, identify potential fraudsters, and deploy proactive measures to prevent fraud before it occurs.
- Machine Learning and AI: Machine learning algorithms and AI-powered models play a crucial role in fraud detection and prevention. By learning from historical data, these models can identify fraudulent patterns, recognise outlier behavior, and improve risk assessment capabilities. Insurers can use these technologies to automate fraud detection processes, reduce false positives, and enhance the accuracy of fraud identification.
- Social Network Analysis: Social network analysis techniques analyze the relationships between policyholders, claimants, and other entities to uncover hidden connections and identify organised fraud networks. By mapping social networks and analyzing transactional data, insurers can

identify suspicious networks and take appropriate measures to prevent fraud.

- Predictive Modelling: Predictive modelling utilizes historical data and statistical techniques to predict the likelihood of fraudulent claims. By developing predictive models, insurers can identify high-risk cases, prioritize investigations, and allocate resources effectively. These models can also be updated in real-time, adapting to evolving fraud patterns and enhancing fraud prevention efforts.

*b. Improving trust in insurance coverage*

- Transparent and Clear Policies: Insurance policies should be transparent, written in clear language, and easily understandable by policyholders. Ambiguous or complex policy wordings can lead to misunderstandings and disputes, eroding trust. Insurers should focus on simplifying policy language, providing comprehensive explanations, and ensuring policyholders have a clear understanding of their coverage.
- Enhanced Customer Communication: Regular and transparent communication with policyholders is crucial for building trust. Insurers should proactively communicate policy details, updates, and changes to their customers. This can include providing information on coverage, claim processes, and any modifications to the policy terms. Timely and open communication fosters trust and ensures policyholders are well-informed.
- Efficient Claims Handling: Efficient and fair claims handling processes are essential for building trust in insurance coverage. Insurers should strive to streamline claims procedures, provide prompt responses, and ensure timely settlements. Leveraging automation and digital technologies can expedite claims processing, reducing delays and enhancing customer satisfaction.
- Customer Feedback and Satisfaction Surveys: Regular feedback from policyholders is valuable for understanding their experiences and improving service quality. Insurers can conduct customer satisfaction surveys, gather feedback through online platforms, and actively listen to policyholders' concerns. This information can be used to address gaps, improve processes, and demonstrate a commitment to customer-centricity.
- Collaboration and Partnerships: Building trust in insurance coverage requires collaboration between insurers, regulators, and other stakeholders. Insurers should actively engage in partnerships with law enforcement agencies, industry associations, and technology providers to share information, exchange best practices, and collectively combat fraud.

Collaborative efforts strengthen the industry's ability to detect and prevent fraudulent activities effectively.

### 4.4.5. Mitigating risks and improving resilience in agriculture

Strategies and approaches to mitigating risks and improving resilience in agriculture

a. Diversification of Crops and Livestock: Diversification is a key strategy for mitigating risks in agriculture. By cultivating a variety of crops and raising multiple livestock species, farmers reduce their vulnerability to specific risks. Diversification can help balance the impacts of adverse weather conditions, diseases, and market fluctuations. Additionally, promoting resilient and climate-adaptive crop varieties and livestock breeds further enhances agricultural resilience.

b. Integrated Pest Management (IPM): Pests and diseases pose significant threats to agricultural productivity. Implementing Integrated Pest Management practices involves a combination of preventive measures, monitoring, and targeted interventions. This approach reduces reliance on chemical pesticides and encourages the use of biological control methods, crop rotation, and resistant varieties. IPM enhances pest and disease management while minimizing environmental risks and preserving natural ecosystems.

c. Sustainable Water Management: Water scarcity and inefficient water management practices pose substantial risks to agriculture, particularly in regions prone to drought. Implementing sustainable water management techniques, such as precision irrigation, rainwater harvesting, and water-efficient practices, can mitigate these risks. Investing in water storage infrastructure and promoting water-saving technologies ensure optimal water usage and enhance agricultural resilience in water-stressed areas.

d. Climate-resilient Farming Practices: Climate change poses a significant challenge to agriculture, affecting weather patterns, precipitation, and temperature regimes. Adapting farming practices to changing climatic conditions is crucial for mitigating risks and ensuring agricultural productivity. This includes adopting climate-smart practices such as conservation agriculture, agroforestry, and soil conservation techniques. These practices improve soil health, water retention, and overall ecosystem resilience.

e. Access to Weather Information and Early Warning Systems: Timely and accurate weather information is critical for farmers to make informed decisions and manage risks effectively. Access to weather forecasts, climate data, and early warning systems enables farmers to prepare for extreme weather events, disease outbreaks, and market fluctuations. Establishing robust weather monitoring networks and disseminating information through mobile apps, SMS alerts, and local extension services empowers farmers with real-time information and improves their capacity to respond to risks.

f. Crop Insurance and Risk Transfer Mechanisms: Crop insurance and risk transfer mechanisms play a vital role in mitigating risks and protecting farmers against unforeseen events. Crop insurance provides financial support in case of crop failure due to natural disasters, pests, or diseases. Innovative insurance products, such as index-based insurance, use weather or yield indices to determine payouts, offering more affordable and accessible coverage. Integrating technology, such as remote sensing and satellite data, improves the accuracy and efficiency of insurance assessments and payouts.

g. Capacity Building and Knowledge Sharing: Enhancing resilience in agriculture requires continuous capacity building and knowledge sharing among farmers, extension services, and researchers. Training programs, workshops, and farmer field schools can educate farmers on best practices, risk management strategies, and technological advancements. Facilitating information exchange and networking platforms enables farmers to learn from each other's experiences, share innovations, and collectively strengthen agricultural resilience.

# Chapter 5

# Livestock management and animal welfare

*Cultivating Compassion: Upholding Integrity, Welfare, and Quality in Livestock Management* - Unknown

**Abstract**

This chapter focuses on the integration of blockchain technology in livestock management and the assurance of animal welfare. It explores the potential of blockchain to improve the monitoring and tracking of livestock health and conditions. Additionally, the chapter examines how blockchain can contribute to ensuring the authenticity and traceability of animal products. The chapter highlights the use of Internet of Things (IoT) devices, sensors, and blockchain in enhancing livestock management practices. The chapter begins by examining the potential benefits of blockchain technology in improving livestock management and ensuring animal welfare. Next, the chapter explores the use of IoT devices, sensors, and blockchain in monitoring and tracking livestock health and conditions. Furthermore, the chapter presents examples of blockchain solutions that ensure the authenticity and traceability of animal products.

**Keywords**: Blockchain technology, Livestock management, Animal welfare, Monitoring, Tracking, Authenticity, Traceability, Internet of Things, Sensors, Livestock health

**Chapter contents**

- Ensuring Transparency and Authenticity in Livestock Supply Chains
- Tracking and Monitoring Animal Health Records Using Blockchain
- Authentication of Organic and Sustainable Farming Practices

***

## 5.1. Ensuring transparency and authenticity in livestock supply chains

Livestock supply chains encounter multifaceted challenges, including a fragmented and intricate structure, the imperative of quality and safety assurance, persistent animal welfare concerns, and the pressing issue of environmental impact. Blockchain technology presents promising solutions to mitigate these hurdles by enabling seamless traceability of livestock from farm to consumer, facilitating verification of origin and breed authenticity, thwarting the illicit sale of stolen or trafficked animals, and bolstering consumer confidence in meat and dairy products. Researchers such as Agrawal, Kumar, Pal, Wang, and Chen (2021) and Blokdyk (2021) have underscored the transformative potential of blockchain in revolutionizing livestock supply chains.

### 5.1.1. Challenges in livestock supply chains

*a. Fragmented and complex structure*

One of the primary challenges in livestock supply chains is their fragmented and complex structure. These supply chains involve multiple stakeholders, including farmers, ranchers, processors, distributors, wholesalers, retailers, and consumers. The coordination and integration of these diverse entities present significant logistical and communication challenges (Cottrill et al., 2018). Fragmentation often leads to information asymmetry, where key data and insights are not effectively shared among stakeholders. This lack of transparency can result in inefficiencies, delays, and increased costs throughout the supply chain. Moreover, it hampers traceability efforts, making it difficult to identify and respond to issues such as food safety concerns or disease outbreaks.

*b. Quality and safety assurance*

Ensuring the quality and safety of livestock products is crucial to protect consumer health and maintain consumer trust. However, guaranteeing consistent quality and safety standards across the entire supply chain is a complex task. Challenges arise at various stages, from on-farm practices to transportation, processing, and storage. Livestock production involves various factors that can affect product quality, including animal health, genetics, nutrition, and handling practices. Maintaining rigorous hygiene standards during transportation and processing is equally important to prevent contamination and ensure food safety. Additionally, the emergence of new pathogens and antimicrobial resistance further complicates the task of ensuring safe and high-quality products.

*c. Animal welfare concerns*

The welfare of animals within the livestock supply chain is an area of increasing concern for consumers, regulatory bodies, and advocacy groups. Addressing animal welfare concerns is not only an ethical obligation but also crucial for maintaining the reputation and sustainability of the industry. However, ensuring high standards of animal welfare throughout the supply chain presents significant challenges. Factors such as crowded housing, long transportation times, and inadequate handling practices can lead to stress, injuries, and compromised welfare. Implementing and enforcing consistent welfare standards across different production systems and geographical regions is a complex endeavour.

*d. Environmental impact*

Livestock production has a substantial environmental footprint, contributing to greenhouse gas emissions, deforestation, water pollution, and biodiversity loss. The challenge for livestock supply chains is to minimize their impact on the environment while meeting the growing demand for animal-based products (Galvez, Mejuto, & Simal-Gandara, 2018). Reducing the environmental impact requires sustainable practices throughout the supply chain. This includes optimizing feed efficiency, managing manure and waste effectively, adopting climate-smart farming techniques, and exploring alternative protein sources. However, implementing these changes at scale, particularly in developing countries where livestock production is expanding rapidly, is a formidable challenge.

*e. Market volatility and global trade*

Livestock supply chains are highly susceptible to market volatility and global trade dynamics. Fluctuating input prices, currency exchange rates, and trade policies can significantly impact the profitability and stability of the entire supply chain. Changes in consumer preferences, dietary patterns, and emerging market trends also pose challenges. For example, the rising demand for plant-based alternatives and the increasing scrutiny of livestock production's environmental impact can disrupt traditional supply chains. Adapting to these changes requires agility, market intelligence, and effective risk management strategies.

### 5.1.2. Traceability of livestock from farm to consumer

Traceability plays a crucial role in ensuring the safety, quality, and transparency of livestock products throughout the supply chain. By tracing the journey of livestock from farm to consumer, stakeholders can identify and address issues such as disease outbreaks, food safety concerns, and ethical considerations. Implementing robust traceability systems not only enhances consumer confidence but also enables swift responses to emergencies, ultimately promoting the integrity and sustainability of the livestock industry (Galvez et al., 2018).

*a. The importance of traceability*

- The Significance of Traceability

  Traceability refers to the ability to track and trace the movement of livestock and their products at every stage of the supply chain. It involves capturing and recording key information, such as the origin of animals, their health records, feed sources, and any treatments or interventions they have received. This information provides a comprehensive overview of the livestock's history and enables stakeholders to make informed decisions regarding their handling, processing, and distribution.
- The significance of traceability can be summarised in three key aspects:
- Food Safety: Traceability allows for swift identification and containment of foodborne illnesses or contamination incidents. In the event of a recall or outbreak, traceability systems enable targeted recalls, reducing the impact on both public health and the industry's reputation.
- Quality Assurance: Traceability provides assurance to consumers about the quality, authenticity, and ethical considerations related to the livestock products they purchase. It helps build trust and allows consumers to make informed choices aligned with their values and preferences.
- Supply Chain Efficiency: Traceability enhances the efficiency of supply chain operations by facilitating inventory management, reducing waste, and optimizing logistics. It enables stakeholders to identify bottlenecks, improve processes, and respond promptly to market demands.

*b. Challenges in implementing traceability systems*

Implementing effective traceability systems in livestock supply chains presents several challenges (Galvez et al., 2018; Kamilaris et al., 2019). These challenges include:

- Data Management: Managing and maintaining a vast amount of data throughout the supply chain can be daunting. It requires standardized data formats, interoperability between different stakeholders' systems, and robust data storage and retrieval mechanisms.
- Information Asymmetry: Information sharing among stakeholders is crucial for traceability. However, stakeholders often operate with different record-keeping practices and technologies, leading to information asymmetry. This hampers the accuracy and completeness of traceability data.
- Infrastructure and Technology: Implementing traceability systems requires appropriate infrastructure and technology, including data capture devices,

communication networks, and information systems. In some regions, especially in developing countries, the lack of infrastructure can pose significant challenges.

- Cost and Adoption: Implementing traceability systems can be costly, particularly for small-scale producers or those with limited resources. The cost of technology, training, and compliance with traceability standards can be prohibitive, hindering widespread adoption.

*c. Solutions and best practices*

To overcome the challenges associated with traceability, several solutions and best practices have emerged (Astill et al., 2019; Kamilaris et al., 2019):

- Standardization: Developing and adopting standardized protocols, data formats, and terminology across the supply chain enhances interoperability and data consistency. Standards such as unique identification codes, barcodes, or radio frequency identification (RFID) tags facilitate accurate and efficient traceability.
- Technology Integration: Leveraging advanced technologies like blockchain, Internet of Things (IoT), and cloud computing can streamline data collection, storage, and sharing. These technologies improve real-time data visibility, enabling stakeholders to access accurate information promptly.
- Collaboration and Partnerships: Collaboration among stakeholders is vital for successful traceability implementation. Governments, producers, processors, retailers, and technology providers need to work together to define traceability standards, share data, and ensure compliance.
- Capacity Building: Providing training and technical assistance to stakeholders, particularly small-scale producers, is crucial for effective traceability. Capacity-building programs can enhance understanding, adoption, and compliance with traceability requirements.
- Regulatory Frameworks: Governments play a crucial role in establishing regulatory frameworks that mandate traceability standards and enforce compliance. Clear regulations provide a level playing field and incentivize stakeholders to invest in traceability systems.

### 5.1.3. Blockchain for verifying the origin and breed of livestock

*a. The need for verifying origin and breed*

Verifying the origin and breed of livestock is essential for several reasons:

- Consumer Trust: Consumers are increasingly interested in knowing the origin of the food they consume and the specific breed of the animals involved. Verifying origin and breed instills confidence, assures quality, and enables consumers to make informed choices based on their preferences and values.
- Regulatory Compliance: Governments and regulatory bodies often require accurate documentation of the origin and breed of animals for various purposes, including disease control, market access, and certification programs. Verifying origin and breed ensures compliance with relevant regulations and standards.
- Intellectual Property Protection: Certain livestock breeds may have unique genetic traits or specific breeding programs associated with them. Verifying breed authenticity helps protect intellectual property rights and ensures fair trade practices.

*b. How blockchain works for verifying origin and breed*

Blockchain is a distributed ledger technology that allows multiple parties to maintain a shared, transparent, and tamper-proof record of transactions or information (Cao et al., 2021; Galvez et al., 2018; Jannat et al., 2021). Here's how blockchain can be used to verify the origin and breed of livestock:

- Data Capture and Registration: Relevant information about the livestock, such as its origin, breed, genetic data, health records, and ownership details, is captured and registered on the blockchain. This information can be entered by various stakeholders involved in the supply chain, including farmers, breeders, veterinarians, and regulatory authorities.
- Immutable Records: Once recorded on the blockchain, the data becomes immutable, meaning it cannot be altered or tampered with without leaving a trace. This ensures the integrity and reliability of the information throughout the supply chain.
- Verification and Consensus: Blockchain technology enables consensus among multiple parties regarding the accuracy of the recorded information. Validators, such as breed associations, certification bodies, or regulatory agencies, can verify and validate the authenticity of the recorded origin and breed information.

- Transparency and Accessibility: Blockchain provides transparency by allowing authorized stakeholders to access the information recorded on the blockchain. This transparency enhances trust among stakeholders, as they can independently verify the origin and breed of the livestock they handle.
- Smart Contracts and Automation: Smart contracts, which are self-executing contracts with predefined rules, can be utilized on the blockchain. They can automate certain processes related to origin and breed verification, such as issuing certificates or triggering alerts when discrepancies are identified.

*c. Benefits and challenges of using blockchain for verification*

Using blockchain for verifying the origin and breed of livestock offers several benefits (Galvez et al., 2018):

- Enhanced Trust and Transparency: Blockchain's transparency and immutability enhance trust among stakeholders by providing a verifiable and tamper-proof record of origin and breed information.
- Improved Efficiency and Accuracy: Blockchain streamlines the verification process by eliminating manual paperwork and reducing the need for intermediaries. This improves efficiency, reduces administrative burden, and minimizes the chances of human errors or fraud.
- Facilitates Market Access and Trade: Verified origin and breed information documented on the blockchain can facilitate market access for livestock products, especially in cases where specific breed requirements or geographical indications are involved. It enables buyers and importers to make informed decisions based on reliable information.
- Intellectual Property Protection: Blockchain can help protect intellectual property rights associated with specific livestock breeds by recording ownership, breeding programs, and genetic information in a secure and tamper-proof manner.

*d. Challenges of implementing blockchain for verifying origin and breed of livestock*

- Data Accuracy and Integrity: The reliability of blockchain-based verification depends on the accuracy and integrity of the data entered into the system. Ensuring accurate and standardized data entry is essential for maintaining the credibility of blockchain records.
- Interoperability and Integration: Integrating blockchain with existing systems and ensuring interoperability among different stakeholders' databases can

be challenging. Effective data sharing and collaboration among multiple parties are crucial for a successful implementation.

- Cost and Infrastructure: Implementing blockchain technology requires investment in infrastructure, technical expertise, and ongoing maintenance. The costs associated with adopting and maintaining a blockchain system can be a barrier, particularly for smaller stakeholders in the livestock supply chain.

### 5.1.4. Preventing the sale of stolen or illegally traded animals

*a. The consequences of illegal animal trade*

Illegal animal trade encompasses various activities, including poaching, smuggling, and trafficking of endangered or protected species, as well as the theft of livestock. The consequences of this illicit trade are far-reaching:

- Biodiversity Loss: The illegal trade of wildlife contributes to the decline of endangered species, disrupts ecosystems, and threatens biodiversity conservation efforts. Similarly, the theft of livestock can result in genetic erosion and the loss of valuable breeding stock.
- Animal Welfare Concerns: Animals involved in illegal trade often endure inhumane treatment, cramped conditions, and long transportation periods. They may be subjected to stress, injuries, or even death during smuggling operations.
- Disease Transmission: Illegal animal trade can facilitate the spread of diseases, including zoonotic diseases, which pose risks to human health. The unregulated movement of animals increases the chances of introducing and spreading pathogens across regions.
- Economic Impact: The theft of livestock and the illegal trade of animals undermine the integrity of legitimate supply chains. It leads to economic losses for farmers, ranchers, and legitimate businesses, disrupts market stability, and compromises consumer confidence.

*b. Strategies to prevent the sale of stolen or illegally traded animals*

Preventing the sale of stolen or illegally traded animals requires a multi-faceted approach involving various stakeholders and strategies:

- Legislation and Enforcement: Governments should enact and enforce robust laws and regulations to combat illegal animal trade. These laws should include stringent penalties for offenders, provisions for asset seizure, and measures to disrupt criminal networks involved in the trade.

- Enhanced Border Controls and Customs: Strengthening border controls and customs procedures is crucial for intercepting illegal animal shipments. This includes investing in technology, training personnel, and improving cooperation among law enforcement agencies at national and international levels.
- Collaboration among Stakeholders: Collaboration among governments, law enforcement agencies, conservation organizations, and industry stakeholders is essential to combat illegal animal trade effectively. Sharing intelligence, resources, and expertise can enhance enforcement efforts and disrupt illicit supply chains.
- Public Awareness and Education: Raising public awareness about the negative impacts of illegal animal trade is crucial for generating support and encouraging responsible consumer behavior. Education programs can help individuals recognise and report suspicious activities, supporting law enforcement efforts.
- Technology and Traceability: Leveraging technology, such as DNA testing, microchipping, and electronic identification, can help establish traceability and verify the legal status of animals. This can assist in identifying stolen or illegally traded animals and deterring their entry into legitimate supply chains.
- Strengthening Market Regulations: Implementing robust market regulations and certification programs can help ensure that animals traded within legitimate supply chains are legally sourced. These regulations can include requirements for documentation, licensing, and proof of origin for traded animals.
- International Cooperation: Illegal animal trade is often transnational in nature, requiring international cooperation and collaboration. Governments should work together through multilateral agreements, information-sharing networks, and joint enforcement operations to combat cross-border illegal trade effectively.

*c. Promoting responsible consumer choices*

Consumer awareness and demand for ethically sourced animals play a significant role in preventing the sale of stolen or illegally traded animals. Consumers can:

- Choose Certified and Traceable Products: Opting for certified and traceable animal products, such as those bearing quality assurance labels or verifiable origin information, can contribute to responsible consumer choices.

- Support Sustainable and Ethical Practices: Consumers can support farmers, ranchers, and businesses that prioritize animal welfare, biodiversity conservation, and sustainability in their operations.
- Report Suspicious Activities: Reporting any suspicions of stolen or illegally traded animals to relevant authorities can aid in investigations and enforcement efforts.

### 5.1.5. Enhancing consumer confidence in meat and dairy products

*a. Understanding consumer concerns*

Consumers today have growing concerns about the meat and dairy products they consume. Some common areas of concern include:

- Food Safety: Consumers expect meat and dairy products to be safe and free from contaminants, pathogens, and harmful residues. They want reassurance that the products they purchase have undergone rigorous quality control measures.
- Animal Welfare: Increasingly, consumers are concerned about the welfare of the animals involved in meat and dairy production. They expect animals to be raised in humane conditions, free from cruelty, and with access to appropriate care and living environments.
- Environmental Impact: Consumers are becoming more aware of the environmental impact associated with meat and dairy production. They seek products that are produced in a sustainable manner, with consideration for resource conservation, greenhouse gas emissions, and land use.
- Ethical Sourcing: Consumers are interested in the ethical aspects of meat and dairy production. They want to know if animals have been raised without the use of unnecessary antibiotics, hormones, or other potentially harmful substances.

*b. Strategies to enhance consumer confidence*

To address consumer concerns and enhance confidence in meat and dairy products, several strategies can be implemented (Katsikouli et al., 2021; Kayikci, Subramanian, Dora, & Bhatia, 2022; Makkar & Costa, 2020):

- Transparency and Traceability: Providing transparent information about the production processes, sourcing, and supply chains of meat and dairy products is essential. Traceability systems can enable consumers to track the journey of products from farm to fork, giving them confidence in their origins and production methods.

- Certification and Quality Assurance: Third-party certifications and quality assurance programs can play a significant role in assuring consumers about the safety, quality, and sustainability of meat and dairy products. These certifications can cover areas such as organic production, animal welfare standards, and environmentally-friendly practices.
- Animal Welfare Standards: Implementing and adhering to robust animal welfare standards can address consumer concerns about the ethical treatment of animals. This includes providing adequate space, proper nutrition, and access to veterinary care.
- Sustainable Production Practices: Emphasising and implementing sustainable production practices, such as reducing environmental impact, conserving resources, and minimizing waste, can resonate with environmentally-conscious consumers. Communicating these practices can build trust and confidence in the sustainability of meat and dairy products.
- Consumer Education and Engagement: Educating consumers about the production processes, challenges, and efforts undertaken by the industry can help bridge the gap in understanding. Engaging with consumers through public outreach programs, farm visits, and social media can foster transparency and build trust.
- Collaboration with Stakeholders: Collaboration among stakeholders, including farmers, processors, retailers, consumer advocacy groups, and governmental bodies, is crucial in addressing consumer concerns. Working together can lead to the development of industry-wide standards, best practices, and initiatives that prioritize consumer needs.
- Continuous Improvement and Innovation: Embracing continuous improvement and innovation is key to meeting evolving consumer expectations. This includes investing in research and development to improve production methods, enhance product safety, and reduce environmental impact.

*c. Building trust and reassurance*

Building trust and reassurance among consumers requires ongoing commitment and communication from the livestock industry. This can be achieved through:

- Transparent Communication: Engaging in open and honest communication with consumers about product attributes, production practices, and industry initiatives builds trust. Clear labeling and accurate product information play a vital role in providing the necessary information for consumers to make informed choices.

- Monitoring and Compliance: Implementing robust monitoring and compliance measures ensures that industry standards and regulations are followed consistently. Regular audits, inspections, and verification processes contribute to maintaining product integrity and consumer confidence.
- Swift Response to Concerns: Addressing consumer concerns promptly and transparently is essential. Promptly investigating and resolving issues, implementing corrective actions, and communicating these efforts to consumers demonstrate a commitment to their well-being.

## 5.2. Tracking and monitoring animal health records using blockchain

Blockchain presents unparalleled opportunities for tracking and monitoring animal health records through various means, including the establishment of blockchain-based digital identities for livestock. This framework ensures secure and tamper-proof storage of health records, enabling real-time monitoring of vaccination and treatment history. By leveraging blockchain, stakeholders can uphold food safety standards and proactively prevent disease outbreaks. Dutta, Choi, et al. (2020) and Iftekhar, Cui, Hassan, and Afzal (2020) have extensively explored the potential of blockchain technology in revolutionizing animal health management and disease prevention strategies.

### 5.2.1. Importance of animal health records

*a. Monitoring and managing animal health*

Animal health records serve as a vital tool for monitoring and managing the health of individual animals and populations. Here are some key reasons why animal health records are important:

- Disease Surveillance and Control: Animal health records help identify disease outbreaks, track disease prevalence, and support disease control measures. Timely access to accurate health records facilitates early detection, response, and prevention of the spread of diseases, protecting both animal and human health.
- Treatment and Medication Management: Animal health records provide a detailed history of treatments, medications, vaccinations, and preventive measures administered to individual animals. This information helps veterinarians make informed decisions regarding treatment plans, dosage adjustments, and managing withdrawal periods to ensure food safety and compliance with regulations.
- Herd or Flock Management: Animal health records enable farmers to track and manage the health status, reproductive history, and productivity of

their livestock. This information helps identify trends, plan breeding programs, make culling decisions, optimize nutrition, and implement targeted health interventions.

- Genetic Improvement: Animal health records contribute to genetic improvement programs by providing data on heritable traits, disease resistance, and performance indicators. This information assists in selective breeding decisions, helping farmers breed healthier and more productive livestock.

*b. Food safety and traceability*

Animal health records play a critical role in ensuring the safety and traceability of livestock products:

- Food Safety Assurance: Animal health records help establish the health status of individual animals, ensuring that only healthy animals enter the food supply chain. This contributes to food safety by minimizing the risk of transmitting diseases or contaminants through meat, milk, and other livestock products.
- Product Labeling and Certification: Accurate health records support product labeling and certification schemes that provide information to consumers about the health status, welfare practices, and quality attributes of livestock products. This transparency enhances consumer trust and facilitates informed purchasing decisions.
- Disease Outbreak Investigation: In the event of a disease outbreak or food safety incident, animal health records aid in tracing the source and identifying affected animals. This facilitates targeted interventions, containment measures, and recall procedures, minimizing the impact on public health and the industry.

*c. Regulatory compliance and market access*

Maintaining animal health records is often a legal requirement and a prerequisite for market access:

- Regulatory Compliance: Many countries have regulations and standards that mandate the recording and reporting of animal health information. Compliance with these requirements is necessary to meet legal obligations, participate in certification programs, and access domestic and international markets.
- Disease Control Programs: Animal health records support disease control programs implemented by governments and industry organizations. These programs often involve disease testing, vaccination campaigns, movement

restrictions, and disease-free certifications, all of which rely on accurate health records.

- Market Access and Trade: Animal health records play a crucial role in gaining market access and ensuring compliance with import/export requirements. Countries or trading partners may require documentation on the health status, vaccination history, and treatment records of animals to ensure disease-free status and prevent the introduction of pathogens.

### 5.2.2. Blockchain-based digital identity for livestock

*a. Understanding blockchain-based digital identity*

Blockchain-based digital identity for livestock refers to the use of blockchain technology to create and manage unique, verifiable identities for individual animals. These digital identities can contain a wide range of information, including genetic data, health records, ownership history, breeding history, and more. By leveraging blockchain's decentralized and tamper-proof nature, digital identities provide a reliable and transparent system for tracking and verifying crucial information throughout an animal's life cycle.

*b. Benefits of blockchain-based digital identity*

Implementing blockchain-based digital identity for livestock offers several benefits to stakeholders across the supply chain, such as traceability and transparency, animal health and welfare management, and genetic improvement and breeding programs. Through blockchain technology, stakeholders can securely track the entire lifecycle of each animal, ensuring transparency in sourcing and production practices. Moreover, blockchain facilitates market access and certification by providing immutable records of livestock provenance and compliance with regulatory standards, while also bolstering fraud prevention efforts and deterring theft through secure digital identities for each animal (Neethirajan & Kemp, 2021):

- Traceability and Transparency: Digital identities enable the seamless tracing of an animal's history from birth to processing or sale. This enhanced traceability provides transparency and accountability, giving consumers confidence in the origin, health, and welfare of the products they purchase.
- Animal Health and Welfare Management: Digital identities facilitate the management of animal health and welfare by maintaining comprehensive and up-to-date health records, vaccination history, and treatment information. This ensures timely interventions, targeted care, and efficient disease management.

- Genetic Improvement and Breeding Programs: Digital identities enable the tracking of genetic information, including parentage, breeding history, and performance indicators. This data supports genetic improvement programs, helping farmers make informed breeding decisions and optimize the productivity and quality of their livestock.
- Market Access and Certification: Digital identities can streamline the certification process for specific attributes, such as organic production, animal welfare practices, or breed authenticity. By providing verifiable data, digital identities help farmers meet market requirements, gain access to premium markets, and command higher prices for their products.
- Fraud Prevention and Prevention of Stolen Animals: Digital identities make it significantly harder for animals to be stolen or misrepresented in the market. The tamper-proof nature of blockchain ensures the authenticity and accuracy of an animal's identity, reducing the risk of fraud and illegal activities in the livestock trade.

*c. Challenges and considerations*

Implementing blockchain-based digital identity for livestock comes with certain challenges and considerations:

- Data Privacy and Security: Safeguarding sensitive animal data is paramount. Proper measures must be in place to protect the privacy and security of information stored on the blockchain, ensuring that only authorized parties have access to specific data elements.
- Interoperability and Standardization: Achieving interoperability and standardization across different blockchain platforms, databases, and stakeholders' systems is crucial for effective implementation. Collaboration and the establishment of common data standards are necessary to ensure seamless data exchange and utilization.
- Adoption and Infrastructure: Widespread adoption of blockchain technology in the livestock industry requires investment in infrastructure, technical expertise, and training. Overcoming these barriers and ensuring access to blockchain solutions for all stakeholders, including small-scale farmers, is essential for realising the full potential of digital identities.
- Legal and Regulatory Considerations: As digital identities become more prevalent, legal and regulatory frameworks must adapt to address issues related to data ownership, liability, consent, and data sharing agreements. Clear guidelines and regulations are necessary to ensure the responsible and ethical use of animal data.

*d. Potential applications*

Blockchain-based digital identity for livestock has a range of potential applications, such as consumer information and product labeling, supply chain efficiency and fraud prevention, and disease management and outbreak response. By utilizing blockchain technology, stakeholders can securely track and verify the origin and history of livestock, ensuring transparency and authenticity throughout the supply chain. Additionally, blockchain facilitates certification and compliance programs by providing immutable records of animal welfare standards and regulatory compliance, thereby enhancing trust and accountability in the livestock industry (Awotunde, Misra, Ayoade, Ogundokun, & Abiodun, 2022; Nayal et al., 2021; Neethirajan & Kemp, 2021):

- Consumer Information and Product Labeling: Digital identities can provide consumers with access to detailed information about the animals involved in meat, dairy, and other livestock products. This information can be presented through product labels, QR codes, or online platforms, empowering consumers to make informed choices.
- Supply Chain Efficiency and Fraud Prevention: Digital identities can streamline supply chain processes by automating data capture, verification, and reconciliation. This improves efficiency, reduces paperwork, and minimizes the risk of fraud, counterfeit products, and misrepresentation.
- Disease Management and Outbreak Response: Digital identities enable swift and targeted responses to disease outbreaks by facilitating the identification, tracking, and isolation of affected animals. This helps contain the spread of diseases and protects both animal and human health.
- Certification and Compliance Programs: Digital identities support certification programs by providing verifiable data on animal welfare practices, organic production, breed authenticity, and more. This simplifies the certification process, reduces administrative burden, and enhances market access for compliant producers.

### 5.2.3. Secure and tamper-proof storage of health records

*a. Importance of secure and tamper-proof storage*

Secure and tamper-proof storage of health records offers several benefits to stakeholders:

- Data Integrity: Ensuring the integrity of health records is essential for making informed decisions about animal health and welfare. Tamper-proof storage prevents unauthorized alterations or modifications to records, maintaining their accuracy and reliability.

- Regulatory Compliance: Many jurisdictions have regulations and standards that mandate the proper storage and retention of health records. Secure storage practices help ensure compliance with these requirements, reducing the risk of penalties, legal issues, and operational disruptions.
- Disease Management: Timely access to accurate health records is critical for effective disease management. Secure storage ensures that the records are readily available when needed, facilitating quick and appropriate responses to disease outbreaks and enabling the tracing of affected animals.
- Treatment and Medication Management: Accurate health records are essential for proper treatment and medication management. Secure storage prevents unauthorized access or tampering, ensuring that treatment protocols are followed correctly and mitigating the risk of medication errors or non-compliance.
- Research and Analysis: Health records provide valuable data for research, analysis, and improving animal health outcomes. Secure storage protects the confidentiality of sensitive information, enabling researchers to analyze trends, identify patterns, and develop evidence-based interventions.

*b. Technologies for secure storage*

Various technologies can be employed for the secure storage of livestock health records:

- Blockchain Technology: Blockchain provides a decentralized and tamper-proof ledger that can securely store and track health records. Each record is time-stamped, cryptographically sealed, and stored across multiple nodes, ensuring transparency, immutability, and data integrity.
- Encrypted Databases: Storing health records in encrypted databases adds an extra layer of security. Encrypted data can only be accessed with the appropriate decryption keys, protecting the confidentiality of sensitive information.
- Access Controls: Implementing access controls, such as user authentication, role-based permissions, and audit trails, ensures that only authorized individuals can access and modify health records. This safeguards against unauthorized access, accidental data loss, or malicious activities.
- Digital Signatures: Digital signatures can be used to verify the authenticity and integrity of health records. They provide a mechanism to detect any

unauthorized modifications to the records, ensuring their accuracy and reliability.

*c. Best practices for secure storage*

In addition to utilizing appropriate technologies, implementing best practices is essential for the secure storage of health records:

- Physical Security: Physical security measures, such as restricted access to storage facilities, locked cabinets, and backup systems, protect records from theft, damage, or loss due to natural disasters.
- Regular Data Backups: Regularly backing up health records to off-site locations or cloud-based storage ensures data resilience and recoverability in case of system failures or data breaches.
- Data Encryption: Encrypting health records both in transit and at rest adds an extra layer of protection against unauthorized access. This safeguards the confidentiality of sensitive information, even if the data is intercepted or stolen.
- Regular Audits and Monitoring: Conducting regular audits and monitoring activities helps identify and address any potential vulnerabilities or security breaches promptly. It ensures that security measures are effective and up-to-date.
- Staff Training and Awareness: Training staff on proper data handling practices, security protocols, and the importance of maintaining data confidentiality is crucial. Staff members should be aware of their responsibilities in safeguarding health records.

*d. Compliance with data protection regulations*

Compliance with data protection regulations, such as the General Data Protection Regulation (GDPR) or relevant local regulations, is essential when storing health records. Understanding the legal requirements, obtaining necessary consent, and implementing appropriate security measures will help ensure compliance and protect the privacy rights of individuals.

### 5.2.4. Real-time monitoring of vaccination and treatment history

*a. Importance of real-time monitoring*

Real-time monitoring of vaccination and treatment history offers several advantages to stakeholders:

- Disease Management and Prevention: Real-time monitoring allows for proactive disease management and prevention. Timely identification of vaccinated and treated animals helps track disease outbreaks, target interventions, and prevent the spread of infectious diseases within the herd or flock.

- Biosecurity Measures: Real-time monitoring facilitates effective biosecurity measures by providing accurate information on the vaccination and treatment status of animals. This information enables better control of disease transmission, reduces the risk of introducing pathogens into the herd or flock, and enhances overall biosecurity protocols.
- Regulatory Compliance: Many countries have regulations that require livestock owners to maintain updated vaccination and treatment records. Real-time monitoring ensures compliance with these regulations, reducing the risk of penalties, legal issues, and trade disruptions.
- Treatment Efficiency: Real-time monitoring allows for better treatment efficiency and effectiveness. By having immediate access to an animal's treatment history, veterinarians can avoid duplication of treatments, accurately assess the response to previous treatments, and make informed decisions regarding subsequent treatments.
- Consumer Confidence: Real-time monitoring enhances consumer confidence in the safety and quality of livestock products. The ability to provide transparent information on vaccination and treatment history assures consumers that animals have received appropriate care, reducing concerns about residues or contaminants in food products.

*b. Challenges in real-time monitoring*

Implementing real-time monitoring of vaccination and treatment history comes with certain challenges:

- Data Accuracy and Completeness: Ensuring the accuracy and completeness of data is critical for reliable monitoring. It requires diligent data entry, verification processes, and training for personnel responsible for recording vaccination and treatment information.
- Data Integration and Interoperability: Integrating data from various sources, such as veterinary clinics, farmers, and laboratories, poses a challenge in achieving comprehensive real-time monitoring. Establishing interoperability and data-sharing mechanisms among different stakeholders is necessary for seamless integration and access to up-to-date information.
- Technology Infrastructure: Adequate technology infrastructure is essential for real-time monitoring. Reliable connectivity, data storage capacity, and user-friendly interfaces are needed to collect, store, and analyze vaccination and treatment data efficiently.

- Data Privacy and Security: Protecting the privacy and security of vaccination and treatment data is crucial. Robust data encryption, access controls, and adherence to data protection regulations are necessary to safeguard sensitive information and prevent unauthorized access or misuse.

*c. Technologies for real-time monitoring*

Several technologies, such as electronic health records, internet of things (IoT), mobile applications, and blockchain technology, can facilitate real-time monitoring of vaccination and treatment history. Electronic health records centralize patient information, including vaccination records, making them easily accessible to healthcare providers. IoT devices can track and transmit data on vaccine administration and patient adherence, while mobile applications provide convenient platforms for patients to access their vaccination history. Blockchain technology enhances data security and immutability, ensuring the integrity of vaccination records and fostering trust in the healthcare system's ability to monitor and manage immunization efforts effectively (Ratta, Kaur, Sharma, Shabaz, & Dhiman, 2021; Sander et al., 2018; Singh & Sharma, 2023; Srivastava, Parizi, & Dehghantanha, 2020):

- Electronic Health Records (EHR): Implementing electronic health records allows for efficient data entry, storage, and retrieval of vaccination and treatment information. EHR systems can be integrated with other data sources and accessed by authorized stakeholders in real time.
- Internet of Things (IoT): IoT devices, such as smart tags or wearable sensors, can be used to monitor and record vaccination and treatment events. These devices can automatically capture data, transmit it wirelessly, and update records in real time.
- Mobile Applications: Mobile applications provide a convenient and user-friendly interface for recording and accessing vaccination and treatment data. They allow farmers, veterinarians, and other stakeholders to update records in the field, ensuring real-time data entry.
- Blockchain Technology: Blockchain offers a secure and transparent platform for recording vaccination and treatment history. Each event can be cryptographically sealed, time-stamped, and stored in an immutable ledger, ensuring data integrity and traceability.

*d. Future directions*

The future of real-time monitoring of vaccination and treatment history holds promising developments:

- Artificial Intelligence (AI) and Machine Learning: AI and machine learning algorithms can analyze large datasets, identify patterns, and provide insights to improve disease management strategies. These technologies can help identify risk factors, predict disease outbreaks, and optimize vaccination and treatment protocols.
- Integration with Livestock Management Systems: Integrating real-time monitoring with livestock management systems allows for comprehensive data analysis, herd-level health assessments, and personalized treatment recommendations. This integration enhances overall herd management and health outcomes.
- Remote Monitoring and Telemedicine: Remote monitoring technologies, such as telemedicine and teleconsultation, enable veterinarians to remotely assess and monitor animals' health, provide guidance, and update vaccination and treatment records in real time. This approach improves accessibility to veterinary services, particularly in remote areas.

### 5.2.5. Ensuring food safety and preventing disease outbreaks

*a. Importance of food safety*

Food safety is crucial for protecting consumers from foodborne illnesses and maintaining public health (Behnke & Janssen, 2020). In the context of the livestock industry, ensuring food safety involves several key aspects:

- Contamination Prevention: Preventing contamination of livestock products with harmful pathogens, chemicals, or residues is essential. Proper handling, storage, and processing practices at all stages of the supply chain are critical to minimizing the risk of contamination
- Hygiene and Sanitation: Maintaining high standards of hygiene and sanitation in animal housing, handling facilities, and processing plants is necessary to prevent the introduction and spread of pathogens. Regular cleaning and disinfection, as well as adherence to hygiene protocols, contribute to food safety.
- Quality Control and Testing: Rigorous quality control measures and testing protocols help identify potential contaminants or hazards in livestock products. Regular monitoring and analysis of samples for pathogens, chemical residues, and other contaminants ensure that only safe products reach the market.

- Traceability and Recall Systems: Establishing robust traceability systems enables the rapid identification and recall of contaminated products in case of an outbreak or contamination event. Traceability allows for swift action to be taken, preventing further distribution and consumption of unsafe products.

*b. Preventing disease outbreaks*

Disease outbreaks can have significant consequences for animal health, welfare, and the overall livestock industry. Preventing disease outbreaks requires a proactive and multifaceted approach:

- Biosecurity Measures: Implementing stringent biosecurity measures is crucial for preventing the introduction and spread of diseases. Measures may include controlling access to farms, proper sanitation practices, isolation of sick animals, and monitoring visitors and vehicles entering the premises.
- Vaccination Programs: Vaccination plays a vital role in preventing and controlling infectious diseases in livestock. Effective vaccination programs tailored to specific disease risks and animal populations are essential for reducing disease incidence and severity.
- Disease Surveillance: Surveillance systems that actively monitor and detect disease occurrences and trends help identify potential outbreaks early on. Rapid detection allows for timely intervention, such as quarantine measures or targeted vaccination, to prevent further spread.
- Education and Training: Providing education and training to livestock producers, veterinarians, and farm workers on disease prevention and control measures is crucial. Knowledge about disease symptoms, reporting procedures, and proper biosecurity practices empowers stakeholders to take proactive measures and respond effectively to potential threats.

*c. Technologies for food safety and disease prevention*

Several technologies can enhance food safety and disease prevention efforts in the livestock industry:

- Rapid Testing and Diagnostics: Rapid testing and diagnostic technologies enable quick and accurate identification of pathogens or contaminants. These technologies, such as PCR-based tests or immunological assays, facilitate early detection, prompt intervention, and targeted control measures.
- Data Analytics and Predictive Models: Data analytics and predictive modelling techniques can analyze vast amounts of data, including

environmental, climatic, and animal health data, to identify disease patterns and predict disease outbreaks. These tools provide valuable insights for preventive measures and early warning systems.

- Internet of Things (IoT) and Sensor Technologies: IoT and sensor technologies can monitor key parameters, such as temperature, humidity, and air quality, in animal housing, storage facilities, or transport vehicles. Real-time monitoring alerts stakeholders to potential risks or deviations from safe conditions, allowing for timely corrective actions.
- Blockchain Technology: Blockchain offers transparent and immutable record-keeping, enhancing traceability and accountability in the supply chain. By providing a decentralized and tamper-proof ledger, blockchain technology facilitates rapid traceability and targeted recalls in the event of contamination or disease outbreaks.

*d. Collaboration and regulatory frameworks*

Ensuring food safety and preventing disease outbreaks requires collaboration among stakeholders and the establishment of robust regulatory frameworks:

- Stakeholder Collaboration: Collaboration among farmers, veterinarians, food processors, regulatory agencies, and consumers is vital for effective disease prevention and food safety. Information sharing, joint initiatives, and knowledge transfer enhance overall efforts and promote a culture of safety.
- Regulatory Frameworks: Governments and regulatory bodies play a crucial role in establishing and enforcing food safety standards and regulations. Robust regulatory frameworks that cover production, processing, and distribution ensure compliance and protect public health.
- International Cooperation: International cooperation and information sharing are essential for preventing the global spread of diseases. Collaborative efforts in surveillance, research, and emergency response facilitate the exchange of best practices and timely containment of potential outbreaks.

### 5.3. Authentication of organic and sustainable farming practices

Blockchain presents a plethora of avenues for authenticating organic and sustainable farming practices, including the verification of organic certifications, tracking the utilization of organic inputs and farming methodologies, and ensuring adherence to sustainability standards. Additionally, blockchain fosters trust in eco-friendly and humane livestock farming by providing transparent and immutable records of animal welfare practices. By harnessing blockchain technology, stakeholders can enhance transparency, traceability, and confidence in the integrity of agricultural processes, thereby facilitating a more sustainable and ethical food system that meets the demands of conscientious consumers (FAO, 2019; Hu, Huang, Huang, & Su, 2021).

### 5.3.1. Demand for organic and sustainable products

*a. Factors driving the demand*

Several factors contribute to the increasing demand for organic and sustainable products:

- Health and Nutrition: Consumers are becoming more conscious of the impact of their food choices on personal health. Organic products are perceived as healthier due to the absence of synthetic pesticides, hormones, and genetically modified organisms (GMOs). Consumers also appreciate the higher nutritional quality and potential benefits of organic farming practices.
- Environmental Concerns: Growing environmental awareness has led consumers to seek products that minimize environmental impact. Organic production systems prioritize soil and water conservation, biodiversity preservation, and reduced use of chemical inputs. By choosing organic products, consumers can contribute to sustainable agricultural practices and support ecosystems.
- Animal Welfare: Animal welfare concerns have gained significant attention, and consumers are increasingly seeking products from farms that prioritize the humane treatment of animals. Organic farming practices generally prioritize animal welfare, such as providing access to outdoor areas, pasture grazing, and natural feed.
- Ethical and Social Responsibility: Consumers are increasingly concerned about the social and ethical aspects of food production. They value fair trade practices, worker welfare, and support for local economies. Organic and sustainable products often align with these values, as they emphasise responsible farming practices and support local and small-scale producers.

*b. Benefits and challenges of organic production*

Organic production systems offer several benefits, but they also come with challenges:

- Benefits:

+ Environmental Sustainability: Organic farming practices promote soil health, biodiversity, and water conservation. They avoid the use of synthetic pesticides and chemical fertilizers, reducing the impact on ecosystems and supporting long-term sustainability.

+ Health and Food Safety: Organic products are free from synthetic pesticides, hormones, and GMOs. They are produced following strict standards, ensuring the absence of chemical residues and potential health risks.
+ Animal Welfare: Organic farming emphasises animal welfare, providing animals with access to outdoor areas, natural feed, and space to exhibit natural behaviors.
+ Market Opportunities: Meeting the demand for organic and sustainable products opens up market opportunities for farmers and producers. Organic certification allows them to tap into niche markets and gain price premiums for their products.

- Challenges:

+ Transition Period: Converting conventional farms to organic production requires a transition period, during which farmers adhere to organic practices without being able to market their products as organic. This period can pose financial challenges and require significant adjustments in farming methods.
+ Limited Availability: Meeting the growing demand for organic products can be challenging due to the limited availability of organic feed, genetics, and inputs. Developing reliable organic supply chains and infrastructure is essential to meet consumer demand.
+ Cost and Price Premiums: Organic production often incurs higher costs due to stricter regulations, increased labor requirements, and potential yield fluctuations. However, the price premiums for organic products can offset these costs and provide economic incentives for farmers.

*c. Meeting the demand for organic and sustainable products*

To meet the increasing demand for organic and sustainable products, stakeholders employ various strategies:

- Organic Certification: Obtaining organic certification assures consumers that products meet specific organic standards. Farmers and producers can pursue organic certification to access organic markets and build consumer trust.
- Education and Consumer Awareness: Educating consumers about the benefits of organic and sustainable products fosters greater demand. Awareness campaigns, labeling initiatives, and transparent communication help consumers make informed choices and understand the value of organic production systems.

- Research and Development: Investing in research and development supports the growth of organic farming practices. Developing improved organic feed, pest management techniques, and livestock genetics can address challenges and enhance productivity within organic systems.
- Collaboration and Knowledge Sharing: Collaboration among farmers, researchers, policymakers, and industry stakeholders facilitates knowledge sharing and the dissemination of best practices. Collaborative efforts can accelerate the adoption of organic and sustainable production methods and address common challenges.

### 5.3.2. Blockchain for verifying organic certifications

*a. The importance of organic certifications*

Organic certifications play a crucial role in assuring consumers that the products they purchase meet specific organic standards. These certifications validate that the food was produced using organic farming methods, free from synthetic pesticides, GMOs, and other prohibited substances. However, verifying the authenticity of organic certifications can be challenging, as the existing systems rely heavily on paper-based records and can be susceptible to fraud or inaccuracies.

*b. How blockchain technology works*

Blockchain is a decentralized and distributed ledger technology that enables secure and transparent record-keeping. It allows multiple participants to maintain and verify a shared database without the need for intermediaries. Here's how blockchain technology works:

- Distributed Ledger: Blockchain consists of a network of computers (nodes) that collectively maintain a replicated and synchronised ledger, known as a distributed ledger. Each participant has a copy of the entire ledger.
- Blocks and Transactions: Transactions are grouped together in blocks, which are then added to the chain of previously validated blocks. Each block contains a cryptographic hash that links it to the previous block, creating an immutable chain of transactions.
- Consensus Mechanism: Blockchain networks use consensus mechanisms to agree on the validity of transactions and ensure the integrity of the ledger. Popular consensus mechanisms include proof of work (PoW) and proof of stake (PoS).
- Transparency and Security: Blockchain provides transparency by allowing all participants to view and verify the transactions recorded on the ledger.

Additionally, the use of cryptographic algorithms ensures the security and immutability of the data.

*c. Verifying organic certifications with blockchain*

Blockchain technology offers several advantages for verifying organic certifications, such as immutable records, traceability and transparency, and smart contracts. Through blockchain's immutable ledger, certification records become tamper-proof, ensuring the integrity of organic certification data. Additionally, smart contracts can automate verification processes, enhancing efficiency and reducing the risk of errors or fraud. Stakeholder collaboration and consumer empowerment are further facilitated by blockchain, as it enables transparent sharing of certification information, fostering trust and accountability in organic supply chains (Hu, Huang, & Qin, 2022):

- Immutable Records: Organic certifications can be securely recorded on the blockchain, creating an immutable record of the certification process. Once recorded, the data cannot be altered or tampered with, providing assurance of the certification's authenticity.
- Traceability and Transparency: Blockchain enables end-to-end traceability of organic products by recording the entire supply chain journey. Each step, from production to distribution, can be documented and verified, ensuring transparency and accountability.
- Smart Contracts: Smart contracts are self-executing contracts with predefined rules and conditions. They can be utilized in the organic certification process to automate compliance checks, validation procedures, and renewal reminders.
- Stakeholder Collaboration: Blockchain allows multiple stakeholders, such as farmers, certification bodies, regulators, and retailers, to participate in the verification process. This collaboration increases trust and streamlines the certification process by eliminating redundant paperwork and manual checks.
- Consumer Empowerment: Blockchain enables consumers to access transparent information about the organic certification of the products they purchase. They can scan QR codes or utilize mobile applications to verify the certification details and make informed choices.

*d. Challenges and considerations*

While blockchain technology holds immense promise for verifying organic certifications, there are several challenges and considerations such as: data accuracy and integrity, interoperability and standardization, data privacy and security, and adoption and infrastructure, to address. Ensuring the accuracy and integrity of data recorded on the blockchain is crucial to maintaining trust in organic certification processes. Additionally, establishing interoperable standards and robust privacy measures are essential for widespread adoption and acceptance of blockchain-based solutions in verifying organic certifications (Hu et al., 2022):

- Data Accuracy and Integrity: Ensuring the accuracy and integrity of the data recorded on the blockchain is crucial. Reliable data sources, standardized formats, and validation mechanisms are necessary to maintain the integrity of the certification records.
- Interoperability and Standardization: Achieving interoperability among different blockchain platforms and organic certification bodies is essential to create a unified and comprehensive verification system. Standardization of data formats and protocols is vital for seamless integration.
- Data Privacy and Security: Protecting sensitive information, such as farm-specific data or confidential business information, is paramount. Blockchain solutions must implement robust privacy and security measures to safeguard against unauthorized access or data breaches.
- Adoption and Infrastructure: Wide-scale adoption of blockchain technology requires investment in infrastructure, training, and awareness among stakeholders. Developing user-friendly interfaces and integrating blockchain with existing systems are crucial for successful implementation.

### 5.3.3. Tracking the use of organic inputs and farming practices

*a. The importance of tracking organic inputs and farming practices*

- Maintaining Organic Integrity: Tracking the use of organic inputs and farming practices is crucial for maintaining the integrity of organic products. Organic standards prohibit the use of synthetic pesticides, genetically modified organisms (GMOs), and certain fertilizers. Accurate tracking ensures that farmers adhere to these standards and that consumers can trust the organic label.
- Regulatory Compliance: Organic certification programs have specific requirements for the use of inputs and farming practices. Tracking allows farmers to demonstrate compliance with these regulations, making it easier to obtain and maintain organic certification.
- Consumer Confidence: Tracking organic inputs and farming practices enhances consumer confidence in the authenticity and sustainability of organic products. When consumers can trace the origin of the inputs used

and understand the sustainable practices employed, they can make informed choices aligned with their values.

- Continuous Improvement: Tracking provides valuable data for farmers to assess the effectiveness of their organic practices. It allows them to monitor inputs, evaluate their impact on soil health and crop yields, and make informed decisions for continuous improvement.

*b. Challenges in tracking organic inputs and farming practices*

- Record-keeping: Accurate record-keeping is crucial for tracking organic inputs and farming practices. Farmers need to document the types and quantities of inputs used, their sources, application methods, and timing. Maintaining comprehensive and up-to-date records can be time-consuming and challenging, particularly for farmers managing large-scale operations.
- Supply Chain Complexity: Organic supply chains often involve multiple actors, including farmers, input suppliers, processors, and retailers. Tracking inputs and practices throughout the supply chain requires collaboration and standardized processes to ensure data accuracy and consistency.
- Data Verification: Verifying the accuracy and authenticity of recorded data can be challenging. Without robust mechanisms in place, there is a risk of inaccurate or falsified information entering the tracking system, compromising its integrity.
- Technological Limitations: Traditional paper-based systems for record-keeping and tracking can be inefficient and prone to errors. Leveraging technology can streamline the process, but implementation challenges, such as limited access to technology and varying levels of digital literacy among farmers, need to be addressed.

*c. Technological solutions for tracking organic inputs and farming practices*

Blockchain offers a variety of technological solutions for tracking organic inputs and farming practices. These solutions enhance transparency and traceability, contributing significantly to the overall sustainability of agricultural processes. By leveraging blockchain technology, farmers and consumers alike can gain confidence in the integrity and authenticity of organic products (Hu et al., 2022).

- Digital Platforms and Mobile Applications: Digital platforms and mobile applications provide farmers with tools to record and track inputs and farming practices. These tools often include features such as digital record-keeping, automated data entry, reminders for application timings, and data analysis capabilities.

- Internet of Things (IoT) Sensors: IoT sensors can be deployed in organic farms to collect real-time data on soil moisture, temperature, and other environmental variables. This data can be integrated into tracking systems, providing insights into the impact of inputs and practices on crop health and overall farm performance.
- QR Codes and Blockchain Technology: QR codes linked to blockchain-based systems can be used to track the origin of organic inputs, their certification status, and the corresponding farming practices. Blockchain ensures transparency, immutability, and traceability of the recorded data, enhancing trust in the tracking process.
- Data Analytics and Artificial Intelligence (AI): Data analytics and AI technologies can be employed to analyze the collected data and provide farmers with valuable insights. By leveraging machine learning algorithms, patterns and correlations can be identified, aiding in decision-making for optimizing organic farming practices.

*d. Collaboration and education*

To overcome the challenges associated with tracking organic inputs and farming practices, collaboration among farmers, certification bodies, researchers, and technology providers is crucial. Education and training programs can enhance farmers' understanding of tracking requirements and the use of technological tools. Knowledge sharing and the establishment of best practices contribute to the development of standardized tracking systems and ensure consistent data management across the organic industry.

### 5.3.4. Ensuring compliance with sustainability standards

*a. The importance of compliance with sustainability standards*

- Environmental Stewardship: Sustainability standards aim to minimize the negative environmental impacts of businesses. Compliance ensures that organizations adopt sustainable practices, such as reducing greenhouse gas emissions, conserving natural resources, protecting biodiversity, and minimizing pollution.
- Social Responsibility: Compliance with sustainability standards encompasses social aspects, including fair labor practices, respect for human rights, and community engagement. Adhering to these standards fosters ethical business conduct, promotes worker well-being, and contributes to the overall welfare of society.

- Market Access and Reputation: Compliance with sustainability standards is often a prerequisite for market access, especially in sectors where sustainability is a key concern for consumers. Meeting these standards enhances an organization's reputation, builds trust with stakeholders, and opens doors to new markets and business opportunities.
- Risk Management: Non-compliance with sustainability standards can expose organizations to reputational risks, legal liabilities, and financial consequences. Ensuring compliance helps mitigate these risks, safeguarding the long-term viability and sustainability of the business.

*b. Challenges in ensuring compliance with sustainability standards*

Ensuring compliance with sustainability standards presents a myriad of challenges, ranging from the complexity and diversity of standards to supply chain intricacies, data collection, verification processes, and the imperative for continuous improvement. These hurdles underscore the multifaceted nature of sustainability initiatives, requiring comprehensive strategies and collaborative efforts across various stakeholders to address effectively. By implementing robust systems for monitoring, reporting, and adapting to evolving standards, organizations can navigate these challenges and advance their sustainability goals while maintaining transparency and accountability throughout their operations (Hu et al., 2022; van Hilten, Ongena, & Ravesteijn, 2020).

- Complexity and Diversity of Standards: Sustainability standards vary across industries and regions, making compliance a complex endeavour. Organizations must navigate a multitude of standards, certifications, and reporting frameworks, requiring dedicated resources and expertise.
- Supply Chain Complexity: Organizations with global supply chains face challenges in ensuring sustainability standards are upheld throughout the supply chain. Monitoring compliance among suppliers, subcontractors, and vendors can be demanding, especially when dealing with diverse cultures, legal frameworks, and business practices.
- Data Collection and Verification: Gathering and verifying data related to sustainability performance can be challenging. Organizations need robust systems for data collection, accurate reporting, and independent verification to ensure the credibility of sustainability claims.
- Continuous Improvement: Sustainability standards evolve over time as new scientific findings and stakeholder expectations emerge. Organizations must continuously adapt their practices, monitor industry trends, and engage in ongoing improvement efforts to stay compliant.

*c. Strategies for effective compliance*

Achieving effective compliance entails employing a range of strategies, including fostering commitment from leadership, implementing robust policies and procedures, and engaging stakeholders. Managing supply chains effectively, ensuring efficient data management and reporting, and prioritizing continuous learning and improvement are also crucial components. These multifaceted approaches underscore the necessity of comprehensive and proactive measures to navigate the complexities of compliance successfully and drive sustainable outcomes in various industries (Tian, 2016; Tomu, 2020; Vangala, Sutrala, Das, & Jo, 2021).

- Commitment from Leadership: Strong commitment from top management is crucial for driving sustainability compliance. Leaders must prioritize sustainability, set clear goals and targets, allocate resources, and integrate sustainability principles into the organization's strategy and decision-making processes.
- Robust Policies and Procedures: Establishing comprehensive policies and procedures that align with sustainability standards is essential. These policies should outline responsibilities, guidelines for sustainable practices, and mechanisms for monitoring and reporting progress.
- Stakeholder Engagement: Engaging with stakeholders, including employees, customers, suppliers, and communities, fosters collaboration and ensures that diverse perspectives are considered. Engaged stakeholders can provide valuable insights, support compliance efforts, and enhance the organization's sustainability performance.
- Supply Chain Management: Implementing sustainable supply chain management practices is crucial for ensuring compliance throughout the supply chain. Organizations should establish clear expectations, conduct supplier assessments, provide training and support, and encourage transparency and accountability among suppliers.
- Data Management and Reporting: Implementing robust systems for data management, including accurate data collection, regular monitoring, and independent verification, is essential. Transparent reporting of sustainability performance enhances credibility and facilitates benchmarking against industry peers.
- Continuous Learning and Improvement: Embracing a culture of continuous learning and improvement is fundamental for sustainability compliance. Organizations should invest in training programs, engage in industry networks and collaborations, and monitor emerging trends and best practices to stay at the forefront of sustainability performance.

### 5.3.5. Building trust in eco-friendly and humane livestock farming

*a. The importance of building trust in eco-friendly and humane livestock farming*

- Environmental Impact: Livestock farming is associated with greenhouse gas emissions, water pollution, deforestation, and soil degradation. Building trust in eco-friendly farming practices helps address these concerns by promoting sustainable land use, reducing pollution, and mitigating climate change.
- Animal Welfare: The well-being of animals raised for food is a significant concern for consumers. Building trust in humane livestock farming practices involves ensuring appropriate housing, access to outdoor areas, and the use of humane handling and slaughter methods.
- Public Health and Food Safety: Livestock farming practices can have implications for public health and food safety. Trust is built through the implementation of proper hygiene practices, responsible use of antibiotics, and adherence to strict food safety regulations.
- Consumer Demand: Consumers are increasingly seeking eco-friendly and humane food options. Building trust in livestock farming practices that prioritize sustainability, animal welfare, and public health can meet this growing demand, attracting a loyal customer base.

*b. Challenges in building trust*

- Lack of Transparency: The livestock industry has historically faced challenges regarding transparency. Limited access to information about farming practices, supply chain operations, and animal welfare standards can erode consumer trust. Increasing transparency is essential for building trust.
- Perception vs. Reality: Negative perceptions of livestock farming, fueled by media reports and documentaries, can overshadow efforts made by responsible farmers. Educating consumers about the variety of farming practices and showcasing eco-friendly and humane approaches can bridge this gap.
- Industry Fragmentation: The livestock industry is diverse and fragmented, ranging from small-scale family farms to large industrial operations. Coordinating efforts, implementing consistent standards, and building trust across the entire industry can be challenging but is necessary for effective change.

- Resistance to Change: Transitioning to eco-friendly and humane livestock farming practices may face resistance due to economic constraints, traditional practices, or a lack of awareness. Overcoming this resistance requires collaborative efforts, knowledge sharing, and incentives for adopting sustainable practices.

*c. Strategies for building trust*

- Transparency and Communication: Livestock farmers should embrace transparency by providing accurate information about their farming practices, animal welfare standards, and environmental initiatives. Open communication builds trust, fosters dialogue, and helps consumers make informed choices.
- Certification and Standards: Implementing credible certification programs and adhering to recognised standards can build trust among consumers. Third-party certifications that verify eco-friendly and humane practices provide assurance and help differentiate responsible farmers from others.
- Collaboration and Knowledge Sharing: Encouraging collaboration among farmers, industry stakeholders, researchers, and policymakers facilitate the exchange of best practices, innovative approaches, and scientific knowledge. Sharing successes and challenges can accelerate the adoption of eco-friendly and humane farming practices.
- Consumer Education: Educating consumers about the benefits of eco-friendly and humane livestock farming is crucial. Initiatives such as farm visits, educational campaigns, and transparent labeling help consumers understand the positive impact of their choices and foster trust in responsible farming practices.
- Continuous Improvement: Livestock farmers should strive for continuous improvement by adopting innovative technologies, sustainable farming methods, and animal welfare practices. Embracing research, investing in training, and actively seeking feedback from consumers and stakeholders demonstrate a commitment to progress.

## 5.4. Blockchain-based animal welfare monitoring

Blockchain technology offers a promising solution for monitoring the welfare of animals through a comprehensive procedure involving various inputs and activities (Detwiler, 2020; Katsikouli et al., 2021; Kayikci et al., 2022; Makkar & Costa, 2020; Tian, 2016; Tomu, 2020; van Hilten et al., 2020; Vangala et al., 2021). This procedure encompasses the utilization of Internet of Things sensors and data collection methods for real-time animal monitoring, the establishment of immutable record-keeping systems to track crucial welfare parameters, rigorous auditing processes to ensure compliance with animal welfare regulations, and efforts to raise consumer awareness and encourage ethical choices in animal products. By integrating these components, blockchain technology can revolutionize animal welfare practices, fostering transparency, accountability, and improved standards throughout the agricultural industry.

### 5.4.1. Animal welfare concerns in livestock farming

*a. Animal welfare concerns in livestock farming*

- Housing and Living Conditions: Livestock housing and living conditions play a crucial role in animal welfare. Concerns include overcrowding, inadequate space, lack of environmental enrichment, and unsuitable flooring or bedding materials. Providing animals with sufficient space, appropriate shelter, and environmental stimulation is essential for their well-being.
- Health Management: Proper health management is vital for ensuring animal welfare. This includes disease prevention, timely veterinary care, and responsible use of medications, including antibiotics. Neglecting health management can lead to suffering, reduced productivity, and compromised animal welfare.
- Handling and Transport: The handling and transport of livestock can significantly impact their welfare. Rough handling, excessive use of force, and inadequate training of personnel can cause stress, injuries, and fear in animals. Ensuring proper handling techniques, minimizing transport duration, and providing comfortable transport conditions are crucial for maintaining animal welfare.
- Slaughter Practices: The methods used for animal slaughter are a significant concern for animal welfare. Humane slaughter techniques, such as stunning methods that minimize pain and distress, are essential to ensure a humane end-of-life experience for livestock.
- Breeding Practices: Breeding practices can impact the welfare of livestock. Selective breeding for production traits, such as high milk yield or fast growth, may result in health issues and reduced welfare. Promoting breeding practices that prioritize animal welfare, genetic diversity, and robustness is crucial.

*b. Challenges in ensuring animal welfare*

- Economic Pressures: Economic pressures in the livestock industry, such as cost efficiency and market demand, can create challenges for implementing and maintaining high animal welfare standards. Balancing economic viability with ethical considerations requires innovative approaches and industry-wide commitment.

- Lack of Awareness and Education: Lack of awareness and education among farmers, workers, and consumers about animal welfare issues can hinder progress. Promoting education and training programs on animal welfare practices, ethics, and regulations is essential to drive positive change.
- Compliance and Enforcement: Ensuring compliance with animal welfare regulations and standards can be challenging. Lack of enforcement, varying standards across regions, and inadequate monitoring systems pose difficulties in ensuring consistent and effective animal welfare practices.
- Industry Complexity: The livestock industry is complex, involving various stakeholders, including farmers, processors, retailers, and consumers. Coordinating efforts and achieving consensus on animal welfare practices can be challenging, especially in globally integrated supply chains.

*c. Strategies for promoting improved animal welfare practices*

- Enhanced Standards and Regulations: Developing and implementing robust animal welfare standards and regulations is crucial for promoting improved practices. These standards should address housing, handling, transport, slaughter, breeding, and health management to ensure comprehensive animal welfare protection.
- Training and Education: Providing training and education programs to farmers, workers, and industry stakeholders can improve awareness and understanding of animal welfare issues. Training should cover best practices, handling techniques, and the importance of animal welfare for sustainable livestock farming.
- Stakeholder Collaboration: Collaborating with animal welfare organizations, industry associations, and other stakeholders can help drive positive change. Sharing knowledge, expertise, and resources fosters a collective effort toward improving animal welfare in livestock farming.
- Technology and Innovation: Embracing technological advancements and innovative solutions can enhance animal welfare in livestock farming. This includes implementing precision farming techniques, monitoring systems, and sensor-based technologies that allow for real-time assessment of animal well-being.
- Transparency and Consumer Education: Promoting transparency in livestock farming practices and engaging in open communication with consumers

builds trust and allows consumers to make informed choices. Educational campaigns and labeling initiatives can help consumers understand the importance of animal welfare and make ethical purchasing decisions.

### 5.4.2. Internet of Things sensors and data collection for animal monitoring

*a. IoT sensors for animal monitoring*

- Environmental Monitoring: IoT sensors can monitor environmental conditions, including temperature, humidity, air quality, and lighting. This data helps ensure optimal living conditions for livestock, promoting their health and productivity.
- Health and Behavior Monitoring: IoT sensors can track vital signs, such as heart rate, respiration, and rumination, allowing for early detection of health issues. They can also monitor behavior patterns, feeding habits, and movement, providing valuable insights into animal well-being and potential distress.
- Reproduction and Fertility Tracking: IoT sensors can aid in monitoring reproductive cycles, estrus detection, and fertility management in livestock. This data helps optimize breeding programs, improve breeding success rates, and enhance overall reproductive efficiency.
- Feeding and Nutrition Management: IoT sensors can track feed consumption, analyze eating patterns, and monitor nutritional intake. This information allows for precise feeding management, ensuring that animals receive the appropriate nutrition for optimal growth and health.
- Location and Tracking: IoT-enabled tracking devices can monitor the location and movement of livestock, preventing theft, enabling efficient herding, and enhancing overall herd management.

*b. Benefits of IoT sensors in animal monitoring*

- Real-time Monitoring: IoT sensors provide real-time data, allowing farmers to promptly identify and address potential issues. Early detection of health problems, deviations in behavior, or changes in environmental conditions enables proactive interventions and improves animal welfare.
- Data-driven Decision-making: IoT sensors generate large amounts of data, which, when properly analyzed, provide valuable insights for informed decision-making. Data-driven decisions help optimize animal management practices, resource allocation, and overall farm operations.
- Improved Efficiency and Productivity: IoT sensors enable precise monitoring of animal health, behavior, and environmental conditions. This optimization

leads to enhanced efficiency, productivity, and cost savings in livestock farming operations.

- Enhanced Animal Welfare: By continuously monitoring vital signs, behavior, and environmental factors, IoT sensors contribute to better animal welfare. Early detection of health issues, timely interventions, and tailored care improve overall animal well-being.
- Sustainability and Resource Management: IoT sensors aid in optimizing resource utilization, such as feed, water, and energy, based on precise monitoring and data analysis. This leads to more sustainable farming practices, reducing waste and environmental impact.

*c. Challenges and considerations*

- Data Privacy and Security: IoT sensors collect and transmit sensitive data, making data privacy and security crucial considerations. Farmers must ensure robust data encryption, secure networks, and adherence to privacy regulations to protect the integrity and confidentiality of the collected data.
- Integration and Interoperability: Integrating IoT sensor technologies with existing farm management systems and ensuring interoperability across different devices can be challenging. Standardization efforts and compatible platforms are necessary to facilitate seamless data integration and analysis.
- Cost and Infrastructure: Implementing IoT sensors and data collection infrastructure may require initial investments in equipment, connectivity, and data storage. Farmers must carefully evaluate costs, potential returns on investment, and scalability before implementing IoT solutions.
- Data Management and Analysis: Managing and analyzing the large volume of data generated by IoT sensors can be overwhelming. Farmers may need to invest in data management tools, analytics capabilities, and skilled personnel to effectively utilize the collected data for decision-making.
- Farmer Education and Adoption: Familiarity with IoT technologies, data analytics, and data-driven decision-making is essential for farmers to fully harness the potential of IoT sensors. Educational programs, training, and support are necessary to facilitate farmer education and promote widespread adoption.

### 5.4.3. Immutable record-keeping of animal welfare parameters

*a. Understanding immutable record-keeping*

- Blockchain Technology: Blockchain is a decentralized and distributed ledger technology that enables secure and transparent record-keeping. It consists of a chain of blocks, where each block contains a set of verified and encrypted transactions. Once recorded, the data in a blockchain cannot be altered or tampered with, ensuring its immutability.
- Animal Welfare Parameters: Animal welfare parameters encompass various aspects of livestock well-being, including health records, feeding routines, environmental conditions, veterinary interventions, and behavior observations. Immutable record-keeping allows for the transparent and unchangeable storage of these parameters throughout an animal's life.

*b. Benefits of immutable record-keeping for animal welfare parameters*

- Data Integrity and Trust: Immutable record-keeping ensures that animal welfare parameters are stored securely and cannot be manipulated or falsified. This enhances data integrity, instills trust in the recorded information, and facilitates accurate decision-making processes.
- Traceability and Transparency: Blockchain-based record-keeping provides a transparent trail of information, allowing stakeholders to trace the history and journey of livestock. This includes details about the animal's health, origin, feeding practices, and overall well-being. Increased transparency builds trust among consumers, regulatory bodies, and other stakeholders.
- Enhanced Animal Welfare Monitoring: Immutable record-keeping enables continuous monitoring of animal welfare parameters throughout an animal's life. This longitudinal data collection allows for the identification of trends, patterns, and potential issues, leading to timely interventions and improved animal welfare outcomes.
- Compliance with Standards and Regulations: Blockchain-based record-keeping can aid in demonstrating compliance with animal welfare standards and regulations. The transparent nature of the technology allows for easy verification and auditing of the recorded parameters, ensuring adherence to industry guidelines.
- Research and Development: Immutable records of animal welfare parameters can serve as a valuable resource for research and development purposes. Researchers can access a wealth of data to analyze trends, identify correlations, and develop evidence-based interventions to further improve animal welfare practices.

*c. Challenges and considerations*

When considering the immutable record-keeping of animal welfare parameters, several challenges must be addressed, including ensuring data input accuracy, maintaining data privacy and confidentiality, integrating with existing systems, and promoting adoption and education (Alshehri, 2023). These hurdles underscore the complexity of implementing robust systems for tracking and documenting animal welfare standards effectively. By addressing these challenges thoughtfully, stakeholders can work towards establishing reliable frameworks that uphold the welfare of animals while also meeting the needs of various stakeholders in the agricultural sector.

- Data Input Accuracy: Immutable record-keeping relies on accurate and reliable data input. Ensuring that the recorded parameters are based on objective measurements, standardized protocols, and proper documentation is crucial to maintain data integrity.
- Data Privacy and Confidentiality: While blockchain technology ensures data immutability, it also raises concerns about data privacy and confidentiality. Protecting sensitive information, such as veterinary records or proprietary farm practices, requires careful consideration of access controls and encryption methods.
- Integration with Existing Systems: Integrating blockchain-based record-keeping systems with existing farm management software or data collection tools can present technical challenges. Compatibility and interoperability between different platforms may require coordination and collaboration with technology providers.
- Adoption and Education: The adoption of blockchain technology in livestock farming requires education and awareness among farmers and industry stakeholders. Training programs, support networks, and user-friendly interfaces are necessary to facilitate widespread adoption and maximize the benefits of immutable record-keeping.

### 5.4.4. Auditing and compliance with animal welfare regulations

*a. The role of auditing in animal welfare*

- Monitoring Compliance: Auditing involves the systematic review and assessment of farm practices, procedures, and records to ensure compliance with animal welfare regulations. It provides an objective evaluation of the extent to which farmers adhere to the prescribed standards and guidelines.
- Identifying Non-compliance: Auditing helps identify areas of non-compliance or potential shortcomings in animal welfare practices. It highlights specific

issues that need attention and corrective actions, leading to continuous improvement in farm management and animal welfare outcomes.

- Building Trust and Transparency: Auditing promotes transparency and accountability in livestock farming. It assures consumers, regulatory bodies, and other stakeholders that farmers are committed to upholding animal welfare standards, fostering trust and confidence in the industry.
- Driving Continuous Improvement: Auditing programs provide feedback and recommendations for improving animal welfare practices. By identifying areas of improvement, farmers can implement necessary changes, adopt best practices, and continuously enhance their compliance with animal welfare regulations.

*b. Key components of effective auditing programs*

- Clear and Robust Standards: Effective auditing programs require clear, well-defined, and scientifically based animal welfare standards and regulations. These standards should be comprehensive, covering all relevant aspects of livestock farming, including housing, handling, transport, slaughter, and health management.
- Competent and Independent Auditors: Auditors should possess the necessary expertise, training, and qualifications to conduct thorough assessments of animal welfare practices. Independence and impartiality are crucial to ensure unbiased evaluations and reliable audit results.
- Risk-based Approach: Auditing programs should adopt a risk-based approach, focusing resources on high-risk areas or sectors where animal welfare concerns are more prevalent. This targeted approach allows for the efficient allocation of auditing resources and effective mitigation of potential welfare risks.
- Regular and Unannounced Audits: Regular auditing, including unannounced visits, helps ensure the integrity of the audit process and provides a comprehensive evaluation of day-to-day farm practices. Unannounced audits offer a realistic assessment of animal welfare conditions and discourage the manipulation of practices solely for audit purposes.
- Data Collection and Documentation: Auditing programs should include robust data collection and documentation processes. Comprehensive records of audit findings, corrective actions, and follow-up activities contribute to transparency, accountability, and continuous improvement in animal welfare.
- Collaboration and Knowledge Sharing: Collaboration between auditors, farmers, industry organizations, and regulatory bodies fosters collective

learning and promotes best practices in animal welfare. Sharing audit results, case studies, and success stories encourages knowledge exchange and drives industry-wide improvement.

*c. Strategies for promoting and achieving compliance*

- Education and Training: Providing education and training programs on animal welfare regulations, best practices, and auditing requirements is essential for promoting compliance. Farmers and farm workers should be equipped with the necessary knowledge and skills to implement effective animal welfare measures.
- Incentives and Recognition: Offering incentives, certifications, and recognition for exemplary animal welfare practices motivates farmers to strive for compliance and continuous improvement. Recognition programs highlight industry leaders and encourage others to adopt similar practices.
- Collaboration with Stakeholders: Collaborating with stakeholders, including farmers, industry associations, animal welfare organizations, and regulatory bodies, is crucial for fostering compliance with animal welfare regulations. Sharing information, resources, and expertise strengthens collective efforts toward promoting and achieving better animal welfare outcomes.
- Enforcement and Penalties: Effective enforcement mechanisms, including penalties for non-compliance, serve as deterrents and encourage farmers to meet animal welfare requirements. Regular inspections, follow-up audits, and strict enforcement actions send a clear message that non-compliance will not be tolerated.
- Public Awareness and Consumer Demand: Increasing public awareness about animal welfare issues and encouraging consumer demand for ethically produced products create market pressures that drive compliance. Consumer preferences for animal-friendly products incentivize farmers to prioritize animal welfare and comply with regulations.

### 5.4.5. Consumer awareness and ethical choices in animal products

*a. The rise of consumer awareness*

- Animal Welfare Concerns: Increasingly, consumers are concerned about the welfare of animals raised for food production. They want to ensure that animals are treated humanely, have access to appropriate living conditions, and are free from unnecessary suffering.
- Environmental Impact: Consumer awareness extends beyond animal welfare to environmental considerations. Many consumers are concerned

about the ecological footprint of animal agriculture and seek sustainable and environmentally friendly options.

- Health and Nutrition: Consumers are becoming more mindful of the health implications of their dietary choices. They prioritize products that align with their personal health goals, such as organic, hormone-free, or antibiotic-free animal products.

*b. Factors influencing ethical choices*

Several factors contribute to influencing ethical choices, encompassing a variety of considerations such as adherence to animal welfare standards, the promotion of transparency and traceability in supply chains, and the adoption of sustainable farming practices. Exploration of alternative production methods and the presence of animal-friendly labeling initiatives also play significant roles in shaping ethical decisions. These diverse influences highlight the multifaceted nature of ethical decision-making in agriculture, emphasizing the importance of considering various perspectives and principles to promote responsible and conscientious practices throughout the food production chain (Ratta et al., 2021; Sander et al., 2018; Singh & Sharma, 2023; Srivastava et al., 2020).

- Animal Welfare Standards: Consumers look for animal products that come from farms adhering to high animal welfare standards. They consider certifications, labeling, and third-party verification systems that provide assurance of ethical treatment.
- Transparency and Traceability: Consumers value transparency in the supply chain and seek information about the origin, production practices, and handling of animals. They are more likely to support brands that provide detailed information and demonstrate a commitment to transparency.
- Sustainable Farming Practices: Ethical consumers prefer products from farms that implement sustainable farming practices, such as regenerative agriculture, reduced use of chemicals, and conservation of natural resources.
- Alternative Production Methods: The rise of alternative production methods, such as organic farming, free-range systems, and plant-based alternatives, provides consumers with ethical choices that align with their values.
- Animal-friendly Labeling: Clear and informative labeling helps consumers make ethical choices. Labels such as “cage-free,” “grass-fed,” or “Certified Humane” provide guidance and assurance of better animal welfare.

*c. The power of consumer preferences*

- Market Forces: Consumer preferences have the power to shape market dynamics. As more consumers prioritize ethical considerations, the demand for ethically produced animal products increases, leading to market shifts and encouraging producers to adopt more humane practices.
- Influencing Industry Practices: Consumer demand for ethical choices can drive industry-wide changes. As producers recognise the value of meeting ethical standards, they may invest in improved animal welfare practices, sustainable production methods, and transparency in response to consumer expectations.
- Collaboration and Advocacy: Consumer awareness can fuel collaboration between consumers, animal welfare organizations, and industry stakeholders. Consumer advocacy and support for animal-friendly initiatives can influence policy changes and encourage collective efforts to improve animal welfare standards.
- Education and Information Sharing: Providing consumers with accurate and accessible information about animal welfare practices, sustainable farming methods, and ethical choices empowers them to make informed decisions and support positive changes in the industry.

*d. Challenges and considerations*

- Price Considerations: Ethically produced animal products may have a higher price point due to the costs associated with meeting higher welfare standards. Balancing consumer demand for ethical choices with affordability is a challenge that requires innovative approaches and market strategies.
- Greenwashing and Misleading Claims: Some companies may make misleading claims or engage in "greenwashing" to capitalize on consumer demand for ethical products. Consumer education and reliable certification systems are crucial to differentiate genuine ethical choices from deceptive marketing tactics.
- Cultural and Societal Factors: Ethical choices can be influenced by cultural norms, dietary preferences, and societal norms. Overcoming these barriers requires a combination of education, awareness campaigns, and the promotion of alternative ethical options.

# Chapter 6

# Food safety and quality assurance

*Preserving Trust, One Bite at a Time: Safeguarding Food Safety and Quality with Blockchain* - Unknown

**Abstract**

This chapter explores the integration of blockchain technology in ensuring food safety and quality assurance. It discusses the potential of blockchain to enhance traceability and transparency throughout the food supply chain. The chapter focuses on the implementation of blockchain to track and verify food origins, handling, and storage conditions. The chapter begins by discussing the potential benefits of blockchain technology in enhancing food safety and quality assurance. It highlights how blockchain's decentralized and immutable nature can enable the secure recording of critical information related to food production, processing, and distribution. Next, the chapter explores the implementation of blockchain to track and verify food origins, handling, and storage conditions. It examines how blockchain, combined with technologies such as Internet of Things (IoT) sensors and smart contracts, can create a comprehensive record of each stage of the food supply chain.

**Keywords**: Blockchain technology, Food safety, Quality assurance, Traceability, Transparency, Food supply chain, Food origins, Handling, Storage conditions, Internet of Things

**Chapter contents**

- Blockchain Applications in Food Safety and Quality Control
- Tracking and Preventing Foodborne Illnesses Using Blockchain Technology
- Certifying Organic and Fair Trade Products Through Blockchain Verification

***

## 6.1. Blockchain applications in food safety and quality control

### 6.1.1. The importance of food safety and quality assurance

*a. Protecting public health*

- Prevention of Foodborne Illnesses: Food safety measures aim to prevent foodborne illnesses caused by bacteria, viruses, parasites, or chemical contaminants. These measures focus on reducing the risk of contamination, implementing proper handling and storage practices, and conducting thorough inspections and testing.
- Minimizing Health Risks: Ensuring food safety minimizes health risks associated with food consumption, such as allergic reactions, microbial infections, or exposure to harmful substances. Comprehensive quality assurance programs help identify and mitigate potential hazards throughout the food production and supply chain.
- Public Health Responsibilities: Governments and regulatory agencies have a responsibility to establish and enforce food safety regulations to protect public health. Robust monitoring, surveillance, and enforcement measures are implemented to ensure compliance with these regulations.

*b. Ensuring quality and consumer confidence*

- Consistency and Uniformity: Quality assurance programs aim to maintain consistency and uniformity in food products, ensuring that consumers receive products that meet their expectations in terms of taste, texture, appearance, and nutritional value.
- Meeting Standards and Specifications: Quality assurance includes adherence to established standards and specifications for food products. This ensures that the products meet the required quality parameters, such as moisture content, size, weight, shelf life, and nutritional composition.
- Consumer Trust and Satisfaction: Consistently delivering safe and high-quality food products fosters consumer trust and satisfaction. When consumers have confidence in the safety and quality of the food they purchase, it strengthens their loyalty to brands and promotes positive word-of-mouth recommendations.

*c. Key aspects of food safety and quality assurance*

- Good Agricultural Practices (GAP): Implementing GAP ensures safe and sustainable agricultural practices, including proper use of fertilizers and pesticides, water management, and appropriate handling of crops, minimizing contamination risks from farm to fork.

- Hazard Analysis and Critical Control Points (HACCP): HACCP is a systematic approach to identify, evaluate, and control food safety hazards. This science-based preventive system helps identify critical control points in the production process and implement effective control measures.
- Quality Control and Testing: Quality control measures involve regular testing and inspection of food products to assess their compliance with safety and quality standards. This includes microbial testing, chemical analysis, sensory evaluation, and physical inspections.
- Traceability and Recall Systems: Effective traceability systems track the movement of food products throughout the supply chain, enabling quick identification and targeted recall of potentially unsafe products in case of contamination or quality issues.
- Compliance with Regulatory Standards: Compliance with national and international food safety and quality standards is essential to ensure consistent adherence to best practices. Regulatory agencies establish guidelines and perform inspections to enforce compliance and ensure public safety.

*d. Advancements in food safety and quality assurance*

Recent advancements in food safety and quality assurance are driven by several key factors, notably technological innovation, enhanced supply chain transparency, and increased international cooperation (Hong, Mao, Wu, & Pu, 2021; Singh & Sharma, 2023; Tripoli & Schmidhuber, 2018). These developments signify a shift towards more proactive and collaborative approaches to ensuring the integrity and safety of the global food supply. By leveraging emerging technologies, fostering transparency across supply chains, and strengthening international partnerships, stakeholders can collectively address complex challenges and work towards safeguarding public health and consumer trust in the food industry.

- Technology and Innovation: Technological advancements, such as automated monitoring systems, DNA-based testing methods, and data analytics, enhance the efficiency and accuracy of food safety and quality assurance processes.
- Supply Chain Transparency: The implementation of advanced tracking and monitoring technologies, such as blockchain, enhances supply chain transparency, enabling faster identification of potential safety or quality issues and facilitating targeted interventions.

- International Cooperation: Global collaboration and information-sharing among regulatory agencies, industry associations, and research institutions promote harmonisation of food safety standards, exchange of best practices, and joint efforts to address emerging food safety challenges.

### 6.1.2. Enhancing transparency and traceability in the food supply chain

*a. The importance of transparency and traceability*

Transparency and traceability play crucial roles in cultivating consumer trust, safeguarding the integrity of the food supply chain, and facilitating compliance with food safety regulations and labeling requirements (Blokdyk, 2021; Yiannas, 2018). By providing clear visibility into the journey of food products from farm to table, transparency instills confidence in consumers regarding the safety, authenticity, and ethical standards upheld throughout the supply chain. Moreover, traceability enables swift and accurate identification of potential issues, thereby enhancing responsiveness to food safety incidents and bolstering overall confidence in the reliability and accountability of the food industry.

- Consumer Trust and Confidence: Transparency and traceability foster consumer trust by providing them with the information they need to make informed decisions. When consumers have access to details about the origin, production methods, and handling of food products, it instills confidence in their safety and quality.
- Supply Chain Integrity: Enhancing transparency and traceability helps ensure the integrity of the food supply chain. It enables stakeholders to identify and address potential risks, such as fraud, adulteration, or contamination, promptly and effectively, safeguarding public health and industry reputation.
- Regulatory Compliance: Transparency and traceability support compliance with food safety regulations and labeling requirements. By accurately documenting and tracking product information throughout the supply chain, stakeholders can demonstrate adherence to regulatory standards.

*b. Technologies enabling transparency and traceability*

The implementation of technologies facilitating transparency and traceability encompasses a diverse range of inputs, activities, and integrations, including blockchain technology, Internet of Things (IoT), data analytics, and mobile applications (Awwad et al., 2018; Blokdyk, 2021; Casino, Kanakaris, Dasaklis, Moschuris, & Rachaniotis, 2019; Chen, Chen, Lin, & Zhuang, 2021; Cottrill et al., 2018; Creydt & Fischer, 2019; Feng et al., 2020; Lin, Shen, Zhang, & Chai, 2018; Tian, 2016; Tsang, Choy, Wu, Ho, & Lam, 2019). These technologies serve as fundamental pillars in revolutionizing supply chain management, offering real-time monitoring, data-driven insights, and seamless communication channels throughout the entire food production and distribution process. By leveraging these innovations, stakeholders can enhance transparency, bolster trust, and mitigate risks associated with food safety and quality.

- Blockchain Technology: Blockchain provides a decentralized and immutable ledger that records and verifies transactions across the supply chain. It enables transparent and tamper-proof tracking of products, facilitating real-time visibility, and enhancing trust among stakeholders.
- Internet of Things (IoT): IoT devices, such as sensors and RFID tags, enable real-time monitoring of temperature, humidity, location, and other vital parameters. This data can be seamlessly integrated into traceability systems, ensuring the accurate and timely tracking of products.
- Data Analytics: Advanced analytics tools allow stakeholders to extract valuable insights from vast amounts of data collected throughout the supply chain. This enables proactive decision-making, early detection of anomalies, and continuous improvement of processes.
- Mobile Applications: Mobile applications provide a convenient means for consumers to access information about food products. QR codes or barcodes on product packaging can be scanned to retrieve detailed information, such as origin, certifications, and production practices.

### *c. Benefits for stakeholders*

Stakeholders across the food supply chain, whose transparency and traceability are enhanced by blockchain technology, encompass a wide spectrum, including producers and farmers, retailers and distributors, regulatory authorities, and consumers. Each of these stakeholders stands to benefit in distinct ways from the implementation of blockchain solutions. Producers and farmers can improve accountability and market access, retailers and distributors can enhance inventory management and consumer trust, regulatory authorities can streamline compliance and enforcement efforts, and consumers can gain access to reliable information about the products they purchase (Blokdyk, 2021; Cottrill et al., 2018).

- Producers and Farmers: Enhanced transparency and traceability empower producers and farmers to showcase their responsible farming practices and differentiate their products in the market. It provides an opportunity to highlight sustainability efforts, ethical considerations, and compliance with quality standards.
- Retailers and Distributors: Transparent supply chains allow retailers and distributors to build stronger relationships with consumers by providing

accurate product information and responding to their concerns. It also enables effective inventory management, reducing waste and ensuring product availability.

- Regulatory Authorities: Transparency and traceability assist regulatory authorities in monitoring and enforcing compliance with food safety regulations. They enable efficient recall management, targeted interventions in case of safety issues, and data-driven policymaking.
- Consumers: Transparent and traceable supply chains empower consumers to make informed choices based on their preferences, dietary needs, and ethical considerations. It strengthens their confidence in the safety, quality, and sustainability of the products they purchase.

*d. Collaboration and industry initiatives*

Augmenting transparency and traceability within the food supply chain not only paves the way for enhanced accountability but also fosters collaboration and industry-driven initiatives (Blokdyk, 2021; Feng et al., 2020). Such initiatives encompass a variety of efforts, including standardization and certification processes to establish uniformity and trust in supply chain practices, partnerships and data sharing arrangements among stakeholders to facilitate seamless information exchange, and targeted consumer education and engagement programs aimed at promoting awareness and empowering informed choices. By embracing these collaborative endeavors, stakeholders can collectively work towards building a more resilient, efficient, and trustworthy food ecosystem.

- Standardization and Certification: Industry-wide standardization and certification programs help streamline transparency and traceability practices. These initiatives establish common guidelines, requirements, and auditing processes, promoting consistency and interoperability across the supply chain.
- Partnerships and Data Sharing: Collaboration among supply chain stakeholders, including producers, retailers, distributors, and technology providers, facilitates data sharing and integration. By working together, they can create a holistic view of the supply chain and ensure seamless traceability and transparency.
- Consumer Education and Engagement: Educating consumers about the significance of transparency and traceability cultivates a more informed and engaged customer base. Initiatives such as campaigns, labels, and digital platforms can effectively communicate the benefits of transparent supply chains and empower consumers to make conscious choices.

### 6.1.3. Blockchain for secure and decentralized data storage

*a. Understanding blockchain technology*

- Decentralization: Blockchain operates on a decentralized network, where multiple participants called nodes maintain copies of the blockchain ledger. This eliminates the need for a central authority, making it resistant to single points of failure or control.
- Immutability: Data stored on a blockchain is secured through cryptographic hashing and linked together in blocks. Once a block is added to the chain, it becomes virtually impossible to alter or tamper with the information contained within it.
- Transparency: Blockchain provides transparency by allowing all participants to access and validate the information recorded on the ledger. Each transaction or data entry is visible to all network participants, promoting accountability and trust.
- Consensus Mechanism: Blockchain utilizes a consensus mechanism, such as proof-of-work or proof-of-stake, to ensure agreement among network participants on the validity of transactions and the order in which they are added to the blockchain.

*b. Benefits of blockchain for data storage*

- Security: Blockchain offers robust security features, such as encryption, digital signatures, and consensus mechanisms, which make it highly resistant to data breaches and unauthorized access. The decentralized nature of the technology reduces the risk of single-point failures and enhances data integrity.
- Immutability and Auditability: Once data is recorded on the blockchain, it becomes practically immutable. This immutability provides a tamper-proof audit trail, ensuring that historical data remains intact and can be verified at any point in the future.
- Data Integrity and Authenticity: Blockchain enables the verification of data authenticity and provenance. Each transaction or data entry is linked to a cryptographic hash, allowing users to verify the integrity of the information and confirm its source.
- Transparency and Trust: Blockchain's transparent nature promotes trust among participants. By allowing all stakeholders to access and verify the data, blockchain ensures transparency and accountability, reducing the need for intermediaries and enhancing trust in the system.

*c. Applications of blockchain in data storage*

The applications of blockchain technology in data storage are diverse and far-reaching, spanning across various sectors such as supply chain management, healthcare and medical records, financial services, intellectual property management, and government and identity management (Iftekhar et al., 2020; Tomu, 2020; Xu, Li, Zeng, Cao, & Jiang, 2022). By leveraging blockchain's decentralized and immutable ledger capabilities, organizations in these industries can enhance data security, transparency, and integrity. Whether it's ensuring the authenticity of products along the supply chain, safeguarding sensitive medical information, or protecting intellectual property rights, blockchain offers a robust solution for secure and efficient data storage across a multitude of domains.

- Supply Chain Management: Blockchain can revolutionize supply chain management by securely storing data related to product provenance, certifications, and logistics. This enables traceability, reduces counterfeiting, and enhances transparency throughout the supply chain.
- Healthcare and Medical Records: Blockchain can securely store and manage sensitive medical records, providing patients with control over their data while enabling interoperability and facilitating secure data sharing between healthcare providers.
- Financial Services: Blockchain has disrupted the financial industry by providing secure storage of transaction records, simplifying cross-border payments, and enabling the issuance of digital assets and smart contracts with self-executing capabilities.
- Intellectual Property Management: Blockchain can assist in the protection of intellectual property rights by securely storing records of creations, patents, and copyrights. This facilitates proof of ownership and reduces disputes over intellectual property.
- Government and Identity Management: Blockchain can enhance identity management systems by providing secure storage of personal data, enabling self-sovereign identity, and reducing the risks associated with centralized identity databases.

*d. Challenges and considerations*

Blockchain technology encounters numerous challenges that warrant careful consideration. These challenges encompass scalability, interoperability, and regulatory and legal considerations. Addressing these hurdles is essential for unlocking the full potential of blockchain technology across various industries (Cappiello, Comuzzi, Daniel, & Meroni, 2019).

- Scalability: Blockchain technology faces scalability challenges due to the increasing volume of data stored on the chain and the consensus mechanisms employed. However, advancements like sharding, layer-2 solutions, and improved protocols are addressing these scalability concerns.
- Interoperability: Ensuring interoperability between different blockchain platforms and existing systems remains a challenge. Standards and protocols are being developed to facilitate seamless integration and data exchange across multiple blockchain networks.
- Regulatory and Legal Considerations: As blockchain adoption expands, regulatory frameworks need to be established to address data privacy, ownership, and compliance with existing laws. These considerations are crucial to ensure the responsible and ethical use of blockchain technology.

### 6.1.4. Immutable record-keeping of food safety practices and quality parameters

*a. Importance of immutable record-keeping*

- Regulatory Compliance: Immutable record-keeping is essential for complying with food safety regulations and industry standards. It enables food businesses to demonstrate their adherence to safety protocols, inspections, and quality control measures required by regulatory authorities.
- Traceability: The ability to trace the origin and movement of food products throughout the supply chain is crucial for rapid and efficient recalls, investigations, and risk assessments. Immutable record-keeping ensures an accurate and reliable audit trail, facilitating effective traceability.
- Quality Assurance: Maintaining detailed records of quality parameters, such as temperature monitoring, storage conditions, and testing results, allows businesses to identify and address potential quality issues promptly. Immutable records provide evidence of quality control practices and aid in maintaining consistent product quality.
- Consumer Confidence: Immutable record-keeping enhances consumer confidence by providing transparency and accountability. When consumers have access to comprehensive and accurate information about food safety practices and quality parameters, they can make informed choices and trust in the products they purchase.

*b. Technologies facilitating immutable record-keeping*

Technologies enabling immutable record-keeping encompass a variety of inputs and activities essential for ensuring the integrity and permanence of data. These include blockchain technology, Internet of Things (IoT) sensors, and digital documentation systems (FAO, 2019; Hong et al., 2021; Jin et al., 2020; Singh & Sharma, 2023; Tao, Cui, Huang, Leigh, & Gu, 2019; Tse et al., 2017). By leveraging Blockchain's decentralized ledger, IoT's real-time data collection capabilities, and digital documentation systems' secure storage mechanisms, stakeholders can establish reliable and tamper-proof records across diverse industries. These technologies play a pivotal role in enhancing transparency, accountability, and trustworthiness in data management practices, driving innovation and efficiency in various sectors.

- Blockchain Technology: Blockchain offers a decentralized and immutable ledger that records and verifies transactions or data entries. By storing food safety practices and quality parameters on a blockchain, businesses can ensure the integrity and immutability of their records while allowing authorized stakeholders to access and verify the information.
- Internet of Things (IoT): IoT devices, such as temperature sensors, humidity monitors, and RFID tags, can collect real-time data on food storage conditions, transportation, and processing. This data can be integrated with blockchain or other record-keeping systems, providing an immutable and tamper-proof record of environmental parameters.
- Digital Documentation Systems: Digital documentation systems, such as electronic record-keeping software or cloud-based platforms, offer secure and centralized repositories for storing food safety practices and quality parameters. These systems often include features for data entry, storage, retrieval, and audit trails, ensuring the integrity and accessibility of records.

*c. Benefits of immutable record-keeping*

- Data Integrity: Immutable record-keeping ensures that food safety practices and quality parameters remain intact and unaltered over time. This eliminates the risk of unauthorized modifications, falsification, or data tampering, preserving the accuracy and reliability of records.
- Efficient Auditing and Inspections: Immutable records simplify the auditing and inspection processes. Regulatory authorities and third-party auditors can access and verify records efficiently, reducing administrative burdens and enhancing the speed and accuracy of compliance assessments.
- Rapid Response to Incidents: In the event of a food safety incident or quality issue, immutable record-keeping enables swift identification of the root cause and targeted response. The ability to trace the history of products and pinpoint the affected batches accelerates recalls, containment measures, and corrective actions.

- Continuous Improvement: Immutable record-keeping facilitates data analysis and trend identification, leading to continuous improvement in food safety practices and product quality. Analyzing historical records allows businesses to identify patterns, areas for enhancement, and opportunities to optimize processes.

*d. Best practices for implementing immutable record-keeping*

Optimal strategies for implementing immutable record-keeping encompass a range of best practices aimed at ensuring data integrity and security. These include standardized data entry protocols to maintain consistency and accuracy, robust data encryption measures and stringent access controls to safeguard against unauthorized access or tampering, regular data backup procedures to mitigate the risk of data loss or corruption, and comprehensive training programs coupled with thorough documentation to empower personnel with the requisite knowledge and skills (Detwiler, 2020). By adhering to these best practices, organizations can establish a robust framework for immutable record-keeping, fostering trust, reliability, and compliance with regulatory requirements.

- Standardized Data Entry: Establishing clear guidelines and standardizing data entry practices ensure consistency and accuracy in record-keeping. This includes capturing relevant information, using standardized formats, and implementing validation checks.
- Data Encryption and Access Controls: Protecting the confidentiality and security of immutable records is essential. Implementing robust encryption methods and access controls prevents unauthorized access, safeguards sensitive information, and maintains the integrity of the records.
- Regular Data Backup: Regular backups of immutable records safeguard against data loss due to technical failures, cyberattacks, or system errors. Backup procedures should be well-documented, and data restoration processes should be periodically tested to ensure their effectiveness.
- Training and Documentation: Training employees on the importance of immutable record-keeping, data entry protocols, and privacy requirements fosters a culture of compliance and accountability. Documenting procedures and guidelines aids in consistency and provides a reference for employees to follow.

### 6.1.5. Ensuring compliance with regulatory standards and certifications

*a. Importance of compliance with regulatory standards and certifications*

- Legal Compliance: Regulatory standards and certifications are established to protect the health and safety of consumers, workers, and the environment. Compliance with these standards ensures that organizations operate within the legal framework and meet their obligations.
- Consumer Confidence: Compliance demonstrates an organization's commitment to quality, safety, and ethical practices. It instills confidence in consumers, who can trust that products and services meet recognised standards and have undergone rigorous assessments.
- Risk Mitigation: Compliance helps identify and address potential risks, reducing the likelihood of accidents, product recalls, fines, and legal liabilities. By implementing robust compliance measures, organizations can proactively manage risks and protect their stakeholders.
- Market Access: Many industries require compliance with specific regulatory standards and certifications to access certain markets or participate in government procurement. Meeting these requirements opens up new opportunities and ensures competitiveness in the marketplace.

*b. Challenges in achieving compliance*

Navigating compliance with regulatory standards and certifications presents a myriad of challenges for organizations. These challenges encompass the dynamic nature of the regulatory landscape, characterized by continual evolution and adaptation to emerging threats and technologies, as well as the inherent complexity and variability inherent in compliance requirements (Hong et al., 2021; Singh & Sharma, 2023). Additionally, resource constraints, such as limited budgets and staffing, pose significant hurdles for organizations striving to achieve and maintain compliance. Furthermore, fostering a robust compliance culture across all levels of the organization is essential to ensuring sustained adherence to regulatory standards. Addressing these challenges demands a proactive and holistic approach, involving strategic planning, resource allocation, and a commitment to cultivating a culture of compliance.

- Evolving Regulatory Landscape: Regulatory standards are continually evolving to address emerging risks and advancements in technology. Staying abreast of these changes and adapting internal processes and procedures accordingly can be challenging for organizations.

- Complexity and Variability: Regulatory standards and certifications can be complex, with multiple requirements and varying interpretations. Understanding the specific requirements, applying them consistently, and navigating different jurisdictional regulations pose challenges for organizations.
- Resource Constraints: Compliance efforts often require significant resources, including financial investments, skilled personnel, and technology infrastructure. Small and medium-sized enterprises, in particular, may face resource constraints in achieving and maintaining compliance.
- Compliance Culture: Establishing a culture of compliance within an organization is crucial. Ensuring that employees are aware of and adhere to regulatory standards, implementing effective training programs, and fostering a culture of accountability can be challenging but essential for long-term compliance success.

*c. Strategies for achieving and maintaining compliance*

- Regulatory Intelligence: Organizations should stay informed about the regulatory landscape, actively monitoring changes, updates, and new requirements. Establishing processes to assess the impact of regulatory changes and proactively adapting internal procedures help maintain compliance.
- Robust Compliance Management Systems: Implementing compliance management systems, such as standard operating procedures, policies, and document control processes, ensures consistent adherence to regulatory standards. These systems should include mechanisms for monitoring, reporting, and addressing non-compliance issues.
- Training and Education: Providing comprehensive training programs for employees at all levels enhances awareness of regulatory requirements and fosters a culture of compliance. Regular training sessions, workshops, and knowledge-sharing initiatives keep employees updated and engaged in compliance efforts.
- Internal Audits and Reviews: Conducting regular internal audits and reviews helps identify gaps, assess the effectiveness of compliance measures, and proactively address issues. These audits should be objective, documented, and involve key stakeholders to ensure thorough evaluations.
- Collaboration and Partnerships: Engaging with industry associations, regulatory bodies, and certification agencies fosters collaboration, knowledge sharing, and alignment with best practices. Seeking guidance

from experts and leveraging external resources can support compliance efforts.

- Continuous Improvement: Compliance is an ongoing process. Organizations should strive for continuous improvement by evaluating feedback, conducting root-cause analysis, implementing corrective actions, and regularly reviewing and updating compliance strategies.

## 6.2. Tracking and preventing foodborne illnesses using blockchain technology

### 6.2.1. Challenges in foodborne illness tracking and prevention

*a. Complexity of foodborne illnesses*

- Wide Range of Pathogens: Foodborne illnesses can be caused by various pathogens, including bacteria, viruses, parasites, and toxins. Each pathogen has unique characteristics, transmission routes, and incubation periods, making it challenging to identify and control specific sources of contamination.
- Diverse Food Supply Chain: The globalised nature of the food supply chain presents challenges in tracing the origin and movement of contaminated food. Complex distribution networks, multiple intermediaries, and international trade increase the difficulty of identifying the exact source of contamination and implementing effective preventive measures.
- Underreporting and Delayed Reporting: Many cases of foodborne illness go unreported or are reported with a delay, hindering timely investigations and preventive actions. Factors such as mild symptoms, lack of awareness, and reluctance to seek medical attention contribute to underreporting, making it difficult to accurately track the prevalence and causes of foodborne illnesses.

*b. Foodborne illness outbreak investigations*

Effective management of foodborne illness outbreaks demands three critical components: prompt detection and response, seamless collaboration among stakeholders, and comprehensive traceability and data integration. Prompt detection and response are crucial to containing the spread of illnesses and minimizing their impact on public health. Seamless collaboration among stakeholders ensures coordinated efforts in identifying the source of contamination and implementing preventive measures to mitigate further risks. Comprehensive traceability and data integration facilitate the tracking of affected products throughout the supply chain, enabling swift removal from shelves and accurate communication with consumers to prevent additional cases (Balzarova, 2021).

- Timely Detection and Response: Rapid identification and response to foodborne illness outbreaks are crucial to prevent further contamination and protect public health. However, the detection of outbreaks can be challenging due to the time lag between consuming contaminated food and experiencing symptoms, as well as the difficulty of linking individual cases to a common source.
- Coordinated Collaboration: Investigating foodborne illness outbreaks requires collaboration among multiple agencies, including public health departments, food safety authorities, laboratories, and healthcare providers. Ensuring effective coordination, information sharing, and communication among these stakeholders can be challenging, especially during cross-jurisdictional outbreaks.
- Traceability and Data Integration: Tracing contaminated food back to its source is essential for identifying the root cause of outbreaks and implementing appropriate control measures. However, the lack of standardized traceability systems and data integration across the food supply chain hinders efficient and accurate traceback investigations.

*c. Prevention strategies*

- Risk Assessment and Management: Assessing and managing risks associated with foodborne illnesses require a comprehensive understanding of potential hazards, vulnerability assessments, and preventive controls. Developing and implementing risk-based strategies can help prioritize resources and preventive measures where they are most needed.
- Education and Awareness: Promoting consumer education and awareness about safe food handling practices, proper cooking temperatures, and recognising symptoms of foodborne illnesses are vital in preventing contamination and reducing the incidence of illnesses. Public health campaigns, educational materials, and outreach programs play a crucial role in empowering consumers to make informed choices.
- Regulatory Framework and Enforcement: Strong regulatory frameworks and effective enforcement of food safety standards are essential to prevent and control foodborne illnesses. Continuous monitoring, inspections, and enforcement actions help ensure compliance with regulations and hold accountable those who violate food safety practices.
- Advances in Technology: Leveraging technological advancements, such as rapid testing methods, DNA sequencing, and data analytics, can significantly improve the tracking and prevention of foodborne illnesses. These

technologies enable quicker detection of pathogens, more accurate identification of sources of contamination, and targeted interventions.

### 6.2.2. Rapid response to food recalls and crises

*a. The significance of rapid response*

- Consumer Safety: The primary objective of a rapid response is to protect consumer safety. By promptly identifying and removing contaminated or hazardous food from the market, the risk of illnesses and adverse health effects can be minimized.
- Risk Mitigation: Swift action helps mitigate potential risks associated with contaminated food, such as further distribution, consumption, and spread of pathogens. This prevents additional cases of foodborne illnesses and reduces the scope and impact of the crisis.
- Reputation Management: A well-executed and timely response demonstrates an organization's commitment to consumer safety and transparency. It helps protect the reputation and brand image by showing stakeholders that appropriate measures are taken to address food safety concerns.
- Regulatory Compliance: Complying with regulatory requirements and cooperating with food safety authorities is crucial during recalls and crises. A rapid response ensures adherence to legal obligations, fosters collaboration with regulatory agencies, and helps maintain compliance with applicable laws and regulations.

*b. Challenges in rapid response*

- Detection and Identification: Detecting and identifying the source of contamination or crisis swiftly is essential for an effective response. However, determining the cause of the issue, whether it's a pathogen, chemical contaminant, or other factors, can be challenging and time-consuming, delaying the initiation of necessary actions.
- Communication and Information Sharing: Timely and accurate communication is crucial during a food recall or crisis. Coordinating internal and external communication, ensuring consistent messaging, and disseminating information to the public, stakeholders, and regulatory agencies can be complex, particularly during high-pressure situations.
- Supply Chain Complexity: The complexity of modern food supply chains presents challenges in quickly identifying and recalling affected products. Tracing the flow of goods from production to distribution and retail can be

intricate, requiring efficient traceability systems and collaboration among multiple stakeholders.

- Consumer Awareness and Response: Promptly reaching and informing consumers about the recall or crisis is vital. However, ensuring that the information reaches the intended audience, addressing public concerns, and managing consumer reactions can be challenging, particularly in the age of social media and rapid information dissemination.

*c. Strategies for effective response*

- Preparedness Planning: Developing comprehensive recall and crisis management plans in advance is crucial. These plans should outline roles and responsibilities, communication protocols, and escalation procedures to enable a rapid and coordinated response when an incident occurs.
- Collaboration and Coordination: Effective collaboration among internal teams, supply chain partners, regulatory agencies, and other relevant stakeholders is essential for a swift response. Building relationships, establishing communication channels, and conducting regular exercises and drills enhance coordination and preparedness.
- Rapid Decision-making: Empowering designated individuals or teams with the authority to make critical decisions quickly is vital during a food recall or crisis. Streamlining decision-making processes, ensuring access to accurate data, and providing clear guidance enable timely actions to protect consumers and mitigate risks.
- Clear Communication Channels: Establishing clear communication channels internally and externally is crucial for effective information sharing. Ensuring that designated spokespersons are trained in crisis communication and providing timely and accurate updates to the public and stakeholders maintains transparency and builds trust.
- Continuous Improvement: Conducting post-incident reviews and analysis helps identify areas for improvement and lessons learned. Incorporating these insights into updated recall and crisis management plans, procedures, and training programs enhances response capabilities and overall preparedness.

### 6.2.3. Blockchain-based real-time monitoring of food products

*a. Understanding blockchain technology*

- Distributed Ledger: Blockchain is a decentralized and immutable ledger that records transactions across multiple computers or nodes. Each transaction, or block, is time-stamped, linked to previous blocks, and encrypted, ensuring transparency, security, and immutability of the data.

- Transparency and Traceability: Blockchain enables end-to-end traceability of food products by recording every transaction, transfer of ownership, and relevant information on the blockchain. This transparency enhances supply chain visibility and enables consumers to verify the origin, quality, and authenticity of the products they purchase.
- Smart Contracts: Smart contracts are self-executing agreements coded on the blockchain. They automatically trigger predefined actions based on predetermined conditions. Smart contracts can facilitate compliance with regulations, automate processes, and streamline interactions among supply chain participants.

*b. Real-time monitoring with blockchain*

- Supply Chain Visibility: Blockchain-based real-time monitoring provides real-time visibility into the movement and condition of food products throughout the supply chain. Each participant in the supply chain can record and verify transactions, including temperature, humidity, location, and other relevant data, ensuring transparency and accountability.
- Quality Assurance: By integrating IoT devices and sensors with blockchain technology, real-time data on temperature, humidity, and other quality parameters can be recorded and stored securely. This information can be accessed by relevant stakeholders, enabling early detection of deviations from desired quality standards and timely interventions to maintain product integrity.
- Food Safety and Recall Management: Blockchain enhances the ability to trace and track the origin of food products, enabling swift and targeted recalls in the event of safety concerns. With real-time monitoring, any anomalies or breaches in safety protocols can be identified promptly, allowing for immediate action to prevent potential foodborne illnesses or contamination.
- Consumer Trust and Transparency: Blockchain empowers consumers with the ability to access transparent and trustworthy information about the products they consume. By scanning QR codes or using mobile applications, consumers can retrieve comprehensive details about the product's journey, including its source, processing, and handling, fostering trust and promoting informed purchasing decisions.

*c. Challenges and considerations*

Implementing blockchain-based real-time monitoring for food products presents several challenges that warrant careful consideration. These encompass integration with current systems, ensuring data accuracy and reliability, and addressing scalability and efficiency concerns. Prioritizing privacy and security is also essential to safeguard sensitive information and maintain consumer trust in the integrity of the monitoring process (Salah, Nizamuddin, Jayaraman, & Omar, 2019).

- Integration with Existing Systems: Integrating blockchain technology with existing supply chain infrastructure and information systems can pose technical challenges. Collaborative efforts, standards, and interoperability solutions are necessary to ensure seamless integration and maximize the benefits of blockchain-based monitoring.
- Data Accuracy and Reliability: The accuracy and reliability of data recorded on the blockchain heavily rely on the integrity of the data sources and the trustworthiness of participants in the supply chain. Verification mechanisms, such as IoT sensors and data authentication protocols, are critical to ensure the accuracy and validity of the recorded information.
- Scalability and Efficiency: As the volume of transactions and data increases, blockchain networks may face scalability and performance issues. Implementing solutions such as off-chain storage, sharding, or layer-2 protocols can help address these challenges and ensure the efficient functioning of blockchain-based monitoring systems.
- Privacy and Security: Protecting sensitive data while ensuring transparency and data accessibility is a critical consideration. Blockchain networks must implement robust security measures, encryption protocols, and access controls to safeguard confidential information and prevent unauthorized access or tampering.

*d. Future prospects and opportunities*

- Interconnected Food Ecosystem: Blockchain-based monitoring can pave the way for a more interconnected food ecosystem, where information flows seamlessly among stakeholders. This enables collaborative decision-making, early detection of issues, and efficient supply chain management.
- Enhanced Trust and Collaboration: By providing a decentralized and transparent platform, blockchain fosters trust and collaboration among supply chain participants. It encourages data sharing, establishes clear accountability, and promotes collective responsibility for food safety and quality.

- Integration of Emerging Technologies: Combining blockchain with other emerging technologies, such as artificial intelligence, machine learning, and big data analytics, can unlock new possibilities for real-time monitoring. These technologies can enable predictive analytics, anomaly detection, and intelligent decision-making, further enhancing food safety and supply chain efficiency.

### 6.2.4. Identifying contamination sources and minimizing the spread of illnesses

*a. The significance of identifying contamination sources*

- Rapid Response: Swiftly identifying the source of contamination is crucial for initiating a timely response to prevent further spread of illnesses. Pinpointing the exact source enables targeted actions, such as recalls, containment measures, and corrective actions, to be implemented promptly.
- Preventing Recurrence: Understanding the contamination source helps prevent similar incidents from occurring in the future. By identifying the root cause, corrective measures can be taken to address underlying issues in the production, processing, or distribution stages of the food supply chain.
- Protecting Public Health: Identifying contamination sources minimizes the risk to public health by enabling quick intervention and preventing additional individuals from consuming contaminated products. It helps mitigate the severity and scope of foodborne illness outbreaks and protects vulnerable populations.

*b. Methods for source tracing*

- Epidemiological Investigations: Epidemiological investigations involve analyzing patterns and trends of illness among affected individuals. By conducting interviews, collecting data, and identifying commonalities in food consumption, investigators can narrow down potential sources of contamination.
- Microbial Testing: Microbial testing involves analyzing food samples to detect the presence of pathogens or harmful contaminants. By comparing the genetic fingerprints of pathogens found in the affected individuals and food samples, it becomes possible to trace back the source of contamination.
- Supply Chain Traceability: Establishing a robust traceability system within the food supply chain is vital for source tracing. By recording and tracking the movement of products from farm to fork, including information on

suppliers, production processes, and distribution channels, it becomes easier to identify the source of contamination in the event of an outbreak.

- Advanced Technologies: Advanced technologies, such as whole-genome sequencing, metagenomics, and data analytics, play a crucial role in source tracing. These technologies provide detailed insights into the genetic makeup of pathogens, enabling more accurate identification of contamination sources and facilitating targeted interventions.

*c. Strategies for minimizing the spread of illnesses*

- Prompt Response and Communication: Timely response and effective communication are essential for minimizing the spread of illnesses. Rapidly notifying relevant stakeholders, including regulatory authorities, healthcare professionals, and the public, ensures swift action and preventive measures are taken.
- Product Recall and Removal: When a contamination source is identified, initiating a product recall or removal is necessary to prevent further distribution and consumption of contaminated products. Collaboration among stakeholders and implementing efficient recall protocols is crucial for effectively removing affected products from the market.
- Enhanced Hygiene and Safety Practices: Strengthening hygiene and safety practices throughout the food supply chain helps minimize the risk of contamination. Implementing robust sanitation protocols, regular training programs, and adherence to industry standards and regulations significantly contribute to preventing the spread of illnesses.
- Continuous Monitoring and Surveillance: Implementing comprehensive monitoring and surveillance systems enables the early detection of contamination and illness outbreaks. This includes regular testing of food products, environmental monitoring, and surveillance of disease patterns to proactively identify potential risks and take preventive measures.
- Collaboration and Knowledge Sharing: Collaboration among stakeholders, including government agencies, industry associations, healthcare professionals, and researchers, is crucial for sharing information, best practices, and lessons learned. Sharing knowledge and experiences facilitates more effective prevention and response strategies.

### 6.2.5. Enhancing consumer trust and confidence in food safety

*a. The importance of consumer trust in food safety*

- Public Health Protection: Consumer trust in food safety measures is directly linked to public health. When consumers trust that the food they consume is safe, they can make informed choices and have confidence in the products they purchase.
- Market Stability: A strong foundation of consumer trust contributes to market stability and sustains the growth of the food industry. Trust fosters brand loyalty, promotes consumer satisfaction, and drives economic stability within the food sector.
- Regulatory Compliance: Consumer trust incentivizes food producers, manufacturers, and distributors to comply with food safety regulations and implement rigorous quality assurance practices. It encourages businesses to prioritize consumer safety and continuously improve their food safety protocols.

*b. Challenges in food safety communication*

- Complex Information: Communicating food safety information to consumers can be challenging due to the complexity of scientific and technical terms. Simplifying and translating technical language into accessible and understandable information is crucial for effective communication.
- Information Overload: In today's digital age, consumers are bombarded with vast amounts of information from various sources, which can lead to confusion and skepticism. Cutting through the noise and providing accurate, relevant, and concise information is key to building trust.
- Misinformation and Rumours: The spread of misinformation and rumours can undermine consumer trust in food safety. Addressing misinformation promptly, providing evidence-based information, and transparently correcting misconceptions are vital for maintaining credibility.
- Crisis Management: Effectively communicating during food safety crises, such as recalls or outbreaks, is challenging. Timely, transparent, and consistent communication is essential to minimize panic, ensure public safety, and rebuild trust in the aftermath of a crisis.

*c. Strategies to foster trust and confidence*

- Transparent Supply Chain: Increasing transparency in the food supply chain is crucial for building trust. Implementing traceability systems, sharing information on sourcing, production practices, and safety protocols, and engaging in third-party certifications help consumers understand the journey of their food.

- Educational Initiatives: Educating consumers about food safety practices, including safe handling, storage, and cooking, empowers them to make informed decisions. Collaborating with educational institutions, public health agencies, and community organizations can effectively disseminate knowledge and promote safe food practices.
- Improved Labeling and Packaging: Clear and accurate labeling is essential in conveying information about allergens, ingredients, nutritional content, and safety certifications. Implementing standardized labeling practices and ensuring transparency in labeling builds consumer confidence in the products they purchase.
- Technology and Digital Tools: Leveraging technology, such as mobile applications, QR codes, and online platforms, allows consumers to access detailed information about food safety, certifications, and recalls. Integrating digital tools into food safety communication enhances transparency and engagement with consumers.
- Collaboration and Industry Standards: Collaboration among industry stakeholders, regulatory bodies, and consumer advocacy groups is essential for setting and implementing industry standards. Active participation in industry initiatives, adherence to strict safety protocols, and voluntary certifications demonstrate commitment to consumer safety.
- Continuous Improvement and Monitoring: Implementing robust quality assurance programs, conducting regular audits, and monitoring food safety practices are crucial for continuous improvement. Periodic reviews and transparent reporting on safety measures inspire consumer confidence in the commitment to maintaining high standards.

## 6.3. Certifying organic and fair trade products through blockchain verification

### 6.3.1. Authenticating organic and fair trade claims

*a. The significance of authenticating organic and fair trade claims*

- Consumer Confidence: Authenticating organic and fair trade claims instills confidence in consumers, assuring them that the products they purchase meet specific standards related to environmental sustainability and ethical practices. It allows consumers to make informed choices aligned with their values.

- Market Integrity: Authenticity in organic and fair trade claims protects the integrity of the market by preventing fraudulent practices. Verifying these claims ensures that products labelled as organic or fair trade genuinely adhere to the established guidelines and support the principles they represent.
- Supporting Sustainable Practices: By authenticating organic and fair trade claims, we support sustainable farming methods, environmental conservation, fair wages for producers, and community development. It encourages responsible business practices throughout the supply chain.

*b. Challenges in verifying organic and fair trade claims*

- Complex Supply Chains: The global nature of supply chains in the organic and fair trade sectors presents challenges in verifying claims. Tracking the origin, production methods, and handling of products across multiple stages and geographical locations can be complex and require robust systems.
- Certification Standards and Variations: Different certification bodies and standards exist for organic and fair trade products, which may vary across regions and countries. Harmonising these standards and ensuring consistent verification processes can be challenging.
- Document Integrity: Verifying organic and fair trade claims often involves scrutinising documentation, such as certificates, invoices, and transaction records. Ensuring the authenticity and accuracy of these documents is essential to establish the veracity of the claims.

*c. Strategies and technologies for authentication*

Numerous strategies and technologies exist for authenticating organic and fair trade claims, ranging from third-party certifications to cutting-edge approaches such as blockchain technology, DNA testing, and traceability. Additionally, supply chain audits, on-site inspections, and fostering collaboration and data sharing play pivotal roles in ensuring credibility. By implementing these multifaceted authentication methods and promoting transparency throughout the supply chain, businesses can build trust with consumers and uphold the integrity of their organic and fair trade practices (Manteghi, Arkat, & Mahmoodi, 2023; Tegeltija et al., 2022; van Hilten et al., 2020).

- Third-party Certifications: Engaging reputable third-party certification bodies adds credibility to organic and fair trade claims. Certification processes involve rigorous audits, inspections, and adherence to specific criteria, providing assurance to consumers and stakeholders.

- Blockchain Technology: Blockchain offers a decentralized and immutable platform for verifying organic and fair trade claims. By recording every stage of the supply chain on a transparent and tamper-proof ledger, blockchain enables easy traceability, ensuring the accuracy of claims and enhancing transparency.
- DNA Testing and Traceability: DNA testing can be utilized to verify the authenticity of organic products by comparing the genetic profile of the product with the known genetic profile of the organic variety. Similarly, traceability systems can track the movement of fair trade products, ensuring compliance with ethical standards.
- Supply Chain Audits and On-site Inspections: Conducting regular audits and on-site inspections throughout the supply chain is essential to verify organic and fair trade claims. These measures help ensure compliance with standards, verify production methods, and assess fair trade practices.
- Collaboration and Data Sharing: Collaboration among stakeholders, including producers, certification bodies, retailers, and consumers, promotes information sharing and strengthens the verification process. Creating platforms for transparent data sharing facilitates verification and builds trust.

*d. Consumer education and transparency*

- Consumer Education: Educating consumers about organic and fair trade practices, certification labels, and verification processes empowers them to make informed choices. Providing accessible information through labeling, websites, and educational campaigns enhances consumer understanding and trust.
- Transparent Labeling and Packaging: Clear and standardized labeling is crucial for authenticating organic and fair trade claims. Including certification logos, product information, and traceability details on packaging helps consumers identify genuine organic and fair trade products.

### 6.3.2. Verifying the origin and production methods of organic products

*a. The significance of verifying organic product claims*

- Consumer Confidence: Verifying the origin and production methods of organic products instills confidence in consumers. It assures them that the products they purchase meet specific organic standards, such as being grown without synthetic pesticides, genetically modified organisms (GMOs), or harmful chemicals.

- Authenticity and Integrity: Verification of organic product claims protects the authenticity and integrity of the organic industry. It ensures that products labelled as organic truly adhere to the established guidelines and have undergone the necessary processes and certifications.
- Market Credibility: Accurate verification processes contribute to the credibility of the organic market. When consumers trust that organic products are genuinely organic, it fosters a positive reputation for the industry, stimulates market growth, and encourages sustainable agricultural practices.

*b. Challenges in verifying organic product claims*

- Complex Supply Chains: Organic products often traverse complex supply chains involving multiple intermediaries and geographical locations. Ensuring traceability and verifying organic claims throughout these intricate networks pose significant challenges.
- Certification Standards and Variations: Organic certification standards can vary across regions and countries. The diverse requirements and certification bodies make it challenging to harmonise standards and establish a consistent verification process.
- Document Integrity: Verifying the origin and production methods of organic products often relies on the scrutiny of documents, including certificates, invoices, and production records. Ensuring the authenticity and accuracy of these documents is essential for reliable verification.

*c. Strategies and technologies for verification*

- Third-party Certification: Third-party certification bodies play a vital role in verifying organic product claims. Engaging reputable certification agencies that conduct thorough audits, inspections, and compliance assessments provides assurance to consumers and stakeholders.
- Traceability Systems: Implementing traceability systems can track the journey of organic products from farm to shelf. Technologies such as barcodes, RFID (Radio Frequency Identification), or blockchain enable the recording and transparent sharing of information, ensuring traceability and verifying organic claims.
- On-site Inspections and Audits: Regular on-site inspections and audits at organic farms and production facilities are essential for verification. These

inspections assess compliance with organic standards, verify production methods, and ensure that organic farming practices are followed.

- Laboratory Testing: Laboratory testing can complement verification processes by analyzing organic products for the presence of pesticides, GMOs, or chemical contaminants. Testing methods such as mass spectrometry, DNA analysis, or residue analysis provide objective data for verification.
- Collaboration and Data Sharing: Collaboration among stakeholders, including farmers, certification bodies, retailers, and consumers, strengthens the verification process. Establishing platforms for transparent data sharing facilitates information exchange, increases accountability, and enhances trust.

*d. Consumer education and labeling*

- Consumer Education: Educating consumers about organic standards, certification labels, and the verification process empowers them to make informed choices. Promoting awareness through educational campaigns, websites, and clear labeling enhances consumer understanding and trust.
- Clear Labeling and Packaging: Transparent and standardized labeling is essential for verifying organic product claims. Including organic certification logos, product information, and traceability details on packaging helps consumers identify genuine organic products.

### 6.3.3. Blockchain-based auditing of supply chain practices

*a. The significance of auditing supply chain practices*

- Transparency and Accountability: Auditing enables transparency by verifying compliance with established standards and regulations throughout the supply chain. It holds stakeholders accountable for their actions and ensures adherence to ethical, social, and environmental practices.
- Risk Mitigation: Auditing identifies potential risks and vulnerabilities within the supply chain, such as labor violations, environmental hazards, or product quality issues. It helps mitigate these risks, protecting consumers, businesses, and the reputation of the supply chain.
- Quality Assurance: Auditing ensures the consistency and quality of products and services by examining production processes, handling procedures, and storage conditions. It helps maintain high standards and builds consumer trust.

*b. Challenges in traditional auditing methods*

- Lack of Transparency: Traditional auditing methods often rely on manual processes and fragmented record-keeping systems, leading to limited visibility and difficulty in obtaining accurate and up-to-date information.
- Data Integrity: Verifying the accuracy and integrity of data collected during audits can be challenging. Data manipulation or tampering is a risk, compromising the reliability of audit findings.
- Time and Cost Constraints: Traditional auditing practices can be time-consuming and costly due to the need for physical inspections, document verification, and data reconciliation.

*c. Blockchain technology for auditing supply chain practices*

Blockchain technology offers a diverse array of applications for auditing supply chain practices, including the utilization of transparent and immutable ledgers, enhanced traceability mechanisms, and implementation of smart contracts for automation. These features ensure data integrity and security, enabling stakeholders to verify the authenticity of information and track the movement of goods throughout the supply chain with unprecedented accuracy. Furthermore, blockchain fosters collaborative auditing efforts by providing a shared platform where all parties can securely access and contribute to the audit trail, enhancing transparency and accountability across the supply chain ecosystem (Galvez et al., 2018; Gupta & Shankar, 2023; Manteghi et al., 2023; van Hilten et al., 2020).

- Transparent and Immutable Ledger: Blockchain technology provides a decentralized and transparent ledger where all transactions and activities within the supply chain are recorded. This enables auditors to access real-time, tamper-proof data for verification and auditing purposes.
- Enhanced Traceability: Blockchain allows for the seamless traceability of products and their associated information, including origins, certifications, and handling practices. Auditors can easily track the movement of goods and verify their compliance with regulations and standards.
- Smart Contracts and Automation: Smart contracts on the blockchain can automate auditing processes by setting predefined rules and conditions. These contracts execute predetermined actions, streamlining auditing procedures and reducing the reliance on manual interventions.
- Data Integrity and Security: Blockchain's cryptographic features ensure data integrity and security. Once data is recorded on the blockchain, it cannot be altered or deleted, providing auditors with a trusted source of information for verification.

- Collaborative Auditing: Blockchain facilitates collaboration among auditors, supply chain participants, and regulators. It enables shared access to data while maintaining confidentiality and privacy, fostering a cooperative approach to auditing practices.

*d. Implementation and adoption challenges*

- Standardization and Interoperability: Establishing common standards and protocols for blockchain implementation across supply chains is crucial for seamless integration and interoperability.
- Integration with Existing Systems: Integrating blockchain with existing supply chain systems and legacy infrastructure requires careful planning and coordination to ensure compatibility and smooth data flow.
- Education and Awareness: Educating stakeholders about the benefits and functionalities of blockchain technology is essential for its widespread adoption in auditing supply chain practices.
- Scalability and Performance: As blockchain networks grow, scalability and performance become crucial considerations. Developing solutions that can handle large volumes of data and process transactions efficiently is necessary for successful implementation.

### 6.3.4. Ensuring fair trade practices and ethical sourcing

*a. The importance of fair trade practices and ethical sourcing*

- Social Responsibility: Fair trade practices and ethical sourcing prioritize the welfare and rights of workers, promoting fair wages, safe working conditions, and respect for human rights. It emphasises social responsibility and sustainable livelihoods for producers and workers.
- Environmental Sustainability: Ethical sourcing considers environmental impacts, encourages sustainable production methods, and reduces the carbon footprint. It aims to minimize harm to ecosystems, protect biodiversity, and promote responsible resource management.
- Consumer Demand: Increasingly, consumers are seeking products that align with their values. Ethical sourcing and fair trade practices meet the growing demand for products that are produced sustainably, support communities, and uphold ethical standards.

*b. Challenges in ensuring fair trade practices and ethical sourcing*

- Supply Chain Complexity: Global supply chains often involve multiple intermediaries, making it challenging to trace and monitor every stage of production. This complexity can create opacity and hinder the verification of fair trade practices and ethical sourcing.

- Verification and Compliance: Verifying fair trade practices and ethical sourcing requires robust systems to ensure compliance with established standards. Monitoring supplier behavior, enforcing regulations, and addressing non-compliance pose significant challenges.
- Lack of Transparency: Transparency gaps within supply chains make it difficult to ascertain the true origin and conditions under which products are sourced. This lack of visibility increases the risk of unethical practices going undetected.

*c. Strategies for promoting fair trade practices and ethical sourcing*

A multitude of strategies exists for promoting fair trade practices and ethical sourcing within industries, encompassing the establishment of certifications and standards, implementation of robust supply chain traceability measures, and fostering engagement and collaboration with suppliers. Educating consumers about ethical considerations and fostering partnerships alongside industry-wide initiatives further contribute to creating a more sustainable and responsible supply chain ecosystem. By adopting these strategies, businesses can uphold ethical values, build trust with stakeholders, and drive positive social and environmental impacts throughout their supply chains (Galvez et al., 2018; Gupta & Shankar, 2023).

- Certification and Standards: Utilizing established fair trade certifications and ethical sourcing standards can provide a framework for assessing compliance and promoting transparency. Certifications such as Fairtrade, Rainforest Alliance, and Fair trade USA enable consumers to identify products that meet specific criteria.
- Supply Chain Traceability: Implementing traceability systems and technologies like blockchain, RFID, and barcoding enhances supply chain transparency. These tools enable the tracking of product origins, labor conditions, and environmental practices, ensuring fair trade and ethical sourcing.
- Supplier Engagement and Collaboration: Engaging with suppliers and fostering collaborative relationships is vital for promoting fair trade practices. Regular audits, supplier assessments, and capacity-building programs can encourage continuous improvement and adherence to ethical standards.
- Consumer Education and Awareness: Educating consumers about fair trade practices, ethical sourcing, and the impact of their purchasing decisions is crucial. Transparent labeling, consumer campaigns, and

information sharing help consumers make informed choices that support ethical businesses.

- Partnerships and Industry Initiatives: Collaborating with industry associations, NGOs, and governmental organizations can drive collective action towards fair trade practices and ethical sourcing. Partnerships can establish industry-wide guidelines, share best practices, and collectively address challenges.

*d. Monitoring and evaluation*

- Performance Metrics: Establishing performance metrics and key performance indicators (KPIs) helps measure progress and assess the impact of fair trade practices and ethical sourcing initiatives. These metrics can include labor conditions, environmental indicators, and social impact assessments.
- Auditing and Verification: Regular auditing and verification processes play a vital role in ensuring compliance and identifying areas for improvement. Independent audits, self-assessments, and supplier reporting mechanisms can enhance transparency and accountability.

### 6.3.5. Building consumer trust in sustainable and socially responsible products

*a. The importance of consumer trust*

- Brand Reputation: Consumer trust is closely linked to a brand's reputation. When consumers trust a brand to deliver sustainable and socially responsible products, it strengthens the brand's image and fosters customer loyalty.
- Ethical Considerations: Consumers are increasingly concerned about the environmental and social impact of their purchasing decisions. Building trust in sustainable and socially responsible products enables consumers to align their values with their consumption choices.
- Competitive Advantage: Brands that prioritize sustainability and social responsibility differentiate themselves in the market. Building consumer trust in such products can create a competitive advantage and attract a growing segment of conscious consumers.

*b. Challenges in building consumer trust*

- Greenwashing: Greenwashing refers to misleading marketing practices that create a false impression of sustainability or social responsibility. It erodes consumer trust and undermines genuine efforts in the industry.

- Lack of Transparency: Limited transparency about a product's supply chain, sourcing practices, and environmental impact can create skepticism among consumers. Transparent communication is essential to address this challenge.
- Verification and Certification: Establishing credible verification processes and certifications that demonstrate a product's sustainability and social responsibility can be complex. Consumers may question the reliability and integrity of these certifications.

*c. Strategies to foster consumer trust*

- Transparent Communication: Open and transparent communication about a product's sustainability practices, supply chain transparency, and social impact is vital. Brands should provide clear information to consumers, backed by verifiable data.
- Collaboration with Certifications and Standards: Collaborating with reputable sustainability certifications and standards organizations can enhance credibility. Brands can seek certifications that align with their values and actively communicate these certifications to consumers.
- Supply Chain Traceability: Implementing traceability technologies, such as blockchain or RFID, allows consumers to trace a product's journey from sourcing to production. Transparently sharing this information helps build trust in sustainable and socially responsible practices.
- Ethical Sourcing and Manufacturing: Brands should prioritize ethical sourcing and manufacturing practices, ensuring fair wages, safe working conditions, and responsible resource management. Sharing these practices with consumers helps build trust in the brand's commitment to social responsibility.
- Stakeholder Engagement: Engaging with stakeholders, including consumers, employees, NGOs, and community organizations, fosters transparency and accountability. Brands can actively involve stakeholders in decision-making processes, listen to feedback, and address concerns.
- Third-party Audits and Impact Assessments: Conducting third-party audits and impact assessments provides independent validation of a brand's sustainability and social responsibility claims. Sharing audit reports and impact assessment results demonstrates a commitment to transparency and accountability.

*d. Continuous improvement and innovation*

- Setting Ambitious Goals: Brands should set ambitious sustainability goals and communicate progress transparently. This demonstrates a commitment to continuous improvement and fosters consumer trust in long-term sustainability initiatives.

- Innovation for Sustainability: Embracing innovation to develop environmentally friendly packaging, eco-friendly materials, and sustainable production processes showcases a brand's commitment to sustainability and can enhance consumer trust.
- Education and Awareness: Educating consumers about sustainability issues, the importance of social responsibility, and the impact of their purchasing decisions deepens their understanding and trust. Brands can use marketing campaigns and educational content to raise awareness.

## 6.4. Blockchain-based supply chain analytics for quality control

Blockchain-based supply chain analytics stand as imperative tools in modern industries, encompassing various facets such as supply chain analytics, real-time monitoring of quality parameters, and predictive analytics for quality control and risk management. Leveraging data-driven insights to enhance product quality and streamline operations, these tools play a pivotal role in boosting efficiency and minimizing waste throughout the supply chain. By integrating blockchain technology into supply chain analytics, businesses can achieve greater transparency, trust, and optimization across their operations, ultimately driving sustainable growth and competitive advantage in today's dynamic marketplace (Jabbar, Lloyd, Hammoudeh, Adebisi, & Raza, 2021).

### 6.4.1. Leveraging blockchain data for supply chain analytics

*a. The promise of blockchain for supply chain analytics*

The potential of blockchain for supply chain analytics encompasses several key advantages, including immutable data storage, heightened traceability capabilities, and real-time data visibility. By leveraging these features, blockchain enables stakeholders to securely track and verify every stage of the supply chain process, enhancing transparency and accountability. Additionally, the facilitation of trust and collaboration among stakeholders further strengthens supply chain integrity and resilience, paving the way for more efficient and sustainable operations (Kamath, 2018; Kamble, Gunasekaran, & Sharma, 2020; Wong, RK Sinha, & Chui, 2021).

- Immutable Data Storage: Blockchain technology provides a tamper-proof and decentralized ledger, ensuring the integrity and immutability of supply chain data. This enables reliable and trustworthy data for analytics purposes.

- Enhanced Traceability: Blockchain enables end-to-end traceability of products throughout the supply chain. Each transaction recorded on the blockchain creates a transparent and auditable trail, facilitating the analysis of product movements and provenance.
- Real-time Data Visibility: With blockchain, supply chain stakeholders can access real-time data, allowing for faster and more accurate analysis. This timely information enables proactive decision-making and facilitates agile responses to supply chain disruptions.
- Trust and Collaboration: Blockchain establishes trust among supply chain participants by eliminating the need for intermediaries and providing a shared, immutable source of truth. This trust fosters collaboration and data sharing, leading to enhanced analytics capabilities.

*b. Leveraging blockchain data for supply chain analytics*

Utilizing blockchain data for supply chain analytics offers diverse avenues for exploration and implementation. These encompass leveraging blockchain for enhanced supply chain visibility, conducting in-depth analyses of supply chain performance, and undertaking comprehensive risk assessments and mitigation strategies. Additionally, organizations can utilize blockchain data to evaluate supplier performance and derive valuable consumer insights, thereby enabling more informed decision-making and fostering greater efficiency and resilience within supply chain operations (Gu, Zhao, Yue, Arshad, & Mohamad, 2023; Kamath, 2018).

- Enhanced Supply Chain Visibility: Blockchain data can provide comprehensive visibility into supply chain activities, including inventory levels, transportation routes, and production processes. Analyzing this data enables better inventory management, optimized logistics, and improved forecasting.
- Supply Chain Performance Analysis: By analyzing blockchain data, supply chain managers can assess performance metrics, such as delivery times, order fulfillment rates, and product quality. These insights help identify bottlenecks, inefficiencies, and areas for improvement.
- Risk Assessment and Mitigation: Blockchain data can be leveraged to identify and analyze supply chain risks, such as disruptions, counterfeit products, or non-compliance with regulations. This allows proactive risk management and the implementation of mitigation strategies.
- Supplier Performance Evaluation: Analyzing blockchain data can help evaluate supplier performance based on criteria such as delivery reliability, product quality, and adherence to ethical standards. This enables data-driven supplier selection and ongoing performance monitoring.

- Consumer Insights: Blockchain data can provide valuable insights into consumer preferences, product authenticity, and purchasing patterns. Analyzing this data helps businesses understand consumer behavior, tailor marketing strategies, and drive product innovation.

*c. Challenges and considerations*

- Data Integration: Integrating blockchain data with existing supply chain systems and analytics platforms can be challenging. Effective data integration strategies and interoperability protocols are crucial to unlock the full potential of blockchain analytics.
- Data Privacy and Security: While blockchain ensures data integrity, privacy and security of sensitive information should be carefully managed. Implementing robust encryption techniques and access controls is essential to protect sensitive supply chain data.
- Scalability and Performance: As blockchain data grows exponentially, scalability and performance become key considerations. Optimizing blockchain networks, implementing efficient consensus mechanisms, and leveraging off-chain storage solutions are essential for smooth data analysis.
- Standardization and Interoperability: To fully leverage blockchain data for supply chain analytics, industry-wide standards and interoperability frameworks need to be established. This ensures seamless data exchange and collaboration among diverse supply chain stakeholders.

*d. The future of supply chain analytics with blockchain*

The outlook for supply chain analytics integrated with blockchain is highly promising, poised to embrace advancements such as artificial intelligence and machine learning, the utilization of smart contracts for automated analytics, and the formation of blockchain consortiums. These developments signify an exciting evolution in the intersection of technology and supply chain management, offering unprecedented opportunities for enhanced transparency, efficiency, and trustworthiness across global supply networks. As organizations increasingly recognize the potential of these integrated solutions, we can anticipate significant transformations in how supply chains are managed and optimized in the coming years (Kshetri, 2019).

- Artificial Intelligence and Machine-learning: Integrating blockchain data with artificial intelligence (AI) and machine-learning (ML) algorithms can unlock advanced analytics capabilities. AI and ML can help identify patterns, predict demand fluctuations, and automate decision-making processes.

- Smart Contracts and Automated Analytics: Smart contracts on the blockchain can facilitate automated analytics processes, triggering real-time data analysis, alert notifications, and performance evaluations based on predefined conditions and rules.
- Blockchain Consortiums and Collaborative Analytics: Collaborative analytics platforms built on blockchain consortiums enable supply chain stakeholders to pool and analyze shared data. This collaborative approach enhances collective intelligence, drives innovation, and improves supply chain performance.

### 6.4.2. Real-time monitoring of quality parameters

*a. The significance of real-time monitoring*

- Timely Detection of Quality Issues: Real-time monitoring allows for immediate detection of deviations from quality standards. By continuously monitoring critical parameters, organizations can identify and address quality issues promptly, preventing the production and distribution of defective products.
- Enhanced Product Consistency: Real-time monitoring ensures consistency in product quality across batches or production runs. By tracking and controlling quality parameters in real time, organizations can maintain uniformity and meet customer expectations consistently.
- Risk Mitigation: Real-time monitoring helps mitigate risks associated with product quality. By promptly identifying deviations or anomalies, organizations can prevent potential safety concerns, product recalls, and reputational damage.
- Improved Customer Satisfaction: Real-time monitoring allows organizations to provide customers with products that meet or exceed their expectations. By ensuring high quality consistently, organizations can build trust, loyalty, and long-term relationships with their customers.

*b. Parameters monitored in real-time*

- Physical Attributes: Real-time monitoring can capture physical attributes such as dimensions, weight, color, texture, and appearance. This ensures that products meet specific visual or physical criteria and conform to quality standards.
- Chemical Composition: Monitoring chemical composition is crucial for industries such as food and pharmaceuticals. Real-time analysis of

ingredients, additives, contaminants, and formulation ratios helps maintain product integrity and safety.

- Environmental Conditions: Real-time monitoring of environmental conditions, such as temperature, humidity, pressure, and light exposure, ensures that products are stored and transported within the specified ranges. This is particularly important for perishable goods and sensitive materials.
- Performance Metrics: In certain industries, real-time monitoring focuses on performance metrics such as efficiency, output quality, reliability, and durability. This allows organizations to optimize production processes, identify bottlenecks, and continuously improve performance.

*c. Technologies enabling real-time monitoring*

- Internet of Things (IoT): IoT devices equipped with sensors and actuators enable real-time data collection and monitoring of quality parameters. These devices can transmit data wirelessly, allowing for remote monitoring and analysis.
- Advanced Analytics: Real-time data collected from sensors and IoT devices can be analyzed using advanced analytics techniques. Machine-learning algorithms can detect patterns, anomalies, and deviations from quality standards, enabling proactive decision-making.
- Cloud Computing: Cloud-based platforms provide scalable storage and processing capabilities for real-time monitoring data. Organizations can leverage cloud computing to store, analyze, and share quality-related data across the supply chain.
- Blockchain Technology: Blockchain offers secure and immutable data storage, ensuring the integrity and traceability of quality-related information. It enables transparent sharing of quality data among supply chain participants and facilitates trust and collaboration.

*d. Benefits of real-time monitoring*

- Proactive Quality Management: Real-time monitoring allows organizations to proactively manage quality by identifying issues in real time and taking immediate corrective actions. This reduces the likelihood of quality-related incidents and improves overall process efficiency.
- Cost Reduction: Detecting quality issues early in the production process minimizes waste, rework, and product rejection. Real-time monitoring helps organizations optimize resource allocation, reduce costs associated with quality control, and streamline operations.

- Regulatory Compliance: Real-time monitoring helps organizations comply with industry-specific regulations and quality standards. By continuously monitoring and documenting quality parameters, organizations can demonstrate compliance and avoid penalties or legal consequences.
- Competitive Advantage: Organizations that excel in real-time monitoring of quality parameters gain a competitive edge. By consistently delivering high-quality products, they differentiate themselves in the market, attract customers, and enhance their brand reputation.

### 6.4.3. Predictive analytics for quality control and risk management

*a. Understanding predictive analytics*

Predictive analytics encompasses a range of methodologies, including predictive modeling, machine-learning algorithms, and the application of data mining and statistical techniques. These methodologies enable organizations to analyze historical data, identify patterns, and forecast future trends or outcomes with a high degree of accuracy. By leveraging predictive analytics, businesses can make data-driven decisions, mitigate risks, and capitalize on opportunities, thereby gaining a competitive edge in their respective industries (Wong et al., 2021).

- Predictive Modelling: Predictive analytics involves developing mathematical models that leverage historical data to make predictions about future quality outcomes. These models consider various variables, patterns, and trends to identify potential quality issues or risks.
- Machine-learning Algorithms: Machine-learning algorithms play a crucial role in predictive analytics by automatically learning from historical data and making predictions or recommendations. These algorithms can identify hidden patterns, correlations, and anomalies, enabling organizations to anticipate quality-related events.
- Data Mining and Statistical Techniques: Predictive analytics relies on data mining and statistical techniques to extract valuable insights from large datasets. These techniques help identify influential factors, determine relationships, and develop accurate predictive models.

*b. Predictive analytics in quality control*

- Early Detection of Quality Issues: By analyzing historical quality data, organizations can identify patterns and indicators that precede quality issues. Predictive analytics enables early detection, allowing timely interventions to prevent or minimize the impact of potential quality problems.

- Process Optimization: Predictive models can uncover inefficiencies and bottlenecks in production processes that may impact product quality. By identifying areas of improvement, organizations can optimize processes, reduce defects, and enhance overall quality performance.
- Root Cause Analysis: Predictive analytics helps uncover the root causes of quality issues by analyzing historical data and identifying correlations. By understanding the underlying factors contributing to quality problems, organizations can implement targeted corrective actions.
- Supplier Quality Management: Predictive analytics can assess the quality performance of suppliers based on historical data and various parameters. This enables organizations to proactively manage supplier relationships, mitigate risks, and ensure consistent quality across the supply chain.

*c. Predictive analytics in risk management*

- Risk Identification and Assessment: Predictive models can analyze historical data to identify potential risks and assess their likelihood and impact. By quantifying risks, organizations can prioritize mitigation efforts and allocate resources accordingly.
- Supply Chain Disruption Prevention: Predictive analytics helps identify factors that may lead to supply chain disruptions, such as material shortages, transportation delays, or supplier performance issues. By proactively addressing these risks, organizations can prevent disruptions and maintain continuity.
- Demand Forecasting and Inventory Optimization: Predictive analytics enables accurate demand forecasting by considering historical sales data, market trends, and external factors. This helps organizations optimize inventory levels, minimize stockouts, and reduce excess inventory.
- Compliance and Regulatory Risks: Predictive analytics can assess compliance risks by analyzing historical data and identifying non-compliance patterns. This enables organizations to proactively address compliance issues, ensure adherence to regulations, and avoid penalties or reputational damage.

*d. Benefits of predictive analytics*

- Proactive Decision-making: Predictive analytics empowers organizations to make proactive decisions based on anticipated quality issues and risks. This leads to timely interventions, improved operational efficiency, and better customer satisfaction.
- Cost Reduction: By predicting quality issues and risks, organizations can minimize the costs associated with defects, rework, recalls, and supply

chain disruptions. Predictive analytics helps optimize resource allocation and reduce financial losses.

- Enhanced Quality Control: Predictive analytics strengthens quality control processes by enabling organizations to identify potential quality issues in advance. This leads to higher product consistency, improved customer satisfaction, and enhanced brand reputation.
- Competitive Advantage: Organizations that leverage predictive analytics gain a competitive edge by staying ahead of quality-related issues and risks. They can deliver products with higher reliability, meet customer expectations consistently, and outperform competitors in the market.

### 6.4.4. Improving product quality through data-driven insights

*a. The importance of data-driven insights*

- Identifying Quality Patterns: Data-driven insights enable organizations to analyze vast amounts of data and identify patterns related to product quality. By uncovering correlations, trends, and anomalies, organizations can gain a deep understanding of quality issues and their root causes.
- Process Optimization: Through data analysis, organizations can identify bottlenecks and inefficiencies in their processes that impact product quality. Data-driven insights highlight areas for improvement, enabling organizations to streamline operations, reduce defects, and enhance overall quality performance.
- Continuous Improvement: Data-driven insights provide a basis for continuous improvement initiatives. By monitoring quality metrics and analyzing data, organizations can set goals, track progress, and implement targeted strategies to continuously enhance product quality.
- Customer Satisfaction: Understanding customer preferences and expectations is crucial for delivering products of high quality. Data-driven insights help organizations gather customer feedback, analyze purchasing patterns, and identify areas where product quality can be enhanced to meet customer demands.

*b. Methodologies for data-driven insights*

- Statistical Analysis: Statistical analysis techniques such as regression analysis, hypothesis testing, and control charts help organizations extract meaningful insights from data. These methodologies enable organizations to understand relationships between variables, identify significant factors affecting quality, and make data-driven decisions.

- Data Mining: Data mining techniques, such as clustering, classification, and association analysis, uncover hidden patterns and relationships within large datasets. By applying data mining algorithms, organizations can identify factors that contribute to product quality and develop strategies for improvement.
- Machine-learning: Machine-learning algorithms automatically learn from data and make predictions or recommendations. These algorithms can uncover complex patterns, detect anomalies, and provide insights into improving product quality. Supervised and unsupervised learning techniques can be employed to classify products, predict defects, and optimize quality control processes.
- Big Data Analytics: With the increasing availability of large volumes of data, big data analytics techniques are used to extract valuable insights. By leveraging technologies such as distributed computing and parallel processing, organizations can analyze diverse data sources and gain holistic views of product quality.

*c. Tools for data-driven insights*

Instruments for deriving data-driven insights encompass a variety of technologies, including data visualization tools, quality management systems, IoT sensors and connectivity solutions, and enterprise resource planning (ERP) systems. These technologies play crucial roles in collecting, analyzing, and presenting data to facilitate informed decision-making processes within organizations. By integrating these tools and systems into their operations, businesses can gain deeper insights into their processes, improve efficiency, and enhance overall performance across various sectors (Yik, Wong, Wong, & Shaw, 2021).

- Data Visualisation: Data visualisation tools help present complex data in a visual format, making it easier to identify trends, outliers, and patterns. Visual representations, such as charts, graphs, and dashboards, allow stakeholders to interpret data quickly and gain actionable insights.
- Quality Management Systems (QMS): QMS platforms integrate data collection, analysis, and reporting functionalities to support data-driven quality improvement efforts. These systems streamline data management, facilitate collaboration among stakeholders, and enable organizations to track and monitor quality metrics effectively.
- IoT Sensors and Connectivity: IoT sensors embedded in production equipment or products collect real-time data, allowing organizations to monitor quality parameters throughout the manufacturing process. Data captured by IoT sensors can be analyzed to identify quality issues, improve process control, and enhance overall product quality.

- Enterprise Resource Planning (ERP) Systems: ERP systems integrate various business functions and consolidate data from multiple sources. By leveraging ERP systems, organizations can access comprehensive data for analysis, gain cross-functional insights, and drive data-driven improvements in product quality.

*d. Benefits of data-driven insights*

- Proactive Quality Management: Data-driven insights empower organizations to take proactive measures to improve product quality. By identifying patterns and trends, organizations can address quality issues before they escalate, leading to reduced defects and customer complaints.
- Operational Efficiency: Data-driven insights highlight process inefficiencies and enable organizations to optimize operations. By streamlining workflows, reducing waste, and improving resource allocation, organizations can enhance productivity, minimize defects, and improve product quality.
- Decision-making Accuracy: Data-driven insights provide a solid foundation for informed decision-making. Organizations can rely on factual data rather than intuition or assumptions, ensuring that quality improvement initiatives are based on accurate information.
- Competitive Advantage: Organizations that harness data-driven insights to improve product quality gain a competitive edge in the market. By consistently delivering high-quality products, they differentiate themselves from competitors, attract customers, and build a strong brand reputation.

### 6.4.5. Enhancing efficiency and reducing waste in the supply chain

*a. Supply chain visibility and collaboration*

- Real-time Tracking: Implementing technologies such as RFID, GPS, and IoT sensors enables real-time tracking of inventory, shipments, and production processes. This visibility allows organizations to identify bottlenecks, optimize routes, and improve the flow of goods throughout the supply chain.
- Collaboration Platforms: Utilizing collaborative tools and platforms facilitates communication and information sharing among supply chain partners. By fostering collaboration, organizations can streamline processes, eliminate redundancies, and reduce lead times, resulting in increased efficiency.
- Demand Planning: Accurate demand forecasting and collaborative demand planning with suppliers and customers minimize inventory holding costs,

prevent stockouts, and optimize production schedules. By aligning supply with demand, organizations can enhance efficiency and reduce waste.

*b. Lean principles and practices*

- Waste Identification: Applying lean principles, such as value stream mapping, organizations can identify and eliminate various forms of waste, including overproduction, excess inventory, waiting time, transportation inefficiencies, defects, and non-value-added activities.
- Continuous Improvement: Adopting a continuous improvement mindset encourages ongoing evaluation and refinement of processes. Through practices like Kaizen, organizations can identify opportunities for waste reduction, implement small-scale improvements, and foster a culture of efficiency throughout the supply chain.
- Just-in-time (JIT) Production: JIT principles aim to minimize inventory by producing goods in response to customer demand. By synchronising production with demand, organizations can reduce waste associated with excess inventory, storage costs, and obsolescence.

*c. Sustainable packaging and logistics*

- Eco-friendly Packaging: Opting for sustainable packaging materials and designs reduces waste and environmental impact. Lightweight packaging, recyclable materials, and eco-friendly practices in product packaging contribute to waste reduction and demonstrate a commitment to sustainability.
- Efficient Transportation: Optimizing transportation routes, consolidating shipments, and utilizing alternative transportation modes, such as rail or sea, can significantly reduce fuel consumption, emissions, and transportation costs. Efficient logistics operations contribute to both environmental sustainability and cost savings.

*d. Efficiency unleashed*

Efforts to bolster efficiency and minimize waste within the supply chain entail the adoption of warehouse automation, harnessing advanced analytics, and fostering seamless supply chain integration. By implementing warehouse automation technologies such as robotic systems and automated inventory management, businesses can optimize storage space utilization and expedite order fulfillment processes. Additionally, leveraging advanced analytics enables organizations to gain insights into supply chain operations, identify areas for improvement, and make data-driven decisions to enhance overall efficiency and reduce waste (Wong et al., 2021).

- Warehouse Automation: Implementing automation technologies, such as robotics, automated guided vehicles (AGVs), and warehouse management systems (WMS), streamlines warehouse operations, improves order fulfillment accuracy, and reduces labor costs.
- Advanced Analytics: Leveraging big data analytics and machine-learning algorithms enables organizations to analyze vast amounts of data and uncover insights for process optimization. Predictive analytics can identify potential inefficiencies, minimize downtime, and optimize resource allocation.
- Supply Chain Integration: Integrating various systems, such as enterprise resource planning (ERP), customer relationship management (CRM), and supply chain management (SCM), enhances data visibility and enables streamlined information flow. Integrated systems improve coordination, reduce errors, and enhance overall supply chain efficiency.

*e. Performance measurement and KPIs*

- Key Performance Indicators (KPIs): Establishing relevant KPIs allows organizations to monitor and evaluate supply chain performance. KPIs such as on-time delivery, order accuracy, inventory turnover, and order cycle time provide valuable insights for identifying areas of improvement and tracking progress.
- Continuous Monitoring and Analysis: Regularly monitoring and analyzing performance data helps identify inefficiencies and waste. Organizations can identify root causes, implement corrective actions, and drive ongoing improvement to enhance efficiency and reduce waste in the supply chain.

# Chapter 7

# Agricultural finance and insurance

*Seeding the Future: Empowering Agriculture through Financial Innovation and Technological Resilience* - Unknown

**Abstract**

This chapter explores the transformative potential of blockchain technology in agricultural finance and insurance. It discusses how blockchain can revolutionize financial transactions and insurance processes in the agriculture sector. The chapter focuses on the use of smart contracts for secure and automated payment systems in agricultural lending, insurance claims, and risk management. The chapter begins by examining the potential benefits of blockchain technology in transforming financial transactions and insurance processes in agriculture. It highlights how blockchain's decentralized and transparent nature can enable secure and efficient transactions, eliminating the need for intermediaries and reducing costs. Next, the chapter delves into the use of smart contracts, a key feature of blockchain technology, for secure and automated payment systems in agriculture. It explores how smart contracts can facilitate transparent and enforceable agreements between farmers, lenders, insurers, and other stakeholders.

**Keywords**: Blockchain technology, Agricultural finance, Insurance, Smart contracts, Financial transactions, Insurance processes, Agricultural lending, Insurance claims, Risk management, Intermediaries

**Chapter contents**

- Blockchain-based Solutions for Agricultural Financing
- Smart Contracts for Crop Insurance and Risk Management
- Decentralized Lending Platforms for Farmers and Agricultural Businesses

***

## 7.1. Blockchain-based solutions for agricultural financing

Blockchain presents a range of solutions for financing agriculture, including the establishment of blockchain-enabled peer-to-peer lending platforms, enhancing access to credit for small-scale farmers, and implementing smart contracts to ensure transparent loan agreements. By leveraging blockchain technology, farmers can bypass traditional financial institutions, directly connecting with lenders and accessing funds with lower barriers. Moreover, blockchain's immutable ledger enhances trust and accountability, reducing the risk of fraud while simultaneously lowering transaction costs associated with agricultural financing (FAO, 2019; Sylvester, 2019).

### 7.1.1. Challenges in agricultural financing

*a. Limited access to credit*

One of the primary challenges in agricultural financing is the limited access to credit for farmers, especially small-scale and marginalized farmers. Traditional financial institutions are often reluctant to provide loans to farmers due to the perceived high risks associated with agriculture. Farmers, particularly those in developing countries, face challenges in meeting the stringent collateral requirements and providing the necessary documentation to secure loans. Additionally, the unpredictable nature of agriculture, including weather-related risks and market fluctuations, makes lenders hesitant to extend credit.

*b. Lack of financial literacy and business skills*

Many farmers, especially those in rural areas, lack the financial literacy and business skills necessary to manage their finances effectively. They may struggle with basic bookkeeping, financial planning, and understanding interest rates, loan terms, and repayment schedules. This lack of financial knowledge can hinder farmers' ability to access and effectively utilize agricultural financing opportunities. Providing financial education and training programs tailored to the specific needs of farmers can help address this challenge.

*c. Seasonality and cash flow management*

Agricultural production is often seasonal, resulting in irregular cash flows for farmers. Income is typically generated at harvest time, while expenses occur throughout the production cycle. This seasonality creates challenges in managing cash flow and repaying loans on time. Farmers may find it difficult to meet repayment obligations during periods of low income, leading to loan delinquencies. Innovative financing models, such as flexible repayment schedules linked to harvest cycles or income-based repayment options, can help farmers manage their cash flow effectively.

*d. Insufficient infrastructure and technology*

Inadequate infrastructure, including access to transportation, storage facilities, and markets, poses significant challenges for farmers in accessing finance. Without proper infrastructure, farmers may struggle to transport their produce to market or store it effectively, resulting in post-harvest losses. Additionally, the lack of access to technology, such as mobile banking or digital financial services, limits the availability of financial services to farmers in remote areas. Investments in rural infrastructure and the adoption of technology can enhance farmers' access to agricultural financing.

*e. Climate change and environmental risks*

Climate change and environmental risks pose additional challenges in agricultural financing. Changing weather patterns, natural disasters, and environmental degradation can increase the vulnerability of farmers and make agricultural production riskier. Lenders may perceive higher risks associated with climate-sensitive agricultural activities, making it difficult for farmers to secure financing for climate-resilient practices. Developing innovative financial products, such as weather-based insurance and green financing schemes, can help address these risks and encourage sustainable agricultural practices.

### 7.1.2. Blockchain-enabled peer-to-peer lending platforms

Blockchain facilitates peer-to-peer lending platforms by bolstering security and data privacy measures, optimizing operational efficiency and cost-effectiveness, and broadening access to credit opportunities. Through its decentralized nature and cryptographic techniques, blockchain ensures that sensitive financial information remains secure and tamper-proof, fostering trust between lenders and borrowers. Additionally, by eliminating intermediaries and automating processes, blockchain streamlines transactions, reducing overhead costs and enabling more competitive interest rates for borrowers (Bermeo-Almeida et al., 2018; Blokdyk, 2021; Detwiler, 2020).

*a. Decentralization and trust*

Blockchain technology operates on a decentralized network of computers, known as nodes, which collectively validate and record transactions. This decentralized nature eliminates the need for a central authority, such as a bank, to oversee lending activities. Instead, the lending process is managed by smart contracts—self-executing agreements stored on the blockchain—that govern the terms and conditions of loans. This decentralized and automated approach instills trust among participants, as transactions are transparent and tamper-resistant. Borrowers and lenders can engage in direct transactions without relying on a third party, fostering a sense of trust and reducing the potential for fraud (Cottrill et al., 2018).

*b. Enhanced security and data privacy*

Blockchain-enabled P2P lending platforms offer enhanced security and data privacy compared to traditional lending systems. In a blockchain, each transaction is recorded in a time-stamped and immutable manner, ensuring a high level of data integrity. The use of cryptography further protects sensitive borrower information, minimizing the risk of data breaches. Additionally, the decentralized nature of the blockchain eliminates the single point of failure associated with centralized databases, making it more difficult for malicious actors to compromise the system. These security features help protect the interests of both borrowers and lenders, inspiring confidence in the platform (Cottrill et al., 2018).

*c. Efficient and cost-effective operations*

Traditional lending processes often involve multiple intermediaries, lengthy paperwork, and manual verification procedures, leading to delays and increased costs. Blockchain-enabled P2P lending platforms streamline these processes, resulting in increased efficiency and cost savings. Smart contracts automate loan agreements, ensuring accurate and timely execution of loan disbursements, repayments, and interest calculations. Additionally, blockchain's ability to facilitate instant and low-cost cross-border transactions opens up new possibilities for global lending, removing geographical barriers and reducing transaction costs associated with traditional banking systems (Cottrill et al., 2018).

*d. Expanded access to credit*

One of the most significant advantages of blockchain-enabled P2P lending platforms is their potential to expand access to credit for underserved individuals and businesses. Traditional lending institutions often have strict eligibility criteria and may overlook borrowers with limited credit history or in emerging markets. P2P lending platforms powered by blockchain technology can leverage alternative data sources, such as social media profiles or transaction histories, to assess the creditworthiness of borrowers. This enables individuals and businesses with limited or no access to traditional financial services to obtain loans based on their reputation and potential rather than just their credit score (Pufahl, Ohlsson, Weber, Harper, & Weston, 2021).

*e. Regulatory and legal considerations*

While blockchain-enabled P2P lending platforms offer numerous benefits, they also present regulatory and legal challenges. The decentralized nature of blockchain complicates the enforcement of existing financial regulations and consumer protection laws. Governments and regulatory bodies are still grappling with creating appropriate frameworks to govern these platforms and ensure investor protection. Addressing these concerns is crucial to strike a balance between fostering innovation and safeguarding the interests of borrowers and lenders (Dutta, Choi, et al., 2020).

### 7.1.3. Improving access to credit for small-scale farmers

Small-scale farmers form the backbone of agricultural production in many countries, playing a crucial role in ensuring food security and rural development. However, accessing affordable and timely credit remains a significant challenge for these farmers, limiting their ability to invest in inputs, technology, and infrastructure necessary for sustainable agricultural practices. In this section, we will explore strategies and initiatives aimed at improving access to credit for small-scale farmers, addressing the unique challenges they face and fostering their economic empowerment.

*a. Microfinance and rural banking*

Microfinance institutions (MFIs) and rural banking initiatives have been instrumental in expanding financial services to underserved communities, including small-scale farmers. These institutions offer small loans, savings accounts, and other financial products tailored to the specific needs of farmers. By operating with a deep understanding of the local context and utilizing flexible repayment options, MFIs and rural banks can mitigate the risks associated with lending to farmers and promote financial inclusion. Collaborations between these institutions, agricultural extension services, and farmers' organizations can further enhance access to credit and provide training on financial management.

*b. Collateral and credit guarantees*

Traditional lending institutions often require collateral as a security for loans. However, small-scale farmers, particularly those in developing countries, often lack the necessary assets to meet collateral requirements. Implementing innovative solutions such as group lending, where farmers come together as a collective to access credit, or using alternative collateral options such as future harvests or livestock can help address this challenge. Additionally, credit guarantee schemes, backed by governments or donor agencies, can provide assurance to lenders and incentivize them to extend credit to small-scale farmers.

*c. Digital financial services and mobile banking*

The rapid expansion of digital financial services and mobile banking has opened new avenues for improving access to credit for small-scale farmers. Mobile-based platforms allow farmers to access financial services, such as loans, savings, and insurance, through their mobile phones, bypassing the need for physical bank branches. Digital platforms can leverage alternative data sources, such as farmers' transaction histories or farm-related data, to assess creditworthiness. Moreover, mobile banking enables convenient loan disbursements and repayments, reducing transaction costs and increasing efficiency (Chinaka, 2016; Kim & Laskowski, 2018; Sajja et al., 2023).

*d. Value chain financing*

Value chain financing approaches aim to address the financial needs of small-scale farmers within the context of the broader agricultural value chain. By establishing partnerships between farmers, input suppliers, buyers, and financial institutions, value chain financing models provide access to credit based on the anticipated income from the sale of agricultural products. This approach aligns the timing of loan repayments with farmers' cash flows, reducing risks for both farmers and lenders. Value chain financing also encourages market linkages and improves farmers' bargaining power, resulting in more favorable terms and conditions (Kim et al., 2018).

*e. Risk management and agricultural insurance*

Agricultural production is inherently exposed to various risks, including weather-related disasters, pests, and market fluctuations. Small-scale farmers are particularly vulnerable to these risks due to their limited resources and capacity to absorb losses. Agricultural insurance schemes that cover yield losses, price fluctuations, or damage to assets can mitigate the risks associated with lending to farmers. Governments, insurance companies, and development agencies should collaborate to develop affordable and accessible insurance products tailored to the specific needs of small-scale farmers. Additionally, integrating climate-smart agricultural practices and risk-reduction strategies into agricultural financing programs can enhance the resilience of small-scale farmers and reduce risks for lenders.

### 7.1.4. Smart contracts for transparent loan agreements and repayments

*a. Understanding smart contracts*

Smart contracts are self-executing agreements stored on a blockchain. They contain the terms and conditions of a contract and automatically execute actions based on predefined rules and triggers. Smart contracts eliminate the need for intermediaries, as they operate on a decentralized network of computers, ensuring transparency, security, and efficiency.

*b. Transparent and immutable loan agreements*

Smart contracts provide a transparent and immutable record of loan agreements. The terms and conditions of the loan, such as the loan amount, interest rate, repayment schedule, and any associated fees, are encoded in the smart contract. All parties involved have access to the contract, ensuring transparency and reducing the potential for disputes or misunderstandings. The immutability of the blockchain ensures that the terms of the contract cannot be altered or tampered with, enhancing trust between borrowers and lenders (Cottrill et al., 2018).

*c. Automated and efficient repayment processes*

Smart contracts automate the repayment process, streamlining loan repayments and reducing administrative burdens. The contract can be programmed to trigger automatic deductions from the borrower's account on specified dates, ensuring timely repayments. This automation eliminates the need for manual tracking and reminders, reducing the risk of late payments and associated penalties. Additionally, smart contracts can calculate interest and fees automatically, ensuring accurate and transparent repayment calculations.

*d. Conditional and escrow-based disbursements*

Smart contracts allow for conditional disbursements of loans, ensuring that funds are released only when predetermined conditions are met. For example, in an agricultural loan, the contract can be programmed to release funds once certain milestones, such as land preparation or the purchase of inputs, are verified. This feature reduces the risk for lenders and encourages responsible use of funds by borrowers.

Escrow-based disbursements are another feature enabled by smart contracts. In this case, the funds are held in an escrow account until specific conditions are met, providing an additional layer of security for both borrowers and lenders. For example, in a construction loan, disbursements can be made in installments upon the completion of predefined stages of the project.

*e. Enhanced security and trust*

Smart contracts leverage the inherent security features of blockchain technology. The decentralized nature of the blockchain ensures that the contract is stored across multiple nodes, making it highly resistant to tampering or fraud. The use of cryptographic algorithms further enhances the security and privacy of the contract. These features instill trust between borrowers and lenders, mitigating the risks associated with traditional lending processes (Cottrill et al., 2018).

*f. Potential challenges and Considerations*

While smart contracts offer numerous advantages, there are still challenges and considerations to address. The legal enforceability of smart contracts varies across jurisdictions, and regulatory frameworks need to be developed to accommodate this emerging technology. Additionally, the reliance on blockchain technology requires robust infrastructure and connectivity, which may pose challenges in certain regions. Ensuring data privacy and protection is also crucial when implementing smart contracts, as sensitive borrower information is stored on the blockchain (Staples et al., 2017).

### 7.1.5. Mitigating fraud and reducing transaction costs

*a. Implementing identity verification*

Identity verification is a crucial step in mitigating fraud and reducing transaction risks. By implementing robust identity verification processes, businesses can ensure that individuals conducting transactions are who they claim to be. This can be achieved through various methods, including document verification, biometric authentication, and two-factor authentication. Integrating digital identity systems and leveraging technologies like facial recognition and fingerprint scanning can strengthen identity verification, making it more difficult for fraudsters to impersonate others.

*b. Utilizing blockchain technology*

Blockchain technology offers inherent security features that can help mitigate fraud and reduce transaction costs. The decentralized and transparent nature of the blockchain enables the validation and recording of transactions in a tamper-resistant manner. By leveraging blockchain for financial transactions, businesses can establish an immutable audit trail, preventing fraudulent activities like double-spending or unauthorized modifications of transaction records. Additionally, blockchain can streamline cross-border transactions by eliminating intermediaries and reducing associated costs (Cottrill et al., 2018).

*c. Employing artificial intelligence and machine learning*

Artificial Intelligence (AI) and Machine Learning (ML) techniques have proven to be effective tools in fraud detection and prevention. These technologies can analyze large volumes of data, detect patterns, and identify anomalies that may indicate fraudulent activities. By employing AI and ML algorithms, financial institutions and businesses can proactively identify suspicious transactions, mitigate risks, and reduce losses. Continuous learning and adaptation of these systems enable them to stay ahead of evolving fraud techniques.

*d. Enhancing data security and encryption*

Protecting sensitive data is crucial in mitigating fraud and ensuring secure transactions. Encryption techniques can be employed to secure data during transmission and storage, making it difficult for unauthorized individuals to access or manipulate information. Strong data security measures, including encryption protocols, firewalls, and regular security audits, can safeguard against data breaches and unauthorized access. Implementing industry best practices and compliance with relevant data protection regulations are essential components of a robust data security strategy.

*e. Collaboration and information sharing*

Collaboration and information sharing among financial institutions, businesses, and law enforcement agencies are essential in combating fraud. Establishing communication channels and sharing fraud-related intelligence can help identify emerging fraud trends, share best practices, and prevent fraudulent activities across industries. Platforms for collaborative fraud detection, such as shared databases or industry-wide fraud monitoring systems, can enable real-time information exchange and proactive fraud prevention.

*f. Streamlining compliance processes*

Compliance requirements and regulatory frameworks play a significant role in mitigating fraud and reducing transaction risks. However, complex and time-consuming compliance processes can increase transaction costs and impede operational efficiency. Leveraging technology, such as RegTech (Regulatory Technology) solutions, can streamline compliance procedures by automating data collection, analysis, and reporting. These solutions can ensure adherence to regulatory standards while minimizing the associated administrative burden and costs.

## 7.2. Smart contracts for crop insurance and risk management

Blockchain technology presents significant opportunities for the development of smart contracts tailored for crop insurance and risk management. Specifically, it facilitates the implementation of Blockchain-enabled parametric insurance contracts, streamlining the process of claims verification and settlements. Moreover, it enables real-time monitoring of crucial factors such as weather conditions and crop yields, enhancing the precision of risk assessment. By leveraging these capabilities, blockchain technology contributes to improving the resilience of agricultural practices and reducing risks for farmers (FAO, 2019; Pranto et al., 2021; Sylvester, 2019; Tarhini, 2021).

### 7.2.1. Introduction to crop insurance in agriculture

*a. The purpose and benefits of crop insurance*

Crop insurance serves as a safety net for farmers, offering protection against yield losses and ensuring their financial stability. The primary purpose of crop insurance is to provide farmers with a means to recover their production costs and mitigate the financial impact of crop failures. By transferring the risk of crop losses to insurance companies, farmers can safeguard their investments and secure their income. Crop insurance also promotes agricultural sustainability by encouraging farmers to adopt best practices, invest in modern technologies, and manage risks more effectively.

*b. Components of crop insurance*

Crop insurance consists of several key components that determine the coverage and terms of the insurance policy. These components include:

- Coverage Level: Farmers can choose the level of coverage based on their risk tolerance and financial needs. Higher coverage levels offer greater compensation in the event of crop losses but also come with higher premiums.
- Perils Covered: Crop insurance policies can cover various perils such as drought, flood, hail, windstorm, pests, diseases, and other natural disasters. The specific perils covered may vary depending on the insurance policy and geographical location.
- Indemnity Calculation: The calculation of indemnity, or the amount of compensation provided to farmers, depends on the coverage level, actual yield, and the price of the crop. Indemnity can be based on individual farm yield or area yield, depending on the type of insurance policy.
- Premiums: Farmers pay premiums to the insurance company in exchange for coverage. Premium rates are determined based on factors such as historical

yield data, crop type, coverage level, and the level of risk associated with the specific region.

*c. Types of crop insurance*

Various types of crop insurance policies are available to farmers, catering to different needs and circumstances. The common types of crop insurance include:

- Yield-based Insurance: This type of insurance provides coverage based on the actual yield of the insured crop. If the actual yield falls below a predetermined threshold due to an insured peril, the farmer receives compensation.
- Revenue-based Insurance: Revenue-based insurance takes into account both the yield and the market price of the crop. If the revenue (yield multiplied by the market price) falls below a certain level, farmers receive indemnity.
- Area-based Insurance: Area-based insurance, also known as index-based insurance, provides coverage based on predetermined indices such as rainfall, temperature, or satellite imagery. If the index triggers a payout, farmers receive compensation, regardless of their individual yield losses.

*d. Challenges and future directions*

Crop insurance implementation faces several challenges that need to be addressed for its effective and widespread adoption. These challenges include:

- Affordability: The cost of crop insurance premiums can be a significant barrier for small-scale farmers, limiting their access to insurance coverage. Developing affordable insurance products and implementing subsidy programs can help overcome this challenge.
- Data Availability and Accuracy: Reliable and accurate data on historical yield, crop performance, and weather patterns are crucial for determining appropriate coverage levels and premium rates. Enhancing data collection systems and improving data quality are necessary to ensure the effectiveness of crop insurance programs.
- Awareness and Education: Many farmers, particularly in developing regions, have limited knowledge and understanding of crop insurance and its benefits. Promoting awareness and providing education about the importance of crop insurance can encourage its adoption and utilization.
- Climate Change: Climate change poses new risks and uncertainties for agriculture. Insurers need to assess and adapt insurance products to address the evolving challenges associated with changing weather patterns, extreme events, and emerging pests and diseases.

### 7.2.2. Blockchain-enabled parametric insurance contracts

Blockchain presents numerous advantages for parametric insurance contracts. Furthermore, there is substantial potential in agricultural risk management through revenue-based insurance contracts (Borselli, 2020; Khan, Loukil, Ghedira-Guegan, Benkhelifa, & Bani-Hani, 2021).

*a. Understanding parametric insurance contracts*

Parametric insurance contracts are designed to provide fast and transparent compensation based on predefined triggers that are easily measurable and verifiable. These triggers can be related to specific weather conditions, indices, or other quantifiable parameters directly linked to the insured risk. By removing the need for time-consuming claims assessments, parametric insurance enables swift payouts, allowing policyholders to recover quickly from losses and resume their operations.

*b. Leveraging blockchain technology*

Blockchain technology offers several advantages for parametric insurance contracts, making them more efficient, secure, and accessible (Gatteschi, Lamberti, Demartini, Pranteda, & Santamaría, 2018). The key benefits of blockchain in this context include:

- Transparency: The decentralized nature of blockchain ensures transparency in the insurance process. Smart contracts on the blockchain contain predefined triggers and payout conditions, which are accessible to all participants. This transparency eliminates information asymmetry and enhances trust between policyholders and insurers.
- Immutable Records: The immutability of blockchain records ensures that the terms of the parametric insurance contract cannot be altered or tampered with. This feature provides policyholders with a verifiable and auditable record of their insurance coverage and triggers confidence in the system.
- Real-time Data Feeds: Blockchain-enabled parametric insurance can integrate real-time data feeds from trusted sources, such as weather stations or satellite systems, to automatically trigger payouts. This eliminates the need for manual data collection and assessment, reducing delays and potential disputes.
- Smart Contract Automation: Smart contracts on the blockchain can automatically execute payout transactions once the predefined triggers are met. This automation minimizes administrative overhead, streamlines claim processing, and ensures swift compensation for policyholders.

- Security and Privacy: Blockchain's cryptographic protocols ensure the security and privacy of data shared within the network. Confidential information related to policyholders and their claims can be securely stored on the blockchain, protecting it from unauthorized access.

*c. Applications in agriculture*

Revenue-based insurance contracts have significant potential in agricultural risk management (Jha et al., 2021). Some applications include:

- Weather-related Risks: Parametric insurance contracts can be designed to cover losses caused by adverse weather conditions, such as droughts, floods, or extreme temperature fluctuations. Smart contracts can integrate weather data from reliable sources, automatically triggering payouts when the predefined thresholds are met.
- Crop Yield Protection: Parametric insurance can be used to protect farmers against low crop yields. By using historical yield data, satellite imagery, or other relevant indices, smart contracts can calculate compensation based on predetermined yield thresholds, offering farmers financial support in the event of reduced productivity.
- Livestock and Disease Outbreaks: Blockchain-enabled parametric insurance can also address risks related to livestock and disease outbreaks. Payout triggers can be based on data from veterinary reports, disease surveillance systems, or other relevant indicators, ensuring quick compensation for affected farmers.

*d. Advantages and considerations*

- Blockchain-enabled parametric insurance contracts offer several advantages, including:

+ Fast Payouts: Parametric insurance eliminates the need for lengthy claims assessment processes, enabling policyholders to receive compensation quickly. This prompt payout mechanism ensures timely financial support for farmers, allowing them to recover from losses and continue their operations.
+ Lower Administrative Costs: Automation and efficiency in claims processing reduce administrative overhead for insurers, resulting in lower transaction costs. This cost-effectiveness can make insurance more accessible and affordable, particularly for small-scale farmers.
+ Improved Risk Management: Parametric insurance encourages proactive risk management practices by providing farmers with predictable and

reliable coverage. This incentivizes investments in risk mitigation strategies, such as improved irrigation systems or disease prevention measures.

+ Trust and Transparency: The transparency and immutability of blockchain enhance trust between policyholders and insurers. The predefined triggers and automated payout processes reduce the potential for disputes and increase confidence in the insurance system.

- While blockchain-enabled parametric insurance contracts offer numerous advantages, there are considerations to keep in mind:

+ Data Accuracy and Reliability: The accuracy and reliability of data sources used as triggers are crucial. Ensuring high-quality data from trusted providers is essential for the effectiveness of parametric insurance contracts.

+ Smart Contract Design: Well-defined and thoroughly tested smart contracts are essential for accurate trigger calculations and automatic payout execution. Careful design and continuous monitoring of smart contracts are necessary to avoid potential vulnerabilities or errors.

+ Regulatory and Legal Frameworks: As with any insurance product, blockchain-enabled parametric insurance contracts must comply with relevant regulatory and legal frameworks. Policyholders' rights, privacy protection, and dispute-resolution mechanisms should be carefully addressed.

### 7.2.3. Automating claims verification and settlements

*a. Benefits of automating claims verification and settlements*

Automating claims verification and settlements offers several benefits to insurance companies, policyholders, and other stakeholders:

- Speed and Efficiency: Automation eliminates manual tasks and accelerates the claims process, leading to faster claim resolution and reduced turnaround times. Policyholders receive timely compensation, minimizing financial disruptions and enhancing their satisfaction.
- Accuracy and Consistency: Automated systems reduce the potential for human errors and inconsistencies in claims processing. By using predefined rules and algorithms, automation ensures consistent and accurate assessment of claims, improving overall decision-making.
- Cost Savings: Automating claims verification and settlements reduces administrative costs associated with manual processes. Insurers can allocate resources more efficiently, focusing on value-added activities and improving operational efficiency.

- Fraud Detection and Prevention: Automated systems can incorporate fraud detection algorithms, enabling early identification of suspicious claims patterns. Fraudulent activities can be flagged, investigated, and prevented, protecting insurers' bottom line and maintaining trust in the insurance ecosystem.
- Enhanced Customer Experience: Swift and accurate claims settlements enhance customer satisfaction and loyalty. Automated systems provide policyholders with a seamless and transparent experience, ensuring a positive interaction with their insurance provider during a potentially stressful time.

*b. Applications of automation in claims verification and settlements*

Automation can be applied to various stages of the claim's verification and settlement process (Khan, Loukil, et al., 2021; Kshetri, 2021a; Miller, Cao, Foth, Boyen, & Powell, 2023), including:

- Data Collection and Validation: Automated systems can collect and validate claim data from various sources, including digital forms, IoT devices, and external databases. By eliminating manual data entry and validation, insurers can improve data accuracy and reduce processing time.
- Claims Assessment: Automation can assess claims by applying predefined rules, algorithms, and machine learning techniques. Claims can be evaluated based on policy terms, historical data, fraud detection algorithms, and other relevant factors, ensuring consistent and objective decision-making.
- Document Processing: Optical Character Recognition (OCR) and Natural Language Processing (NLP) technologies enable automated extraction and analysis of information from claim documents. This streamlines document processing, reduces manual effort, and improves accuracy in capturing relevant data.
- Settlement and Payment: Automated systems can calculate claim amounts based on policy terms, coverage limits, and predefined rules. Once the assessment is complete, payouts can be initiated automatically, with funds transferred to the policyholder's account electronically.
- Communication and Updates: Automated communication systems can keep policyholders informed about the status of their claims, providing regular updates and notifications. This proactive communication improves transparency, reduces customer inquiries, and enhances the overall claims experience.

*c. Challenges and considerations*

While automation offers significant benefits, there are challenges and considerations to address (Khan, Loukil, et al., 2021):

- Data Quality and Integration: Automation relies on accurate and reliable data. Insurers must ensure data integrity, address data silos, and establish secure integrations with external data sources for effective automation.
- Regulatory Compliance: Compliance with regulatory requirements, including data privacy and protection, is crucial. Insurers must ensure that automated systems adhere to legal and regulatory frameworks governing claims verification and settlements.
- Complex Claims: Some claims may involve complex scenarios that require human intervention or expert assessment. Insurers should identify such cases and establish clear protocols for appropriate escalation and manual review.
- Change Management and Training: Introducing automation requires change management strategies and training programs to familiarise employees with new processes and technologies. Engaging and upskilling the workforce is essential for successful implementation.

*d. Future directions*

The automation of claims verification and settlements is an evolving field, driven by advancements in technology and changing customer expectations. Future directions in this domain include:

- Integration of Emerging Technologies: The integration of artificial intelligence, machine learning, and predictive analytics can further enhance automation capabilities, enabling more accurate claims assessment, fraud detection, and personalized customer experiences.
- Collaboration and Data Sharing: Collaboration between insurers, industry stakeholders, and data providers can lead to shared databases, enabling better data quality, streamlined processes, and improved customer outcomes.
- Continuous Process Improvement: Insurers should adopt a continuous improvement mindset, leveraging data analytics and customer feedback to refine and optimize automated claims processes. Regular evaluation and adjustment of automation workflows ensure ongoing efficiency gains.

### 7.2.4. Real-time monitoring of weather and yield data

*a. Importance of real-time monitoring*

Real-time monitoring of weather and yield data provides farmers with timely and accurate information, allowing them to:

- Weather Monitoring: Understand current and forecasted weather conditions, including temperature, rainfall, humidity, wind speed, and solar radiation. This information helps farmers plan irrigation schedules, assess pest and disease risks, and optimize planting and harvesting activities.
- Crop Growth Monitoring: Track the growth and development of crops, monitor nutrient levels, and identify potential stress factors. Real-time monitoring helps farmers identify growth abnormalities, implement timely interventions, and optimize inputs such as fertilizers and irrigation.
- Yield Estimation: By continuously monitoring crop growth and environmental factors, farmers can estimate yield potential. Real-time yield estimation enables proactive decision-making regarding marketing, storage, and post-harvest management.
- Risk Mitigation: Real-time monitoring allows farmers to identify risks, such as extreme weather events or pest outbreaks, and take timely actions to mitigate their impact. This proactive approach helps reduce crop losses and preserve yields.

*b. Benefits of real-time monitoring*

Real-time monitoring of weather and yield data offers several benefits to farmers:

- Improved Decision-making: Accurate and up-to-date information empowers farmers to make data-driven decisions. By understanding the current conditions and potential future trends, farmers can optimize resource allocation, adjust planting and harvesting schedules, and implement appropriate agronomic practices.
- Resource Optimization: Real-time monitoring enables farmers to optimize the use of resources such as water, fertilizers, and pesticides. By aligning inputs with crop requirements and weather conditions, farmers can minimize waste, reduce environmental impact, and enhance sustainability.
- Early Warning Systems: Real-time monitoring allows for the development of early warning systems. Farmers can receive alerts and notifications about potential risks, enabling them to take preventive measures, such as applying targeted pest control or implementing irrigation strategies, to avoid or mitigate crop losses.

- Increased Efficiency: By leveraging real-time data, farmers can streamline operations and improve overall efficiency. They can allocate resources based on actual crop needs, optimize labor utilization, and reduce unnecessary manual monitoring and interventions.
- Enhanced Market Competitiveness: Real-time monitoring provides farmers with a competitive edge. By staying informed about weather conditions, crop growth, and yield potential, farmers can align their products with market demands, enhance product quality, and potentially access premium markets.

*c. Applications in agriculture*

Real-time monitoring of weather and yield data finds diverse applications in agriculture:

- Weather Stations and Sensors: Weather stations equipped with sensors can provide real-time data on temperature, humidity, rainfall, wind speed, and solar radiation. This data helps farmers make precise irrigation decisions, assess crop water requirements, and predict disease risks.
- Remote Sensing and Satellite Imagery: Satellite-based remote sensing allows for the monitoring of crop growth, vegetation indices, and soil moisture content. This information aids in crop health assessment, yield estimation, and the identification of potential nutrient deficiencies or pest outbreaks.
- IoT and Sensor Networks: Internet of Things (IoT) devices and sensor networks provide real-time data on soil moisture, nutrient levels, and environmental conditions. These data sources assist farmers in implementing precision agriculture practices, optimizing input usage, and managing field variability.
- Mobile Applications and Data Platforms: Mobile applications and data platforms allow farmers to access real-time data and receive personalized recommendations. These platforms often integrate weather forecasts, crop models, and historical data to provide actionable insights for farm management.
- Blockchain Integration: Blockchain technology can enhance the credibility and traceability of real-time weather and yield data. By securely recording data on a decentralized ledger, farmers can ensure data integrity and potentially share information with stakeholders in the supply chain.

*d. Considerations and challenges*

While real-time monitoring offers significant advantages, there are considerations and challenges to address:

- Data Accuracy and Reliability: Ensuring the accuracy and reliability of real-time data is crucial. Calibration, validation, and quality control processes should be in place to ensure data integrity and minimize measurement errors.
- Data Integration and Compatibility: Integrating data from multiple sources and platforms can be challenging. Farmers need compatible systems that allow seamless data integration and provide a comprehensive view of the farm's conditions.
- Data Privacy and Security: Protecting sensitive farm data, including location information and crop performance data, is essential. Farmers should carefully choose data platforms and ensure compliance with data privacy regulations to maintain confidentiality and prevent unauthorized access.
- Accessibility and Affordability: Real-time monitoring technologies should be accessible and affordable to farmers of all scales. The availability of cost-effective monitoring solutions and tailored packages can facilitate wider adoption and ensure inclusivity.
- Farmer Education and Adoption: Farmers need education and support to effectively interpret and utilize real-time data. Training programs and extension services can play a vital role in facilitating farmer adoption and maximizing the benefits of real-time monitoring.

### 7.2.5. Improving resilience and reducing risks for farmers

*a. Building resilience in farming systems*

Resilience is the ability of farming systems to withstand and recover from shocks and stressors while maintaining productivity and livelihoods. Here are key strategies to build resilience:

- Diversification: Diversifying crops, livestock, and income sources can spread risks and provide a buffer against market fluctuations and production challenges. Farmers can explore alternative crops, value-added products, or non-farm income-generating activities.
- Climate-smart Agriculture: Implementing climate-smart practices, such as conservation agriculture, agroforestry, and precision irrigation, can enhance resilience to climate change. These practices promote soil health, water efficiency, and adaptive farming techniques.

- Improved Water Management: Efficient water management techniques, including rainwater harvesting, micro-irrigation systems, and water recycling, help farmers cope with water scarcity and drought conditions.
- Strengthened Financial Capacities: Access to financial services, such as savings, credit, and insurance, is crucial for building resilience. Farmers should explore financial tools tailored to agriculture, including crop insurance, weather-indexed insurance, and savings schemes.
- Knowledge and Information Sharing: Facilitating knowledge exchange and providing access to relevant information, such as weather forecasts, pest and disease alerts, and best practices, empowers farmers to make informed decisions and adapt to changing conditions.

*b. Managing climate and weather risks*

Climate and weather risks pose significant challenges to agriculture. Farmers can adopt various strategies to manage these risks effectively:

- Crop Selection and Rotation: Choosing crops that are resilient to local climate conditions and diversifying crop rotations can help mitigate climate risks. Rotating crops with different growth cycles and water requirements can also optimize resource use and reduce vulnerability to weather extremes.
- Early Warning Systems: Monitoring weather conditions and implementing early warning systems can provide farmers with timely information about potential risks. Weather forecasts, pest and disease alerts, and flood or drought monitoring systems help farmers take proactive measures to protect their crops and livelihoods.
- Irrigation Management: Optimal irrigation practices, such as scheduling irrigation based on crop needs and using water-efficient techniques, minimize water-related risks. Precision irrigation technologies and soil moisture sensors enable farmers to apply water where and when it is most needed.
- Crop Protection Strategies: Implementing integrated pest management (IPM) approaches, including biological controls, crop rotation, and scouting for pests, reduces the reliance on chemical pesticides and minimizes the risk of pest outbreaks.
- Disaster Preparedness and Recovery: Developing contingency plans and disaster response strategies is essential. Farmers should have emergency response plans in place, including measures to protect livestock, secure machinery, and store harvested crops safely.

*c. Market and financial risk management*

Market volatility and financial risks can impact farmers' profitability and stability. To manage these risks, farmers can consider the following:

- Market Diversification: Expanding market access and diversifying marketing channels reduce reliance on a single buyer or market. Direct marketing, farmer cooperatives, and value-added processing can provide alternative avenues for selling agricultural products.
- Forward Contracting and Price Hedging: Engaging in forward contracting or utilizing price hedging instruments, such as futures contracts or options, can help farmers secure stable prices and minimize the impact of market fluctuations.
- Access to Credit and Insurance: Maintaining access to credit and insurance products is crucial for managing financial risks Farmers should explore agricultural lending programs, microfinance institutions, and crop insurance schemes to protect against production and market-related risks.
- Farm Business Planning: Developing robust farm business plans that consider market trends, production costs, and potential risks enables farmers to make informed decisions and respond to changing conditions effectively.
- Market Information Systems: Access to accurate and timely market information helps farmers understand price trends, demand-supply dynamics, and market opportunities. Market information systems and platforms facilitate informed decision-making and enhance market engagement.

*d. Technology and innovation*

Technological advancements and innovation play a pivotal role in improving resilience and reducing risks for farmers (Khan, Loukil, et al., 2021):

- Remote Sensing and Digital Tools: Remote sensing technologies, satellite imagery, and digital tools provide real-time data on weather patterns, crop health, and soil conditions. Farmers can leverage these tools to make data-driven decisions and optimize resource management.
- Smart Farming Technologies: Smart farming technologies, including precision agriculture tools, sensor networks, and Internet of Things (IoT) devices, enable farmers to monitor and manage their farms more efficiently. These technologies optimize resource use, minimize waste, and improve productivity.
- Mobile Applications and Data Platforms: Mobile applications and data platforms provide farmers with access to market information, weather

forecasts, and agronomic advice. These platforms empower farmers to make informed decisions, enhance market participation, and adopt best practices.

- Climate-resilient Crop Varieties: The development and adoption of climate-resilient crop varieties that can withstand extreme weather conditions, pests, and diseases enhance farmers' resilience. Breeding programs and genetic engineering techniques contribute to the availability of resilient crop varieties.
- Collaborative Networks and Extension Services: Collaborative networks, farmer cooperatives, and extension services play a vital role in disseminating knowledge, facilitating technology adoption, and supporting farmers in managing risks. These networks foster information sharing, peer learning, and collective problem-solving.

## 7.3. Decentralized lending platforms for farmers and agricultural businesses

### 7.3.1. Challenges in traditional lending for agriculture

*a. Lack of collateral and credit history*

One of the primary challenges in traditional lending for agriculture is the lack of sufficient collateral and credit history. Agricultural assets, such as land and livestock, may have uncertain values or be subject to external risks, making them less desirable as collateral. Additionally, small-scale farmers often lack formal credit histories, limiting their access to loans from traditional financial institutions (Mazzei et al., 2020; McIntosh, 2018); Tomu (2020). The following solutions can be considered:

- Alternative Collateral Options: Encouraging the acceptance of alternative collateral options, such as warehouse receipts, equipment, or future produce, can help mitigate the collateral challenge. Implementing mechanisms that accurately assess the value and potential risks of agricultural assets can enable lenders to extend loans based on alternative forms of collateral.
- Credit Scoring Models: Developing credit scoring models specifically designed for agriculture can help evaluate the creditworthiness of farmers without extensive credit histories. These models can take into account factors like farm productivity, market potential, and farmers' experience to assess repayment capacity.

*b. Seasonal and cyclical nature of agriculture*

Agriculture is highly influenced by seasonal and cyclical factors, leading to irregular cash flows and repayment schedules. Traditional lending institutions may find it challenging to structure loans that align with the unique cash flow patterns of agricultural activities. The following solutions can be considered:

- Flexible Repayment Structures: Introducing flexible repayment structures that align with the cash flows of agricultural activities can alleviate the challenges associated with irregular income. Seasonal repayment options, grace periods, and customized repayment schedules that consider the timing of harvests and market sales can be implemented.
- Agricultural Credit Lines: Offering agricultural credit lines or revolving credit facilities can provide farmers with access to funds throughout the year. These credit lines can be utilized as per the farmer's cash flow needs, enabling them to address short-term funding requirements during lean periods.

*c. High risk and uncertainty*

Agriculture is inherently exposed to various risks, including weather events, market volatility, and pest outbreaks. These risks make agricultural lending perceived as high risk by traditional financial institutions, leading to reluctance in providing loans to farmers (Kumar et al., 2020). The following solutions can be considered:

- Risk Mitigation Tools: Promoting the use of risk mitigation tools, such as crop insurance and weather-indexed insurance, can provide lenders with greater confidence in extending loans to farmers. These tools help transfer and manage risks associated with unforeseen events, protecting both the farmer and the lender.
- Collaborative Risk Sharing: Encouraging partnerships and collaborations between financial institutions, agricultural stakeholders, and government agencies can facilitate risk sharing. Sharing risk through co-lending arrangements, loan guarantees, or risk-sharing agreements can help mitigate the perceived high risk associated with agricultural lending.

*d. Limited financial literacy and awareness*

Financial literacy and awareness among farmers, particularly small-scale farmers, is often limited. Many farmers may not have a clear understanding of loan requirements, repayment terms, or the benefits of formal financial services. This lack of awareness poses a challenge in accessing and effectively utilizing traditional lending options. The following solutions can be considered:

- Financial Education and Training: Implementing financial education programs and training initiatives targeted at farmers can enhance their understanding of loan processes, financial management, and the benefits of formal financial services. Extension services, farmer organizations, and government agencies can play a crucial role in delivering these programs.
- Farmer-lender Partnerships: Building strong partnerships between lenders and farmer organizations can bridge the gap in financial literacy. Lenders can collaborate with local agricultural cooperatives or farmer associations to provide tailored financial education and support services to their members.

### 7.3.2. Blockchain-based decentralized lending platforms

*a. Understanding blockchain technology*

Blockchain is a decentralized, transparent, and immutable digital ledger that records transactions across multiple computers. It eliminates the need for intermediaries, as transactions are validated and stored in a distributed manner (Pal, Tiwari, & Behl, 2021; Xu et al., 2019). Key features of blockchain technology include:

- Decentralization: Blockchain operates on a peer-to-peer network, removing the need for intermediaries such as banks or lending institutions. Transactions are validated by a network of participants, ensuring transparency and reducing reliance on centralized authorities.
- Transparency: All transactions recorded on the blockchain are transparent and visible to network participants. This transparency enhances trust and accountability, as transaction details can be audited and verified by anyone with access to the blockchain.
- Security: Blockchain employs cryptographic algorithms to secure transaction data. Each transaction is cryptographically linked to previous transactions, forming a chain of blocks that is resistant to tampering or unauthorized modifications.
- Smart Contracts: Smart contracts are self-executing agreements coded on the blockchain. They automatically execute predefined conditions when certain events are met. Smart contracts can automate lending processes, including loan disbursement, repayment, and interest calculations.

*b. Advantages of blockchain-based decentralized lending platforms*

Blockchain-based decentralized lending platforms offer several advantages over traditional lending models:

- Accessibility and Inclusion: These platforms enable farmers, especially those in underserved areas, to access loans without relying on traditional financial institutions. Blockchain technology removes geographical barriers and provides financial services to individuals who may have limited access to traditional banking.
- Enhanced Efficiency: Blockchain-based lending platforms streamline and automate the lending process, eliminating the need for intermediaries and reducing administrative burdens. Smart contracts automate loan agreements, repayment terms, and interest calculations, resulting in faster and more efficient loan processing.
- Transparency and Trust: The transparent nature of blockchain ensures that all loan transactions and details are recorded and visible to all participants. This transparency enhances trust among lenders and borrowers, reducing the risk of fraud and increasing the overall credibility of the lending process.
- Reduced Costs: By removing intermediaries, blockchain-based lending platforms can significantly reduce transaction costs associated with traditional lending. Automated processes, elimination of manual paperwork, and lower operational overheads contribute to cost savings for both lenders and borrowers.
- Security and Data Privacy: Blockchain technology provides enhanced security for sensitive borrower information. The decentralized nature of the blockchain reduces the risk of data breaches and unauthorized access. Additionally, borrowers have control over their own data, deciding what information to share and with whom.

*c. Challenges and considerations*

While blockchain-based decentralized lending platforms offer exciting possibilities for agricultural financing, there are several challenges and considerations to address:

- Regulatory Frameworks: The regulatory landscape surrounding blockchain and decentralized finance is still evolving. Governments and regulatory bodies need to develop frameworks that ensure consumer protection, prevent money laundering, and promote fair lending practices.
- Scalability: Blockchain technology faces scalability limitations in terms of transaction processing speed and capacity. As lending platforms grow,

ensuring that the blockchain infrastructure can handle increasing transaction volumes and maintain efficiency becomes crucial.

- Adoption and User Experience: Encouraging farmers and lenders to adopt blockchain-based lending platforms requires user-friendly interfaces, education, and awareness programs. Simplifying the user experience and providing technical support can facilitate wider adoption.
- Volatility and Cryptocurrency Risks: Blockchain-based lending platforms often involve the use of cryptocurrencies for loan transactions. The volatility of cryptocurrencies poses risks for borrowers and lenders. Implementing mechanisms to mitigate cryptocurrency volatility risks is necessary for stable lending operations.

*d. Future outlook*

Blockchain-based decentralized lending platforms have the potential to transform agricultural financing by increasing access, efficiency, and transparency. As blockchain technology continues to mature and regulatory frameworks evolve, we can expect to see increased adoption of these platforms in agricultural lending. Collaborations between blockchain developers, financial institutions, and agricultural stakeholders will be key to developing user-friendly and robust lending solutions that cater to the specific needs of farmers.

### 7.3.3. Peer-to-peer lending and crowdfunding for farmers

*a. Peer-to-peer lending for farmers*

Peer-to-peer lending platforms enable farmers to access loans directly from individual lenders or investors. These platforms leverage technology to facilitate loan matching, application processing, and repayment management. Key features of peer-to-peer lending for farmers include:

- Direct Borrower-lender Connection: Peer-to-peer lending platforms eliminate the need for traditional financial intermediaries by connecting borrowers and lenders directly. Farmers can present their loan requirements and financial profiles to potential lenders, who can assess the risk and choose to fund the loan.
- Diverse Loan Options: Peer-to-peer lending platforms offer farmers a variety of loan options, including working capital loans, equipment financing, and farm expansion loans. The loan terms and interest rates are often flexible and tailored to meet the specific needs of farmers.
- Risk Assessment and Creditworthiness: Peer-to-peer lending platforms utilize algorithms and credit scoring models to assess the creditworthiness

of farmers. Factors such as farm productivity, financial history, and market potential may be considered in the evaluation process.

- Community Support and Engagement: Peer-to-peer lending creates an opportunity for lenders to directly support farmers and engage with the agricultural community. Lenders can choose to fund projects aligned with their values, fostering a sense of connection and support for farmers.

*b. Crowdfunding for agricultural projects*

Crowdfunding platforms enable farmers to raise funds for specific projects or initiatives by soliciting small contributions from a large number of individuals. These platforms harness the power of collective support and community engagement. Key features of crowdfunding for agricultural projects include:

- Project-specific Funding: Farmers can showcase their agricultural projects or initiatives on crowdfunding platforms and seek financial contributions from interested individuals. Projects may include farm expansions, sustainable practices, technology adoption, or community-oriented initiatives.
- Community Engagement and Marketing: Crowdfunding provides an avenue for farmers to engage with their local communities, share their vision, and generate support for their projects. The crowdfunding campaign acts as a marketing tool, raising awareness about the farm's activities and attracting potential customers or partners.
- Rewards and Incentives: Crowdfunding campaigns often offer rewards or incentives to contributors. For agricultural projects, these may include farm produce, farm visits, or personalized experiences that connect supporters directly with the farm and its activities.
- Validation and Feedback: Crowdfunding campaigns serve as a validation of the viability and desirability of agricultural projects. Feedback and comments from contributors can provide valuable insights and suggestions, enhancing the project's overall quality.

*c. Benefits and considerations*

Peer-to-peer lending and crowdfunding for farmers offer several benefits, but careful considerations are necessary for successful implementation:

- Access to Finance: Peer-to-peer lending and crowdfunding provide farmers with alternative financing options, especially for those who may face challenges in obtaining loans from traditional financial institutions. These models bridge the gap and create opportunities for farmers to secure funds for their operations or projects.

- Community Support and Engagement: These models foster community engagement and support, allowing farmers to connect directly with lenders or contributors who have an interest in agriculture. This engagement builds trust, creates a sense of shared responsibility, and promotes the local agricultural economy.
- Flexibility and Customization: Peer-to-peer lending platforms and crowdfunding campaigns often offer flexible loan terms, interest rates, or project-specific funding options that can be tailored to the unique needs of farmers. This customization increases the likelihood of obtaining funding that aligns with the farmers' requirements.
- Considerations for Success: Implementing peer-to-peer lending and crowdfunding requires careful attention to legal and regulatory compliance, platform security, and risk management. Farmers must present compelling projects or loan requests, engage with potential lenders or contributors, and maintain transparency throughout the process.

### 7.3.4. Transparent credit assessment and scoring

*a. The importance of transparent credit assessment*

Transparent credit assessment is essential in agricultural lending to ensure fairness, accountability, and informed decision-making. By providing farmers with clear insights into the credit assessment process, they can better understand how lenders evaluate their creditworthiness. Transparent credit assessment offers several advantages:

- Trust and Confidence: Transparent credit assessment instills trust and confidence in the lending process. When farmers have visibility into the factors considered during credit evaluation, they are more likely to trust the outcome and have confidence in the lender's decision-making.
- Fairness and Equal Opportunities: Transparency in credit assessment promotes fairness by ensuring that all borrowers are evaluated based on consistent and objective criteria. It reduces the risk of bias or discrimination, providing equal opportunities for all farmers to access financing.
- Empowerment and Knowledge: Transparent credit assessment empowers farmers with knowledge about the factors that influence their creditworthiness. Farmers can identify areas for improvement and take proactive steps to strengthen their financial profile, making them more attractive to lenders.

*b. Components of transparent credit assessment*

Transparent credit assessment incorporates clear and well-defined criteria that are easily understandable to borrowers. Key components of transparent credit assessment include:

- Financial Performance: Financial performance indicators such as income, cash flow, profitability, and debt service coverage ratios are considered in credit assessment. Farmers should have visibility into how these indicators are evaluated and what benchmarks lenders use to assess their financial health.
- Farm Productivity and Management: Factors related to farm productivity, such as crop yields, livestock productivity, land utilization, and sustainable farming practices, may be considered in credit assessment. Farmers should understand how these factors impact their creditworthiness and what improvements can positively influence their evaluation.
- Market Potential and Off-take Arrangements: Lenders may assess the market potential of a farmer's produce and the existence of off-take arrangements with buyers. Farmers should be aware of how these aspects are evaluated and the role they play in credit assessment.
- Repayment Capacity and Credit History: Repayment capacity, including historical repayment records and credit history, is an essential aspect of credit assessment. Transparency should be maintained regarding the significance of repayment capacity and how it affects creditworthiness.

*c. Benefits of transparent credit scoring*

Transparent credit scoring complements transparent credit assessment by providing a clear and understandable credit score to borrowers. Benefits of transparent credit scoring in agricultural lending include:

- Clarity and Accountability: Transparent credit scoring provides borrowers with a numerical representation of their creditworthiness. Farmers can understand how their credit score is calculated, including the weightage given to different factors, enabling them to identify areas for improvement.
- Access to Opportunities: Transparent credit scoring allows farmers to benchmark their creditworthiness against industry standards and eligibility criteria. By knowing their credit score, farmers can proactively seek out lenders who are likely to provide financing based on their score range, increasing their chances of accessing suitable loan products.
- Financial Planning and Decision-making: Transparent credit scoring helps farmers make informed financial decisions. By understanding their credit score, farmers can evaluate the potential impact of borrowing on their

overall financial health and plan for the appropriate loan amount and repayment terms.

- Building Trust and Collaboration: Transparent credit scoring fosters trust between lenders and borrowers. When borrowers have a clear understanding of their credit score and the factors that influence it, they can engage in meaningful discussions with lenders, negotiate loan terms, and build collaborative relationships.

### 7.3.5. Facilitating investments in sustainable agriculture

*a. Importance of investments in sustainable agriculture*

Investments in sustainable agriculture are essential for the following reasons:

- Environmental Stewardship: Sustainable agriculture practices promote environmental conservation, including soil health, water management, biodiversity preservation, and reduced greenhouse gas emissions. Investments in sustainable agriculture enable farmers to adopt these practices and contribute to environmental stewardship.
- Improved Productivity and Efficiency: Sustainable agricultural investments often involve modernizing farming techniques, adopting innovative technologies, and improving resource management. These initiatives enhance productivity and efficiency, leading to higher crop yields, optimized resource utilization, and reduced input costs for farmers.
- Market Demand and Consumer Preferences: Increasingly, consumers are demanding sustainably produced food. Investments in sustainable agriculture enable farmers to meet this demand, access premium markets, and differentiate their products based on sustainability credentials. This, in turn, enhances market opportunities and profitability.
- Resilience and Adaptability: Sustainable agriculture practices help farmers build resilience to climate change, extreme weather events, and other environmental challenges. Investments in sustainable agriculture enable farmers to implement climate-smart practices, diversify their income streams, and mitigate risks associated with changing environmental conditions.

*b. Initiatives to facilitate investments in sustainable agriculture*

Several initiatives and strategies have been developed to facilitate investments in sustainable agriculture:

- Financial Incentives and Support Programs: Governments, development agencies, and financial institutions can provide financial incentives such as grants, low-interest loans, and subsidies to encourage farmers to adopt sustainable practices. These initiatives reduce the financial burden on farmers and incentivize the adoption of sustainable agriculture.

- Impact Investing and Socially Responsible Investing: Impact investors and socially responsible investment funds allocate capital to agricultural enterprises that demonstrate a commitment to sustainable practices. These investments not only provide financial support but also contribute to the growth and development of sustainable agriculture.
- Public-private Partnerships: Collaborations between governments, private sector entities, and non-profit organizations can facilitate investments in sustainable agriculture. These partnerships leverage the resources, expertise, and networks of multiple stakeholders to support farmers in adopting sustainable practices and accessing financing.
- Capacity Building and Technical Assistance: Providing farmers with access to training, technical assistance, and advisory services is crucial for the successful implementation of sustainable agriculture practices. Investments in capacity-building initiatives enhance farmers' knowledge and skills, making them better equipped to adopt and maintain sustainable practices.
- Certification and Labeling Programs: Certification schemes and labeling programs that verify and promote sustainable agricultural practices play a vital role in facilitating investments. These programs provide assurance to consumers and investors that agricultural products meet specific sustainability criteria, enhancing market opportunities for farmers.

*c. Benefits of investments in sustainable agriculture*

Investments in sustainable agriculture yield numerous benefits:

- Environmental Conservation: Sustainable agriculture investments contribute to preserving natural resources, reducing pollution, and protecting ecosystems. These initiatives support biodiversity, promote soil health, conserve water, and mitigate climate change impacts.
- Economic Prosperity: Investments in sustainable agriculture enhance the economic prosperity of farmers and rural communities. By improving productivity, reducing input costs, and accessing premium markets, farmers can achieve higher incomes and contribute to local economic development.
- Resilience and Adaptation: Sustainable agriculture investments strengthen the resilience of farming systems to climate change and other environmental challenges. By implementing climate-smart practices, farmers are better

prepared to cope with extreme weather events, water scarcity, and changing market dynamics.

- Long-term Sustainability: Investments in sustainable agriculture ensure the long-term viability of the agricultural sector. By safeguarding natural resources, diversifying production systems, and promoting sustainable livelihoods, these investments lay the foundation for a resilient and sustainable future.

## 7.4. Tokenization of agricultural assets and investment

Tokenization, stemming from blockchain technology, is a transformative concept reshaping the agricultural landscape. It entails the digitization of real-world agricultural assets, offering a plethora of advantages to the sector. These include the creation of blockchain-based digital assets representing farmland and livestock, the facilitation of fractional ownership and investment avenues, the bolstering of liquidity, and the unlocking of previously untapped value in agricultural assets. Furthermore, tokenization enables seamless global investment in agriculture (Pal et al., 2021; van Wassenaer et al., 2023; Xiong et al., 2020; Zheng et al., 2022).

### 7.4.1. Introduction to tokenization in agriculture

*a. Understanding tokenization in agriculture*

Tokenization involves converting tangible assets or rights into digital tokens that can be stored, transferred, and traded on a blockchain network. In the context of agriculture, tokenization allows for the representation and management of agricultural assets in a digital format. Key aspects of tokenization in agriculture include:

- Digital Representation: Agricultural assets, such as land parcels, crops, or livestock, are represented as unique digital tokens on a blockchain. Each token contains relevant information about the asset, such as ownership, characteristics, location, and history.
- Fractional Ownership: Tokenization enables the division of agricultural assets into fractional ownership units. These units can be traded or sold, allowing investors or individuals to own a portion of an asset rather than the entire asset. Fractional ownership facilitates liquidity and broadens investment opportunities in agriculture.
- Immutable and Transparent Records: Blockchain technology ensures that tokenized assets' ownership records and transaction history are stored in a decentralized and transparent manner. This immutability and transparency enhance trust, mitigate fraud, and simplify asset verification processes.

- Smart Contracts: Tokenization can be coupled with smart contracts, self-executing agreements written on the blockchain. Smart contracts automate and enforce the terms and conditions associated with the tokenized assets, such as revenue sharing, profit distribution, or lease agreements, reducing the need for intermediaries.

*b. Benefits of tokenization in agriculture*

Tokenization in agriculture offers numerous benefits for various stakeholders involved in the agricultural ecosystem:

- Increased Liquidity: Tokenization allows for the fractional ownership and trading of agricultural assets, providing liquidity and investment opportunities to a wider range of investors. Farmers can access capital by tokenizing their assets and attracting investment from interested parties.
- Enhanced Market Efficiency: Tokenization improves market efficiency by streamlining asset transactions, reducing paperwork, and lowering transaction costs. It enables the seamless transfer of ownership and facilitates faster and more transparent settlement processes.
- Access to Capital: Tokenization opens up new avenues for farmers and agricultural enterprises to raise capital. By tokenizing their assets, farmers can unlock the value of their land, crops, or livestock, allowing them to access funding for expansion, technology adoption, or sustainable practices.
- Fractional Ownership and Diversification: Tokenization allows investors to diversify their portfolios by owning fractional shares of multiple agricultural assets. It provides opportunities for smaller investors to participate in the agricultural sector and gain exposure to diverse agricultural activities.
- Improved Traceability and Quality Assurance: Tokenization enables the tracking of agricultural assets throughout their lifecycle, ensuring traceability and quality assurance. For instance, tokenizing a crop can provide verifiable information about its origin, cultivation practices, and compliance with quality standards, enhancing transparency and consumer confidence.

*c. Potential applications of tokenization in agriculture*

Tokenization can be applied to various areas within agriculture, including:

- Land Ownership and Leasing: Tokenizing land parcels enables fractional ownership, simplifies land leasing processes, and facilitates transparent land transactions. It allows individuals to invest in agricultural land while providing farmers with opportunities to access capital or lease out portions of their land.

- Crop and Livestock Management: Tokenizing crops and livestock provides a digital representation of these assets, enabling better inventory management, tracking of production data, and supply chain transparency. It enhances visibility and efficiency in managing agricultural produce.
- Commodity Trading: Tokenization can revolutionize the trading of agricultural commodities by representing them as digital tokens. It facilitates efficient and transparent trading, reduces intermediaries, and enables seamless settlement processes.
- Carbon Credits and Environmental Incentives: Tokenization can be used to tokenize carbon credits and other environmental incentives, promoting sustainable practices in agriculture. It provides a transparent and tradable mechanism for incentivizing and rewarding farmers for their environmental stewardship efforts.

### 7.4.2. Blockchain-based digital assets for farmland and livestock

*a. Understanding blockchain-based digital assets*

Blockchain-based digital assets for farmland and livestock refer to the representation of ownership and associated information as digital tokens on a blockchain network. These digital assets can include:

- Farmland Tokens: Farmland tokens represent ownership or fractional ownership rights to agricultural land. Each token contains relevant information such as the location, size, boundaries, land use, and ownership history of the farmland.
- Livestock Tokens: Livestock tokens represent ownership or fractional ownership rights to livestock, such as cattle, poultry, or sheep. These tokens may include details about the breed, age, health records, genetic information, and ownership history of the livestock.
- Data Tokens: Data tokens represent specific data related to farmland or livestock, such as soil quality reports, weather data, crop yields, or animal health records. These tokens enable secure and transparent access to valuable agricultural data.

*b. Benefits of blockchain-based digital assets*

The use of blockchain-based digital assets for farmland and livestock offers several advantages:

- Ownership and Transfer Transparency: Blockchain-based digital assets provide transparent and immutable records of ownership and transfer history. This transparency reduces disputes, enhances trust, and simplifies the process of buying, selling, or transferring ownership rights.

- Fractional Ownership and Investment Opportunities: Digital assets enable fractional ownership, allowing individuals to invest in farmland or livestock with smaller capital amounts. This fractional ownership model increases investment opportunities, diversification, and liquidity in the agricultural sector.
- Improved Traceability and Provenance: Digital assets on the blockchain provide a transparent and auditable trail of information for farmland and livestock. This traceability enhances supply chain management, enables accurate provenance verification, and supports consumer demands for transparency and sustainability.
- Efficient Transactions and Reduced Intermediaries: Blockchain-based digital assets streamline the transaction process by removing intermediaries and reducing paperwork. Smart contracts facilitate automated and secure transactions, ensuring that the terms and conditions are enforced without the need for intermediaries.
- Data Security and Privacy: Blockchain technology ensures data security and privacy by employing encryption and decentralized storage. Agricultural data associated with digital assets can be securely stored and selectively shared, ensuring confidentiality while allowing authorized access for relevant parties.

*c. Applications of blockchain-based digital assets*

Blockchain-based digital assets can find applications across various areas within the agricultural industry, including:

- Farmland Investments and Crowdfunding: Digital assets enable individuals to invest in farmland by purchasing fractional ownership tokens. This model opens up investment opportunities for a broader range of investors, facilitates crowdfunding initiatives, and supports sustainable land management.
- Livestock Traceability and Supply Chain Management: Digital assets for livestock provide traceability throughout the supply chain. Each token can record key information about the animal's health, vaccinations, breeding history, and movements, enabling improved livestock management and supply chain transparency.
- Access to Agricultural Data: Data tokens associated with digital assets allow stakeholders to access and share specific agricultural data securely. This data can be used for research, analysis, precision farming, or making informed decisions related to land use, crop selection, or livestock management.
- Land Governance and Title Management: Blockchain-based digital assets can revolutionize land governance and title management systems. By

recording land ownership information on the blockchain, it becomes easier to establish and verify ownership, reduce fraud, and streamline land administration processes.

- Agricultural Insurance and Risk Management: Digital assets for farmland and livestock can facilitate efficient insurance and risk management solutions. Blockchain-based records of ownership, asset conditions, and relevant data enable accurate assessment, underwriting, and claims processing in agricultural insurance.

### 7.4.3. Fractional ownership and investment opportunities

*a. Understanding fractional ownership*

Fractional ownership refers to the division of an asset into smaller ownership units, allowing multiple individuals or investors to own a fraction or share of that asset. In the context of agriculture, fractional ownership enables individuals to invest in agricultural assets such as farmland, livestock, or agricultural infrastructure without the need for substantial capital or the responsibility of sole ownership.

*b. Benefits of fractional ownership in agriculture*

Fractional ownership in agriculture offers several benefits for both investors and the agricultural industry:

- Access to Investment Opportunities: Fractional ownership opens up investment opportunities in the agricultural sector to a wider range of individuals. It allows investors with smaller capital amounts to participate in agricultural ventures and gain exposure to the industry's potential returns.
- Diversification: Fractional ownership enables investors to diversify their portfolios by investing in multiple agricultural assets. By spreading their investments across different types of crops, livestock, or geographical regions, investors can mitigate risks associated with any particular asset or region.
- Lower Barrier to Entry: Fractional ownership reduces the barrier to entry for investors interested in agricultural investments. Instead of needing significant capital to purchase an entire farm or livestock herd, investors can acquire fractional shares at a lower cost, making agricultural investments more accessible and inclusive.
- Shared Responsibilities: With fractional ownership, investors share the responsibilities and costs associated with the agricultural asset. Maintenance,

operational expenses, and management tasks can be divided among the co-owners, reducing the burden on individual investors.

- Professional Management: Fractional ownership often involves professional management of the agricultural asset. This allows investors to benefit from the expertise of agricultural professionals who oversee operations, implement best practices, and optimize returns on behalf of the co-owners.
- Liquidity and Exit Options: Fractional ownership can provide greater liquidity compared to traditional forms of ownership. Investors may have the option to sell their fractional shares or exit the investment at any time, subject to the terms and conditions set forth in the investment agreement.

*c. Applications of fractional ownership in agriculture*

Fractional ownership can be applied to various aspects of agriculture, including:

- Farmland Investments: Fractional ownership allows investors to own a portion of farmland, enabling them to participate in the agricultural land market without the need for full ownership. This model provides opportunities for individuals to diversify their investment portfolios and support sustainable land management practices.
- Livestock Investments: Fractional ownership of livestock offers investors the chance to own a share of livestock herds, such as cattle, poultry, or sheep. Investors can benefit from the income generated by the livestock's sale or breeding while sharing the associated costs and risks.
- Agricultural Infrastructure: Fractional ownership can be applied to agricultural infrastructure, such as storage facilities, processing units, or renewable energy installations. Investors can collectively own and benefit from these essential infrastructure assets, supporting agricultural value chains and efficiency.
- Agribusiness Ventures: Fractional ownership can also be utilized for investing in agribusiness ventures, such as agricultural technology startups, organic food companies, or sustainable agriculture initiatives. This allows investors to participate in the growth potential of innovative agricultural enterprises.

### 7.4.4. Enhancing liquidity and unlocking value in agricultural assets

*a. Technology-driven solutions*

- Tokenization: Tokenization involves representing agricultural assets as digital tokens on a blockchain network. By tokenizing assets such as farmland or livestock, ownership rights can be divided into fractional shares, enabling easier transferability and access to liquidity. Tokenization provides a transparent and efficient mechanism for buying, selling, and trading agricultural assets.

- Peer-to-peer Trading Platforms: Online platforms that connect buyers and sellers directly, without intermediaries, have emerged in the agricultural sector. These platforms utilize technology to match supply and demand, facilitating transactions and improving liquidity for agricultural assets. Peer-to-peer trading platforms enable farmers, investors, and other stakeholders to unlock the value of their assets by accessing a wider market.
- Digital Marketplaces: Digital marketplaces provide a platform for farmers to showcase their agricultural products, connecting them with potential buyers. These platforms improve visibility, streamline transactions, and enhance liquidity for agricultural commodities. By digitizing the buying and selling process, farmers can unlock the value of their produce and access a broader customer base.

*b. Financial innovations*

- Agricultural Asset-backed Securities: Agricultural asset-backed securities involve bundling agricultural assets, such as farmland or livestock, into investment products that can be traded on financial markets. These securities allow investors to gain exposure to the agricultural sector and provide farmers with a means to unlock the value of their assets. Agricultural asset-backed securities enhance liquidity by creating a secondary market for agricultural assets.
- Supply Chain Finance: Supply chain finance involves the use of financial instruments and technology to optimize cash flows in the agricultural supply chain. By providing financing options to farmers, suppliers, and buyers, supply chain finance improves liquidity and unlocks value by accelerating payment cycles and reducing working capital constraints. This approach enables farmers to access funds earlier and improves their cash flow management.
- Agricultural Credit Platforms: Online platforms that connect farmers with lenders have emerged as a way to enhance liquidity and access to credit in the agricultural sector. These platforms utilize technology to streamline the loan application and approval process, enabling farmers to access financing more efficiently. By improving access to credit, agricultural credit platforms unlock the value of agricultural assets by providing farmers with the necessary funds to invest and grow their operations.

*c. Value-added services*

- Data Analytics and Risk Assessment: Advanced data analytics and risk assessment models can provide valuable insights into the value and potential risks associated with agricultural assets. By analyzing historical data, market trends, and risk factors, farmers and investors can make informed decisions, unlock value, and mitigate risks. These value-added services enhance liquidity by improving asset valuation and risk management practices.
- Sustainable and Organic Certifications: Certifications for sustainable and organic agricultural practices add value to agricultural assets. These certifications cater to consumer demands for ethically produced, environmentally friendly products. By adopting sustainable and organic practices, farmers can unlock the value of their assets by capturing premium prices and accessing niche markets.

### 7.4.5. Enabling global investment in agriculture

*a. Overcoming regulatory challenges*

- Investment Frameworks: Governments can create investment-friendly frameworks that attract foreign investors to the agricultural sector. By establishing transparent regulations, providing incentives, and simplifying administrative procedures, countries can foster an environment conducive to global investment.
- Bilateral and Multilateral Agreements: Bilateral and multilateral agreements facilitate cross-border investment by harmonizing regulations, protecting investor rights, and promoting cooperation between countries. These agreements can streamline investment processes, reduce barriers, and encourage collaboration in the agricultural sector.
- Investment Promotion Agencies: Investment promotion agencies can play a crucial role in attracting foreign investment in agriculture. These agencies provide information, assistance, and support to potential investors, helping them navigate regulatory requirements and identify investment opportunities.

*b. Leveraging technology for information access*

- Digital Platforms: Online platforms and databases can aggregate agricultural investment opportunities from around the world, providing investors with a centralized source of information. These platforms facilitate access to project details, market insights, and due diligence reports, enabling investors to make informed investment decisions.

- Remote Sensing and Data Analytics: Remote sensing technologies, coupled with data analytics, provide valuable information on land suitability, climate patterns, and agricultural productivity. Investors can leverage these technologies to assess investment opportunities, monitor performance, and mitigate risks in different regions.
- Mobile Applications: Mobile applications offer convenient access to agricultural investment opportunities, market trends, and real-time information. These applications enable investors to stay informed, manage investments, and engage with stakeholders across borders, enhancing global investment in agriculture.

*c. Innovative financial mechanisms*

- Impact Investing: Impact investing channels capital towards projects that generate positive social and environmental impacts alongside financial returns. Impact investors in agriculture seek to address challenges such as sustainable farming practices, poverty alleviation, and food security while achieving financial objectives.
- Agricultural Funds: Agricultural funds pool capital from multiple investors to invest in agriculture. These funds provide opportunities for diversification, professional management, and access to global agricultural assets. They allow investors to participate in the growth potential of the agricultural sector while mitigating risks associated with individual investments.
- Crowdfunding: Crowdfunding platforms provide a means for individuals to invest in agriculture projects globally. By connecting project owners with a larger pool of potential investors, crowdfunding opens up investment opportunities to a broader range of individuals, democratizing access to agricultural investments.

*d. Public-private partnerships*

Public-private partnerships (PPPs) facilitate collaboration between governments, private investors, and other stakeholders in the agricultural sector. These partnerships leverage the strengths of each participant, combining resources, expertise, and networks to drive investment and sustainable development. PPPs can promote knowledge transfer, technology adoption, and inclusive growth in agriculture.

# Chapter 8
# Sustainable agriculture and supply chain sustainability

*Cultivating a Sustainable Future: Transforming Agriculture's Footprint with Blockchain Technology* - Unknown


**Abstract**

This chapter focuses on the integration of blockchain technology in promoting sustainable practices in agriculture and achieving supply chain sustainability goals. It examines how blockchain can contribute to environmental, social, and economic sustainability in the agriculture sector. The chapter explores the integration of blockchain with certifications, carbon credits, and sustainability metrics to drive supply chain sustainability. The chapter begins by examining how blockchain technology can promote sustainable practices in agriculture. It discusses how blockchain's transparency and traceability features can enable stakeholders to track and verify sustainable farming practices, such as organic production, fair trade, and responsible water usage. Next, the chapter explores the integration of blockchain with certifications, carbon credits, and sustainability metrics. It discusses how blockchain can provide a reliable and tamper-proof record of certifications, allowing consumers to make informed choices about sustainable products.



**Keywords**: Blockchain technology, Sustainable practices, Agriculture, Supply chain sustainability, Environmental sustainability, Social sustainability, Economic sustainability, Certifications, Carbon credits, Sustainability metrics


**Chapter contents**

- Using Blockchain to Incentivize and Track Sustainable Farming Practices
- Carbon Footprint Tracking and Reduction through Blockchain
- Promoting Fair Trade and Ethical Sourcing with Blockchain Applications

***

## 8.1. Using blockchain to incentivize and track sustainable farming practices

There is an increasingly widespread acknowledgment of the critical significance of sustainable agriculture in addressing global challenges. Blockchain technology emerges as a pivotal tool offering innovative solutions to incentivize and monitor sustainable farming practices. These solutions encompass the development of blockchain-based systems tailored for verifying sustainable farming practices, the establishment of digital identities for sustainable farms, the meticulous tracking and documentation of sustainable practices on the blockchain ledger, and the implementation of mechanisms to incentivize farmers through token rewards and certifications (Blokdyk, 2021; FAO, 2019; Li, Luo, & Zhou, 2023; Li, Wang, & Li, 2020; Nayal et al., 2021; Tomu, 2020).

### 8.1.1. The importance of sustainable agriculture

*a. Preserving natural resources*

One of the fundamental principles of sustainable agriculture is the conservation of natural resources. By implementing practices such as crop rotation, soil conservation, and integrated pest management, farmers can minimize soil erosion, preserve water quality, and protect biodiversity. For instance, crop rotation helps prevent nutrient depletion and reduces the reliance on chemical fertilizers, thus maintaining the fertility of the land. Additionally, sustainable irrigation techniques, such as drip irrigation and precision agriculture, minimize water waste and promote efficient water use, especially in regions facing water scarcity. By safeguarding these vital resources, sustainable agriculture ensures the long-term viability of food production while mitigating the negative impacts on the environment.

*b. Promoting biodiversity and ecosystem health*

Agricultural systems designed with sustainability in mind prioritize the preservation of biodiversity and the health of ecosystems. Monoculture farming, which relies on the extensive cultivation of a single crop, can lead to the loss of biodiversity and the disruption of natural habitats. In contrast, sustainable agriculture encourages the cultivation of diverse crops, which enhances biodiversity and strengthens the resilience of ecosystems. By creating habitats for beneficial insects, birds, and other wildlife, sustainable farming practices support natural pest control and reduce the need for chemical pesticides. Furthermore, the preservation of natural ecosystems surrounding farmlands contributes to the overall health of the environment, providing clean air, water, and regulating climate patterns. A thriving ecosystem is not only essential for sustaining agriculture but also for maintaining the overall health and balance of our planet.

*c. Ensuring food security*

Sustainable agriculture plays a critical role in ensuring global food security. By employing practices that conserve resources and minimize waste, sustainable farming systems are better equipped to cope with environmental challenges, such as climate change and droughts. Through the adoption of climate-smart agricultural techniques, such as agroforestry and conservation agriculture, farmers can increase their resilience to extreme weather events, improve soil quality, and maintain stable yields. Furthermore, sustainable agriculture emphasises local food production, reducing the dependence on long-distance transportation and lowering carbon emissions associated with food miles. By promoting local, diverse, and resilient food systems, sustainable agriculture enhances food security, reduces food waste, and strengthens local economies (Cottrill et al., 2018).

*d. Fostering rural development and livelihoods*

Sustainable agriculture is not only beneficial for the environment and global food security but also for rural development and livelihoods. By promoting sustainable farming practices, we can create opportunities for small-scale farmers, empower local communities, and improve the socio-economic conditions of rural areas. Sustainable agriculture encourages the adoption of agroecological techniques, which often require fewer external inputs and are more affordable for small farmers. Moreover, by supporting organic and regenerative farming practices, sustainable agriculture opens up new markets for sustainably produced goods, providing farmers with fair prices and economic stability. This, in turn, helps combat rural poverty, reduces migration to urban areas, and preserves cultural heritage and traditional knowledge related to agriculture.

### 8.1.2. Blockchain-based systems for verifying sustainable farming practices

*a. Understanding blockchain technology*

At its core, blockchain is a decentralized and transparent digital ledger that records and verifies transactions across multiple computers. Unlike traditional centralized databases, blockchain allows for secure, immutable, and tamper-proof storage of information. Each transaction, or "block," is cryptographically linked to the previous block, forming a chain of blocks that cannot be altered without consensus from the network participants. This distributed ledger technology provides an unprecedented level of transparency and trust, making it an ideal platform for verifying and validating sustainable farming practices (Detwiler, 2020).

*b. Enhancing transparency and traceability*

Blockchain-based systems offer enhanced transparency and traceability throughout the agricultural supply chain. By recording every transaction and interaction between stakeholders, from farmers and suppliers to processors and retailers, blockchain creates an immutable and auditable trail of information. This transparency enables consumers to track the journey of their food, ensuring that sustainable farming practices were followed at each step. For instance, information such as the origin of the product, farming methods used, certifications obtained, and environmental impact data can be securely stored on the blockchain. This empowers consumers to make informed choices and supports the demand for sustainable and ethically produced food (Kim & Laskowski, 2018).

*c. Verifying certification and compliance*

Certifications and compliance standards play a vital role in ensuring sustainable agricultural practices. However, verifying the authenticity and accuracy of these certifications can be challenging. Blockchain technology can provide a solution by enabling the verification of certifications and compliance in a transparent and efficient manner. By storing certification data on the blockchain, it becomes accessible to all relevant stakeholders, eliminating the need for manual audits and paperwork. Smart contracts, self-executing agreements built on the blockchain, can be utilized to automatically verify compliance based on predefined criteria. This streamlines the certification process, reduces costs, and minimizes the potential for fraud or misrepresentation (Kim & Laskowski, 2018).

*d. Promoting data sharing and collaboration*

Blockchain facilitates secure data sharing and collaboration among stakeholders in the agricultural supply chain. Traditionally, sharing data across different organizations or entities has been complex and prone to issues such as data silos, privacy concerns, and lack of trust. With blockchain, data can be securely shared among authorized participants, ensuring data integrity and confidentiality. This enables collaboration and the exchange of valuable information, such as best practices, research findings, and sustainability metrics. By fostering a culture of data sharing, blockchain-based systems can accelerate the adoption of sustainable farming practices by enabling collective learning, innovation, and continuous improvement.

*e. Mitigating greenwashing and building consumer trust*

Greenwashing, the practice of making misleading or unsubstantiated claims about the environmental or sustainable attributes of a product, remains a concern in the agricultural industry. Blockchain can help combat greenwashing by providing an immutable record of farming practices and certifications. Consumers can easily verify the authenticity of sustainability claims by accessing the transparent and auditable information stored on the blockchain. This builds trust between consumers and producers, encourages responsible practices, and rewards genuine efforts towards sustainability.

### 8.1.3. Creating digital identities for sustainable farms

*a. Defining digital identities for farms*

A digital identity for a sustainable farm can be seen as a comprehensive profile or record that encapsulates various aspects of its sustainability practices. It encompasses information such as the farm's location, size, farming methods, certifications obtained, use of resources, environmental impact data, and other relevant details. This digital identity serves as a digital passport that showcases the farm's commitment to sustainability and provides stakeholders with a reliable and transparent source of information.

*b. Enhancing transparency and trust*

Digital identities for sustainable farms promote transparency and trust within the agricultural supply chain. By establishing a standardized format for documenting and sharing sustainability-related information, these identities enable consumers, retailers, and other stakeholders to access reliable data about a farm's practices. This transparency helps build trust by providing verifiable evidence of a farm's sustainability claims. Consumers can make more informed choices, retailers can confidently source products from sustainable farms, and stakeholders can assess the environmental and social impact of their supply chains.

*c. Enabling traceability and accountability*

Digital identities facilitate traceability and accountability throughout the agricultural value chain. By associating a farm's digital identity with its products, it becomes possible to track the journey of food from farm to fork. This traceability ensures that sustainable farming practices were followed at each stage, including cultivation, harvesting, processing, and distribution. In case of any quality or sustainability issues, the digital identity provides a means to identify and address them swiftly, holding farms accountable for their practices. Additionally, digital identities can enable feedback loops, allowing stakeholders to provide insights, suggestions, or evaluations, further improving accountability and driving continuous improvement.

*d. Facilitating certification and compliance*

Certifications and compliance standards are crucial for sustainable farming, but managing and verifying them can be complex. Digital identities simplify this process by incorporating certification information within the farm's profile. Farms can provide verifiable evidence of their compliance with specific standards, such as organic farming, fair trade practices, or animal welfare regulations. This eliminates the need for manual verification, streamlines certification processes, reduces administrative burdens, and enhances the reliability of sustainability claims.

*e. Promoting data-driven decision making*

Digital identities create a framework for collecting and analyzing data that can drive data-driven decision-making in agriculture. Farms can leverage their digital profiles to capture and store real-time data on key sustainability metrics, such as water usage, energy consumption, greenhouse gas emissions, and biodiversity indicators. This data can be analyzed to identify trends, assess performance, and identify areas for improvement. Moreover, aggregated and anonymized data from multiple farms can provide valuable insights at the industry level, enabling benchmarking, research, and the development of best practices.

*f. Overcoming challenges and ensuring adoption*

While the concept of digital identities for sustainable farms holds great potential, it faces several challenges. These include the need for standardized frameworks and protocols, data interoperability, data security and privacy concerns, and integration with existing systems and technologies. Additionally, ensuring broad adoption and participation from farms of all sizes and regions is crucial for the effectiveness of digital identities. Collaborative efforts among stakeholders, including farmers, industry organizations, technology providers, and policymakers, are necessary to address these challenges and create a supportive ecosystem for the implementation and adoption of digital identities.

### 8.1.4. Tracking and recording sustainable practices on the blockchain

*a. The Power of transparent tracking*

Blockchain provides a transparent and tamper-proof platform for tracking sustainable practices throughout supply chains. By recording each transaction or interaction on the blockchain, stakeholders can trace the journey of products from their origin to the end consumer. This transparency ensures that sustainability standards and practices are adhered to at every stage. Whether it is monitoring the use of environmentally friendly materials, tracking energy consumption, or verifying fair trade practices, blockchain enables a comprehensive and auditable record of sustainable actions.

*b. Immutable records for proof of sustainability*

The immutability of blockchain records offers a robust mechanism for providing proof of sustainability. Sustainable practices and certifications can be securely stored on the blockchain, creating an unalterable audit trail. This eliminates the risk of falsification or manipulation of sustainability claims, enhancing the credibility of the information. Stakeholders, such as consumers, retailers, and regulatory bodies, can easily access and verify the authenticity of sustainability records, promoting trust and confidence in the claims made by businesses.

*c. Smart contracts automating compliance*

Smart contracts, self-executing agreements built on the blockchain, have the potential to automate compliance with sustainability standards. These contracts can be programmed to execute predefined actions when specific conditions are met. For instance, a smart contract can automatically trigger a verification process when a sustainable certification expires, ensuring ongoing compliance. This automation reduces the administrative burden associated with manual compliance checks and fosters a culture of accountability and continuous adherence to sustainability practices.

*d. Enhancing supply chain collaboration*

Blockchain facilitates collaboration among stakeholders in the supply chain to achieve sustainability goals. With a shared and transparent ledger, participants can securely share information and collaborate on sustainability initiatives. This enables the exchange of best practices, identification of areas for improvement, and collective problem-solving. Blockchain-based platforms can foster a collaborative ecosystem, where businesses, suppliers, and consumers work together towards common sustainability objectives, driving positive change at a systemic level.

*e. Real-time data for continuous improvement*

The real-time data recorded on the blockchain provides valuable insights for the continuous improvement of sustainable practices. By capturing and analyzing data related to energy consumption, waste generation, carbon emissions, and other sustainability metrics, businesses can identify inefficiencies and areas of improvement. This data-driven approach enables evidence-based decision-making, allowing organizations to implement targeted strategies that reduce their environmental footprint and improve overall sustainability performance (Dakshayini & Balaji Prabhu, 2020; Song & Li, 2021).

*f. Addressing challenges and ensuring adoption*

While the potential benefits of tracking and recording sustainable practices on the blockchain are immense, certain challenges need to be addressed for widespread adoption. These include scalability issues, interoperability between different blockchain platforms, data privacy concerns, and the need for industry-wide standards and protocols. Overcoming these challenges requires collaboration among stakeholders, including businesses, technology providers, policymakers, and industry organizations. Together, they can create a supportive ecosystem that fosters the adoption and integration of blockchain technology into sustainable supply chains.

### 8.1.5. Incentivising farmers through token rewards and certifications

*a. Token rewards for sustainable actions*

Token rewards offer a novel way to incentivize farmers for adopting sustainable practices. By utilizing blockchain-based tokens, farmers can earn rewards for implementing specific sustainability measures, such as soil conservation, water management, biodiversity preservation, or reduced pesticide use. These tokens can represent a digital asset that holds value within a specific ecosystem, allowing farmers to exchange them for various benefits. Rewards can range from financial incentives, such as access to capital or discounted supplies, to non-monetary benefits, including access to training programs, technical support, or market opportunities. Token rewards create a direct link between sustainable actions and tangible benefits, motivating farmers to embrace sustainable practices and continuously improve their environmental stewardship.

*b. Certifications and verifiable credentials*

Certifications have long been recognized as a means to incentivize sustainable practices in agriculture. However, the traditional certification process can be cumbersome and costly for farmers. Blockchain technology offers a solution by providing a transparent and tamper-proof platform to store and verify certifications. Digital certificates can be securely recorded on the blockchain, ensuring their authenticity and accessibility to relevant stakeholders. Farmers can showcase their certifications as verifiable credentials, earning recognition for their sustainable efforts. These credentials can open doors to market opportunities, better prices for their products, and enhanced reputation within the industry. By streamlining the certification process and providing transparent proof of sustainable practices, blockchain-based certifications incentivize farmers to strive for higher sustainability standards (Mukherjee et al., 2021).

*c. Creating token economies and marketplaces*

Token economies and marketplaces built on blockchain technology can further incentivize farmers by creating additional revenue streams and opportunities. By participating in these economies, farmers can earn tokens not only for their sustainable actions but also for contributing to data sharing, knowledge exchange, or participating in collaborative initiatives. These tokens can be used within the ecosystem to purchase inputs, access services, or even trade with other participants. Token marketplaces enable farmers to monetize their sustainable practices and benefit from their efforts in a dynamic and decentralized manner. This encourages active engagement, collaboration, and the sharing of best practices among farmers, driving a collective push towards sustainability.

*d. Collaboration with stakeholders*

Incentivizing farmers through token rewards and certifications requires collaboration among multiple stakeholders. This includes farmers' associations, industry organizations, supply chain actors, certification bodies, and technology providers. Collaborative efforts can ensure the development of robust token reward systems, establish credible certification frameworks, and foster the adoption of blockchain technology in the agricultural sector. Stakeholders can collaborate to design and implement incentive programs, define sustainability metrics, establish token economies, and create marketplaces that provide tangible benefits to farmers. Such collaboration strengthens the incentive ecosystem, amplifies its impact, and cultivates a supportive environment for farmers' transition to sustainable practices.

## 8.2. Carbon footprint tracking and reduction through blockchain

Agriculture stands as a cornerstone in nourishing the ever-expanding global population, yet its operations often impose significant environmental burdens. The widespread utilization of natural resources, emission of greenhouse gases, contamination of water sources, and the erosion of biodiversity epitomize the challenges entrenched within conventional agricultural practices. However, blockchain technology emerges as a potent enabler of sustainable agricultural paradigms. Through its innovative applications, blockchain fosters sustainability in myriad ways. These include the facilitation of blockchain-enabled carbon footprint tracking, the real-time collection and monitoring of emissions data, the utilization of smart contracts for carbon-offset trading, and the promotion of collaborative initiatives aimed at curbing greenhouse gas emissions. These transformative pathways underscore the potential of blockchain to reshape agricultural practices for the betterment of both humanity and the environment (Esmaeilian, Sarkis, Lewis, & Behdad, 2020; Jabbar et al., 2021; Manupati et al., 2020; Saberi, Kouhizadeh, Sarkis, & Shen, 2019; Yadav & Singh, 2020).

### 8.2.1. Addressing the environmental impact of agriculture

*a. Sustainable farming practices*

Transitioning to sustainable farming practices is a key step in addressing the environmental impact of agriculture. Sustainable agriculture focuses on minimizing resource use, reducing chemical inputs, and enhancing ecosystem health. Practices such as organic farming, agroforestry, conservation agriculture, and precision farming promote soil health, biodiversity conservation, and water and energy efficiency. By adopting these practices, farmers can minimize environmental degradation, protect natural resources, and promote ecological balance.

*b. Precision agriculture and technology*

The integration of precision agriculture and advanced technologies can significantly reduce the environmental footprint of farming. Precision agriculture utilizes data-driven techniques, such as remote sensing, geographic information systems (GIS), and sensors, to optimize inputs and manage resources more efficiently. By precisely targeting irrigation, fertilization, and pesticide application, farmers can minimize waste, improve crop yields, and reduce the release of harmful substances into the environment. Additionally, technologies like artificial intelligence (AI) and machine learning enable farmers to make data-driven decisions, optimize resource allocation, and enhance overall sustainability performance.

*c. Climate-smart agriculture*

Climate change poses a significant threat to agriculture, and addressing its impacts is crucial. Climate-smart agriculture encompasses practices that increase resilience to climate variability, mitigate greenhouse gas emissions, and support carbon sequestration. Strategies such as crop diversification, agroforestry, conservation tillage, and integrated water management can help farmers adapt to changing climate conditions and reduce the sector's contribution to global warming. By implementing climate-smart practices, farmers can not only reduce environmental impact but also enhance their own resilience and productivity (Fernando, Rozuar, & Mergeresa, 2021).

*d. Water management and conservation*

Water scarcity and pollution are pressing challenges in agricultural systems. Implementing efficient water management practices, such as drip irrigation, precision sprinklers, and rainwater harvesting, can minimize water wastage and ensure its judicious use. Additionally, conservation practices like cover cropping, contour plowing, and terracing can prevent soil erosion, improve water infiltration, and protect water quality. By prioritizing water conservation and adopting sustainable irrigation practices, farmers can reduce their impact on freshwater resources and contribute to long-term water security.

*e. Sustainable supply chains and consumer awareness*

Addressing the environmental impact of agriculture requires collaboration across the entire supply chain, from farmers to consumers. Sustainable sourcing, fair trade, and responsible consumption practices can promote environmentally friendly farming methods and support farmers who prioritize sustainability. Educating consumers about the environmental consequences of their food choices and empowering them to make informed decisions can create demand for sustainable products and drive market transformations. By establishing sustainable supply chains and fostering consumer awareness, we can collectively promote sustainable agriculture and reduce its ecological footprint (Shakhbulatov, Arora, Dong, & Rojas-Cessa, 2019).

*f. Policy and financial support*

Government policies and financial incentives play a crucial role in encouraging sustainable agriculture practices. Governments can provide support through subsidies, grants, and favorable regulations that promote sustainable farming methods and reward environmental stewardship. Investing in research and development for sustainable agricultural technologies, promoting knowledge sharing and capacity building, and facilitating market access for sustainable products are also essential. Public-private partnerships and collaborations can further mobilize resources and promote sustainable agriculture on a larger scale.

### 8.2.2. Blockchain-enabled carbon footprint tracking

Blockchain technology empowers the meticulous tracking of carbon footprints through various avenues, including the implementation of smart contracts for verification and accountability, the establishment of carbon credits and token economies, fostering collaboration and seamless data sharing, and safeguarding data integrity and privacy. This multifaceted approach not only facilitates the quantification and verification of carbon emissions but also encourages transparency and responsible environmental stewardship (Al Sadawi, Madani, Saboor, Ndiaye, & Abu-Lebdeh, 2021; Liu, Chang, Huang, & Lu, 2019).

*a. Transparent and trustworthy data*

Blockchain technology enables the transparent and secure recording of carbon emissions data throughout the agricultural supply chain. By using distributed ledger technology, stakeholders can securely store and share data related to energy use, fuel consumption, fertilizer application, and other emissions sources. This transparency enhances trust among participants, as the shared ledger ensures that emission data cannot be tampered with or manipulated. It enables accurate measurement and tracking of carbon footprints, facilitating more informed decision-making and targeted emission reduction strategies.

*b. Smart contracts for verification and accountability*

Smart contracts, self-executing agreements built on the blockchain, can automate the verification and accountability of carbon reduction efforts. These contracts can be programmed to trigger predefined actions when specific emission reduction targets or milestones are met. For example, a smart contract can automatically issue carbon credits to farmers who implement sustainable practices that lead to emissions reductions. This automation streamlines the verification process, eliminates the need for intermediaries, and ensures that emission reductions are accurately accounted for and rewarded (Wang, Wang, & Abareshi, 2020).

*c. Carbon credits and token economies*

Blockchain-enabled systems can facilitate the creation and trading of carbon credits, providing financial incentives for emission reductions in agriculture. Carbon credits represent a measurable reduction in greenhouse gas emissions, and blockchain ensures their transparency and traceability. Farmers can earn carbon credits for adopting sustainable practices, sequestering carbon in soils, or implementing renewable energy solutions. These credits can be tokenized on the blockchain, creating a digital asset that can be bought, sold, and traded within a token economy. Token marketplaces allow participants to monetize their emission reductions, providing a tangible incentive to reduce their carbon footprint and fostering a market for sustainable practices (Woo et al., 2021).

*d. Collaboration and data sharing*

Blockchain technology encourages collaboration and data sharing among stakeholders in the agricultural sector. By securely and transparently sharing emission data on the blockchain, farmers, researchers, policymakers, and supply chain actors can collectively work towards emissions reduction goals. Collaboration enables the identification of best practices, knowledge exchange, and the development of innovative solutions to address the environmental challenges associated with agriculture. Blockchain-enabled platforms can facilitate this collaboration, creating a network where stakeholders can access and contribute to a comprehensive and shared emissions dataset.

*e. Ensuring data integrity and privacy*

While blockchain technology offers transparency, it is crucial to address data integrity and privacy concerns in carbon footprint tracking. Secure data protocols, encryption techniques, and permissioned blockchain networks can safeguard sensitive information while ensuring transparency and traceability. Participants can control access to their data, granting permission only to trusted entities involved in emissions verification or trading. Privacy-preserving techniques can be implemented to protect proprietary information while still enabling accurate emissions accounting and verification (Kouhizadeh, Saberi, & Sarkis, 2021).

### 8.2.3. Real-time data collection and monitoring of emissions

*a. Sensor technology and IoT integration*

Sensor technology plays a crucial role in real-time data collection and monitoring of emissions in agriculture. Sensors can be deployed in fields, livestock facilities, and processing plants to measure various parameters such as air quality, temperature, humidity, methane emissions, and nitrogen oxide levels. These sensors can be connected through IoT networks, allowing data to be collected and transmitted in real time. The integration of sensor technology and IoT provides a comprehensive and continuous monitoring system, capturing emissions data at different stages of agricultural activities.

*b. Emissions modelling and predictive analytics*

Real-time emission data, combined with advanced data analytics and modeling techniques, can provide valuable insights into emissions patterns and trends.

By analyzing the collected data, emissions models can be developed to understand the factors contributing to emissions and their variations over time. Predictive analytics can then be applied to forecast emissions levels based on specific farming practices, weather conditions, or other relevant parameters. These predictive models empower farmers to anticipate emission hotspots, optimize management strategies, and implement targeted mitigation measures.

*c. Remote sensing and satellite imagery*

Remote sensing technologies, such as satellite imagery and aerial drones, offer a bird's-eye view of agricultural landscapes, enabling the assessment of emissions at larger scales. Satellites equipped with specialized sensors can capture data on land use, vegetation health, and atmospheric conditions, providing valuable information on emissions sources, such as deforestation, changes in soil carbon, or agricultural burning. Combining remote sensing data with ground-based measurements enhances the accuracy and reliability of emissions monitoring, supporting evidence-based decision-making for sustainable agricultural practices.

*d. Real-time feedback and decision support systems*

Real-time data collection and monitoring systems can provide farmers with immediate feedback on their emissions performance. Through user-friendly interfaces and decision support systems, farmers can access real-time information about their emissions levels, compare their performance against benchmarks or targets, and receive recommendations for emission reduction strategies. This feedback loop enables farmers to make timely adjustments to their practices, optimize resource use, and adopt more sustainable management approaches.

*e. Stakeholder collaboration and knowledge sharing*

Real-time emissions data collection and monitoring systems foster collaboration among stakeholders in the agricultural sector. By sharing data and insights through platforms or networks, farmers, researchers, policymakers, and industry actors can collectively work towards reducing emissions and improving environmental management. Collaborative platforms enable knowledge sharing, best practice dissemination, and the identification of innovative solutions to address emissions challenges. This collaborative approach amplifies the impact of real-time monitoring, leading to a more coordinated and effective response to environmental issues.

*f. Overcoming challenges and ensuring data integrity*

Real-time data collection and monitoring of emissions face challenges related to data quality, integration, and interoperability. It is crucial to establish data standards, protocols, and interoperable systems that facilitate the seamless integration of data from multiple sources and platforms. Ensuring data integrity, privacy, and security are also paramount. Robust data governance frameworks, encryption techniques, and privacy safeguards should be implemented to protect sensitive information while enabling data sharing and collaboration.

### 8.2.4. Smart contracts for carbon-offset trading

*a. Automated verification and transparency*

Smart contracts, self-executing agreements built on the blockchain, automate the verification process in carbon-offset trading. They can incorporate predefined criteria and algorithms that automatically validate carbon-offset projects and transactions. This automation eliminates the need for intermediaries, reduces administrative overheads, and enhances the speed and efficiency of the verification process. Moreover, the transparency of blockchain ensures that the entire transaction history and carbon-offset details are securely recorded and accessible to all participants, promoting trust and accountability.

*b. Tokenization of carbon offsets*

Blockchain technology enables the tokenization of carbon offsets, creating digital assets that can be securely stored and traded. By representing carbon offsets as tokens on the blockchain, the market for carbon-offset trading becomes more accessible, efficient, and inclusive. Tokens can be easily transferred between parties, facilitating the buying and selling of carbon offsets in a decentralized manner. Tokenization also enhances the traceability and transparency of carbon offsets, as each token represents a specific emission reduction or removal project, allowing buyers to track the origin and impact of their offsets.

*c. Decentralized marketplaces*

Smart contracts enable the creation of decentralized marketplaces for carbon-offset trading. These platforms bring together buyers and sellers of carbon offsets, facilitating direct transactions without the need for intermediaries. Buyers can browse through different projects, assess their environmental credentials, and make informed decisions based on verified data stored on the blockchain. Smart contracts automatically execute the transaction once the agreed-upon conditions are met, ensuring secure and efficient trading while reducing costs and time delays.

*d. Additionality and permanence safeguards*

Smart contracts can embed safeguards for additionality and permanence in carbon-offset projects. Additionality ensures that emission reductions or removals would not have occurred without the carbon-offset project, providing a net reduction in emissions. Permanence safeguards address concerns about the reversal of carbon sequestration efforts over time. Smart contracts can include mechanisms that periodically verify and validate the ongoing additionality and permanence of carbon-offset projects, providing assurance to buyers and maintaining the integrity of the market.

*e. Enhancing trust and accountability*

Blockchain-based smart contracts enhance trust and accountability in carbon-offset trading. The immutable nature of the blockchain ensures that all transactions and emission reductions are recorded transparently and cannot be altered. This transparency, coupled with automated verification, mitigates the risk of fraud or double counting, promoting integrity in the market. Buyers can have confidence in the authenticity and quality of the carbon offsets they purchase, encouraging greater participation and investment in emission reduction projects.

*f. Integration with existing carbon standards*

Smart contracts can be designed to integrate with existing carbon standards and methodologies, ensuring compatibility and interoperability. By aligning with established frameworks such as the Verified Carbon Standard (VCS) or the Gold Standard, smart contracts provide a familiar structure for carbon-offset projects and trading. This integration enables seamless transition and collaboration between traditional carbon-offset mechanisms and blockchain-enabled systems, fostering a harmonized and robust carbon market.

### 8.2.5. Collaborative efforts for reducing greenhouse gas emissions

*a. Multi-stakeholder partnerships*

Multi-stakeholder partnerships are key to driving collective action for emissions reduction. Governments, businesses, civil society organizations, research institutions, and communities must collaborate to set ambitious emission reduction targets, develop sustainable policies, and implement effective strategies. Such partnerships enable the pooling of expertise, resources, and innovative solutions, leading to more comprehensive and impactful emission reduction efforts.

*b. Knowledge sharing and capacity building*

Collaborative efforts facilitate knowledge sharing and capacity building, empowering stakeholders to take informed actions. Sharing best practices, lessons learned, and technological advancements across sectors and regions can accelerate the adoption of sustainable practices. Capacity-building programs can enhance the skills and knowledge of individuals and organizations, equipping them with the tools to implement effective emission reduction measures. By fostering a culture of learning and knowledge exchange, collaborative initiatives maximize the collective potential for emissions mitigation.

*c. Research and development collaborations*

Research and development collaborations play a crucial role in advancing low-carbon technologies, innovative solutions, and sustainable practices. By bringing together researchers, scientists, and industry experts, collaborative research initiatives can tackle complex challenges, explore new avenues, and drive technological breakthroughs. These collaborations enhance our understanding of the causes and impacts of GHG emissions, enabling evidence-based decision-making and the development of targeted mitigation strategies.

*d. Supply chain collaboration*

Collaborative efforts within supply chains have significant potential to reduce emissions. Businesses can collaborate with their suppliers, customers, and logistics partners to optimize processes, reduce waste, and improve energy efficiency. Supply chain collaborations can also promote the adoption of sustainable sourcing practices, encourage the use of renewable energy, and enhance product lifecycle management. By working together, supply chain stakeholders can achieve emissions reductions across the entire value chain, leading to a more sustainable and resilient business ecosystem (Kshetri, 2021b; Yousefi & Tosarkani, 2022).

*e. Public-private partnerships*

Public-private partnerships (PPPs) play a critical role in driving emissions reduction initiatives. Governments and businesses can collaborate to develop and implement sustainable policies, financial incentives, and regulatory frameworks. PPPs can leverage the strengths of both sectors, combining public resources, expertise, and regulatory authority with the agility, innovation, and market influence of the private sector. These partnerships facilitate the scaling up of sustainable technologies, infrastructure development, and the mobilization of financial resources for emissions reduction projects.

*f. Local community engagement*

Engaging local communities is essential for successful emissions reduction efforts. Collaborative initiatives can involve communities in decision-making processes, raise awareness, and empower individuals to take action at the grassroots level. Community-led projects, such as renewable energy cooperatives or sustainable agriculture initiatives, can contribute to emission reductions while promoting local ownership and resilience. By fostering inclusive and participatory approaches, collaborative efforts ensure that emission reduction strategies reflect the needs and aspirations of the communities they aim to serve.

## 8.3. Promoting fair trade and ethical sourcing with blockchain applications

Blockchain presents a robust toolkit for advancing fair trade and ethical sourcing practices across industries. These multifaceted solutions encompass ensuring fair wages and humane working conditions throughout supply chains, utilizing blockchain for transparent verification of fair trade certifications, establishing a clear and accountable tracking system for supply chain practices, combating child labor, and championing ethical sourcing initiatives. By leveraging blockchain, consumers are empowered with the knowledge and tools to make informed decisions that align with their values of sustainability and ethics (Kshetri, 2022b; Rana et al., 2021; Sharma, Samad, Jabbour, & de Queiroz, 2021).

### 8.3.1. Ensuring fair wages and working conditions in supply chains

*a. Understanding the supply chain landscape*

Supply chains can span multiple countries and involve various tiers of suppliers, making it challenging to monitor and enforce fair labor standards. Understanding the complexity and interconnectedness of supply chains is crucial for addressing issues related to wages and working conditions. It involves mapping out the entire supply chain, identifying key actors, and assessing the labor practices at each stage.

*b. Responsible sourcing and supplier engagement*

Responsible sourcing is a critical strategy for ensuring fair wages and working conditions. Organizations can adopt policies and guidelines that prioritize suppliers who demonstrate commitment to labor rights and fair treatment of workers. Engaging with suppliers through audits, assessments, and regular communication can encourage continuous improvement in labor practices. Collaboration between buyers and suppliers fosters a shared responsibility for upholding fair working conditions throughout the supply chain.

*c. Ethical codes of conduct*

Developing and implementing ethical codes of conduct is an effective way to set clear expectations for suppliers and promote fair labor practices. These codes outline standards for wages, working hours, occupational health and safety, and other critical aspects of working conditions. Ethical codes of conduct should be comprehensive, aligned with international labor standards, and enforced through regular monitoring and supplier evaluations.

*d. Collaboration and capacity building*

Collaboration among stakeholders is key to driving positive change in supply chains. Governments, industry associations, non-governmental organizations, and trade unions can work together to establish industry-wide initiatives, share best practices, and develop capacity-building programs. Capacity building aims to enhance suppliers' understanding of labor rights, train workers on their rights and responsibilities, and build the capabilities needed to improve working conditions.

*e. Worker empowerment and voice*

Ensuring fair wages and working conditions requires giving workers a voice and empowering them to assert their rights. Mechanisms such as worker committees, grievance mechanisms, and worker empowerment programs can provide platforms for workers to express concerns, report violations, and collectively negotiate for better conditions. Creating an environment where workers feel safe and empowered enhances their well-being and contributes to sustainable supply chains.

*f. Transparent supply chain reporting*

Transparent reporting on labor practices and working conditions in supply chains promotes accountability and drives positive change. Organizations should disclose information about their supply chain management, including efforts to ensure fair wages and working conditions. Transparent reporting enables stakeholders, including consumers and investors, to make informed choices and hold companies accountable for their commitments to social responsibility.

*g. Supporting local communities and social development*

Ensuring fair wages and working conditions goes beyond the workplace. It also involves supporting local communities and contributing to social development. Companies can engage in community development initiatives, such as providing access to education, healthcare, and essential services. Supporting local communities strengthens the social fabric and creates opportunities for economic empowerment, contributing to sustainable livelihoods beyond the workplace.

### 8.3.2. Blockchain for verifying fair trade certifications

*a. Transparency and traceability*

Blockchain provides a decentralized and transparent platform for verifying fair trade certifications. By recording certification details, supply chain information, and transaction history on the blockchain, stakeholders can access real-time, immutable data about the journey of a product from its origin to the end consumer. This transparency and traceability enable consumers to verify the authenticity of fair trade certifications, ensuring that products meet the required ethical standards.

*b. Immutable certification records*

Blockchain's immutability ensures that fair trade certification records cannot be altered or tampered with. Each certification is securely stored on the blockchain, creating a permanent and auditable record of compliance with fair trade standards. This feature eliminates the risk of fraudulent certifications and provides verifiable proof of ethical sourcing and production practices.

*c. Smart contracts for compliance automation*

Smart contracts can be integrated into the blockchain to automate compliance with fair trade standards. These self-executing contracts can contain predefined rules and conditions that must be met for a certification to be issued or maintained. Smart contracts can facilitate automatic verification of supply chain data, monitor compliance with fair trade requirements, and trigger alerts or penalties in case of non-compliance. This automation streamlines the verification process and ensures that fair trade certifications remain valid and up to date.

*d. Collaborative verification networks*

Blockchain enables the creation of collaborative verification networks, bringing together stakeholders such as producers, certifying bodies, retailers, and consumers. These networks can collectively validate fair trade certifications by cross-referencing information stored on the blockchain and conducting peer-to-peer audits. By leveraging the collective knowledge and expertise of network participants, blockchain-based verification networks enhance the accuracy and reliability of fair trade certifications.

*e. Empowering consumers*

Blockchain empowers consumers by providing them with accessible and verifiable information about fair trade certifications. With a simple scan of a product's QR code or barcode, consumers can access detailed information about the product's journey, including fair trade certifications, origins, and the associated supply chain. This transparency enables consumers to make informed purchasing decisions aligned with their values, supporting companies and producers that prioritize fair trade practices.

*f. Assurance for producers and suppliers*

Blockchain-based verification of fair trade certifications benefits producers and suppliers as well. By providing a secure and immutable record of their adherence to fair trade standards, blockchain instills trust among buyers and consumers. Producers and suppliers can demonstrate their commitment to ethical practices and differentiate themselves in the market, gaining a competitive edge while fostering long-term relationships with responsible buyers.

*g. Continuous improvement and feedback mechanisms*

Blockchain-based verification systems can incorporate feedback mechanisms and channels for continuous improvement. Stakeholders can provide feedback on fair trade practices, report issues or concerns, and suggest improvements directly through the blockchain platform. This feedback loop enables the evolution of fair trade standards, ensures accountability, and drives positive change within the industry.

### 8.3.3. Transparent tracking of supply chain practices

*a. The need for transparent supply chains*

Transparent supply chains are essential for ensuring ethical sourcing, reducing environmental impact, and promoting responsible business practices. By tracking the journey of products from raw materials to the end consumer, stakeholders can assess compliance with social, environmental, and quality standards. Transparent supply chains empower consumers to make informed purchasing decisions and hold companies accountable for their practices.

*b. Blockchain for supply chain transparency*

Blockchain technology provides a decentralized and immutable platform for transparently tracking supply chain practices. Through the use of distributed ledger technology, information about each stage of the supply chain can be securely recorded and shared among stakeholders. Blockchain ensures data integrity, prevents tampering, and enables real-time visibility into product provenance, certifications, and compliance with sustainability standards.

*c. Internet of things (IoT) sensors and data*

The Internet of Things (IoT) complements blockchain technology by enabling the collection and transmission of real-time data from various points in the supply chain. IoT sensors can monitor temperature, humidity, location, and other relevant parameters, providing granular insights into product conditions and handling. This data, when integrated with blockchain, enhances transparency by creating an immutable record of the product's journey and ensuring compliance with quality and sustainability requirements.

*d. Supplier verification and auditing*

Transparent tracking of supply chain practices enables efficient supplier verification and auditing processes. By leveraging blockchain and IoT data, companies can verify the authenticity and compliance of suppliers, ensuring they meet the required social and environmental standards. Smart contracts can automate verification processes, triggering audits when predefined conditions are not met. This ensures that suppliers adhere to ethical practices and fosters a more responsible and accountable supply chain ecosystem.

*e. Consumer engagement and trust*

Transparent supply chains foster consumer engagement and trust. By providing access to detailed information about a product's origins, manufacturing processes, and sustainability practices, companies can empower consumers to make conscious choices aligned with their values. Transparent tracking cultivates trust, allowing consumers to verify claims, support ethical brands, and drive demand for sustainable products.

*f. Collaboration and industry standards*

Transparent supply chains promote collaboration among stakeholders, including suppliers, manufacturers, retailers, and consumers. By adopting common industry standards and sharing supply chain data, stakeholders can collectively work towards sustainable practices and address systemic challenges. Collaborative platforms and initiatives enable the sharing of best practices, lessons learned, and innovative solutions, fostering continuous improvement throughout the supply chain.

*g. Regulatory compliance and risk mitigation*

Transparent tracking of supply chain practices helps companies comply with regulations and mitigate risks associated with non-compliance. By having a clear overview of the entire supply chain, companies can identify potential vulnerabilities, proactively address issues, and demonstrate compliance with legal requirements. Transparent supply chains also reduce the risk of reputational damage by enabling prompt responses to potential controversies or violations.

### 8.3.4. Eliminating child labor and promoting ethical sourcing

*a. Understanding the impact of child labor*

Child labor has severe consequences on the well-being and development of children. It denies them access to education, exposes them to hazardous working conditions, and perpetuates a cycle of poverty. Understanding the root causes, social dynamics, and economic factors that contribute to child labor is crucial for designing effective interventions and promoting ethical sourcing practices.

*b. Legislation and international standards*

Legislation and international standards play a vital role in combating child labor and promoting ethical sourcing. Governments should enact and enforce laws that prohibit child labor, set minimum age requirements for work, and establish penalties for non-compliance. International standards, such as those defined by the International Labor Organization (ILO), provide guidelines for responsible business conduct and supply chain management, including the elimination of child labor.

*c. Supply chain due diligence*

Companies have a responsibility to conduct due diligence throughout their supply chains to identify and address child labor risks. This involves mapping out the supply chain, engaging with suppliers, conducting regular audits, and implementing robust monitoring systems. Supply chain due diligence helps identify areas where child labor may be present and enables companies to take appropriate actions to eliminate it.

*d. Supplier engagement and capacity building*

Engaging with suppliers is essential for promoting ethical sourcing practices and eliminating child labor. Companies should communicate their expectations regarding child labor and work collaboratively with suppliers to ensure compliance. Supplier capacity-building programs, including training on labor rights and responsible practices, can empower suppliers to improve working conditions and eliminate child labor from their operations.

*e. Traceability and certification programs*

Traceability systems and certification programs enhance transparency and accountability in supply chains. By implementing technologies like blockchain and incorporating certification programs, companies can verify the authenticity and integrity of their supply chains. Certifications, such as Fairtrade or Responsible Sourcing, provide assurance to consumers that products are sourced ethically and are free from child labor.

*f. Collaboration and industry initiatives*

Collaboration among companies, governments, non-governmental organizations, and civil society is crucial for addressing the complex issue of child labor. Industry-wide initiatives can foster knowledge sharing, promote best practices, and establish common guidelines for ethical sourcing. By joining forces, stakeholders can leverage their collective influence to create systemic change and eradicate child labor from supply chains.

*g. Empowering communities and supporting education*

Long-term solutions to child labor require addressing the root causes and supporting community development. By investing in education, access to healthcare, and social programs, companies and organizations can empower communities to break the cycle of poverty and provide alternative opportunities for children. Education is a powerful tool for combating child labor, as it equips children with the knowledge and skills they need for a brighter future.

### 8.3.5. Empowering consumers to make sustainable and ethical choices

*a. Raising awareness*

Raising awareness about sustainability and ethical issues is the first step in empowering consumers. Education campaigns, media coverage, and public discourse can highlight the environmental, social, and ethical impacts of consumer choices. By understanding the consequences of their actions, consumers can make more informed decisions and consider the long-term implications of their purchases.

*b. Accessible information*

Access to transparent and reliable information is crucial for consumers to make sustainable and ethical choices. Companies should provide clear and accurate labeling, certifications, and product information that disclose the sourcing, production processes, and environmental footprint of their products. Technology platforms, such as mobile apps and online databases, can enable consumers to access this information easily and compare the sustainability credentials of different brands.

*c. Consumer education*

Consumer education plays a vital role in empowering individuals to make sustainable choices. Educational programs, workshops, and online resources can educate consumers about sustainable consumption patterns, the impact of their choices, and the significance of certifications and labels. By equipping consumers with knowledge and critical thinking skills, they can navigate the marketplace and make informed decisions aligned with their values.

*d. Promoting responsible brands*

Promoting and supporting responsible brands is essential for driving sustainable and ethical practices. Companies that prioritize environmental stewardship, social responsibility, and ethical sourcing should be celebrated and rewarded. Through marketing campaigns, consumer advocacy groups, and social media influence, consumers can amplify the voices of responsible brands and create a demand for sustainable and ethically produced goods.

*e. Conscious consumerism*

Conscious consumerism is about aligning personal values with purchasing decisions. It involves considering the environmental impact, labor conditions, and social implications of products and services. Consumers can prioritize products that are environmentally friendly, produced ethically, and contribute positively to communities. By supporting sustainable and ethical businesses, consumers can send a strong market signal and drive industry-wide change.

*f. Collective action*

Collective action can amplify the impact of individual choices. Consumers can join forces with like-minded individuals, consumer advocacy groups, and non-governmental organizations to advocate for sustainable and ethical practices.

By participating in campaigns, petitions, and boycotts, consumers can leverage their collective power to hold companies' accountable, influence industry standards, and shape policy agendas.

*g. Collaboration across stakeholders*

Empowering consumers requires collaboration among various stakeholders, including businesses, governments, civil society organizations, and educational institutions. By working together, these stakeholders can create a supportive ecosystem that promotes sustainable and ethical consumption. Collaborative efforts can involve sharing best practices, conducting research, developing industry standards, and implementing regulatory frameworks that protect consumers and incentivize responsible practices.

## 8.4. Blockchain-based circular economy in agriculture

Blockchain technology offers a suite of powerful tools to drive the transition towards a circular economy. These encompass leveraging blockchain for precise tracking and tracing of recycled materials, facilitating efficient resource utilization and consequent waste reduction, fostering collaborations within circular supply chains, and championing sustainable practices that culminate in the closure of loops, particularly within agricultural contexts. Through these mechanisms, blockchain not only enhances transparency and accountability but also promotes the sustainable utilization of resources (Chauhan, Parida, & Dhir, 2022; Ramadoss, Alam, & Seeram, 2018).

### 8.4.1. Transitioning to a circular economy model

Blockchain technology serves as a catalyst for the evolution from a linear to a circular economy through various strategic interventions. These include the deliberate design for durability and reusability of products, the implementation of strategies aimed at extending product life cycles, the adoption of resource-efficient practices and robust material recovery mechanisms, the embrace of innovative product-as-a-service and sharing models, the promotion of sustainable supply chain practices, and the provision of policy support and economic incentives. By integrating blockchain into these pivotal aspects, industries can effectively facilitate the transition towards a more sustainable and circular economic framework (Kouhizadeh, Zhu, & Sarkis, 2020).

*a. Understanding the circular economy*

The circular economy aims to decouple economic growth from resource consumption by closing the loop of material flows. It emphasizes the principles of designing out waste and pollution, keeping products and materials in use for as long as possible, and regenerating natural systems. Shifting to a circular economy model requires rethinking the entire lifecycle of products, from design and production to consumption and disposal.

*b. Designing for durability and reusability*

Designing products for durability and reusability is a fundamental principle of the circular economy. By creating products that are long-lasting, repairable, and upgradable, their lifespan can be extended, reducing the need for constant replacements. Emphasizing modular designs and standardized components can facilitate easy repairs and component replacements, further extending the product's life cycle.

*c. Implementing product life extension strategies*

Extending the life of products through repair, refurbishment, and remanufacturing is a key strategy in the circular economy. Repair cafes, refurbishment programs, and take-back schemes enable consumers to repair and upgrade their products rather than discard them. Manufacturers can also play a role by offering repair services, providing spare parts, and promoting the reuse of components.

*d. Adopting resource efficiency and material recovery*

Maximizing resource efficiency is central to the circular economy. This involves optimizing production processes, minimizing waste generation, and adopting innovative recycling and material recovery practices. Advanced recycling technologies can transform waste materials into new resources, reducing the demand for virgin materials and decreasing the environmental impact of extraction and production.

*e. Embracing product-as-a-service and sharing models*

The circular economy encourages the transition from ownership to access-based models. Product-as-a-service (PaaS) and sharing models enable consumers to access products when needed, promoting a shift from product ownership to product utilization. Collaborative consumption platforms, such as car sharing and tool libraries, reduce the overall demand for resources and increase the utilization rate of products.

*f. Promoting sustainable supply chains*

Transitioning to a circular economy requires transforming supply chains to be more sustainable and resilient. Businesses can prioritize sourcing materials from recycled or renewable sources, promote efficient logistics and reverse logistics, and establish partnerships for resource sharing and waste valorization.

Collaboration across industries and sectors is key to creating closed-loop supply chains and promoting circularity at a systemic level (Pakseresht, Ahmadi Kaliji, & Xhakollari, 2022).

*g. Policy support and economic incentives*

Government policies and economic incentives play a crucial role in facilitating the transition to a circular economy. Policies can include extended producer responsibility, waste reduction targets, tax incentives for recycling and resource recovery, and regulatory frameworks that promote sustainable production and consumption. Public-private collaborations can drive innovation, support research and development, and provide financial assistance to businesses transitioning to circular practices.

### 8.4.2. Blockchain for tracking and tracing recycled materials

*a. The challenges of tracking and tracing recycled materials*

Tracking and tracing recycled materials throughout the supply chain is essential for maintaining their value and ensuring proper handling. However, traditional systems often lack transparency and face difficulties in verifying the authenticity and quality of recycled materials. Blockchain technology offers a promising solution by providing a decentralized and tamper-proof platform for recording and validating recycling transactions.

*b. Blockchain technology and its features*

Blockchain is a distributed ledger technology that enables secure and transparent record-keeping of transactions. Its key features, such as decentralization, immutability, and consensus mechanisms, make it ideal for tracking and tracing recycled materials. By storing information in a decentralized network, blockchain ensures that records are tamper-proof and accessible to all relevant stakeholders.

*c. Creating digital identities for recycled materials*

Blockchain enables the creation of digital identities for recycled materials, allowing for the seamless tracking of their origin, composition, and transformation throughout the recycling process. Each material can be assigned a unique digital token or identifier that carries essential information, including its type, weight, quality, and previous processing steps. These digital identities provide a comprehensive and verifiable history of the material's journey, increasing trust and accountability.

*d. Establishing chain of custody*

Blockchain facilitates the establishment of a transparent chain of custody for recycled materials. Each time the material changes hands or undergoes a transformation, the transaction is recorded on the blockchain, creating an immutable trail of ownership and custody. This information allows stakeholders to verify the provenance, authenticity, and quality of recycled materials, ensuring compliance with regulations and standards.

*e. Enhancing supply chain traceability*

Blockchain enhances supply chain traceability by providing real-time visibility into the movement of recycled materials. Through smart contracts and Internet of Things (IoT) devices, data on collection, sorting, transportation, and processing can be automatically recorded on the blockchain. This data can include location, temperature, humidity, and other relevant parameters, enabling stakeholders to monitor and optimize recycling processes.

*f. Certifications and standards*

Blockchain can facilitate the integration of certifications and standards into the tracking and tracing of recycled materials. Certificates of origin, quality, and sustainability can be securely stored on the blockchain, allowing for easy verification by stakeholders. This ensures that recycled materials meet specific criteria, such as recycled content percentages or environmental impact indicators, fostering trust and facilitating market access.

*g. Collaboration and industry adoption*

The successful implementation of blockchain for tracking and tracing recycled materials requires collaboration among stakeholders across the recycling ecosystem. Governments, recycling facilities, manufacturers, retailers, and consumers must work together to establish common standards, share data, and integrate blockchain solutions into existing systems. Industry consortia and partnerships can drive innovation, promote knowledge sharing, and accelerate the adoption of blockchain-enabled recycling processes.

### 8.4.3. Enabling efficient resource utilization and waste reduction

*a. Resource efficiency: designing for a circular economy*

Designing products and processes for resource efficiency is a key strategy for minimizing resource consumption and waste generation. Adopting principles such as eco-design, life cycle assessment, and lean manufacturing can help optimize resource use at every stage, from raw material extraction to end-of-life disposal. By maximizing the use of renewable resources, reducing material inputs, and embracing modular designs, we can enhance resource efficiency and reduce waste.

*b. Industrial symbiosis: turning waste into resources*

Industrial symbiosis promotes the sharing of resources, energy, and waste between different industries and businesses. By identifying synergies and establishing symbiotic relationships, one company's waste can become another company's resource. For example, the byproducts of one manufacturing process can serve as inputs for another, reducing the need for virgin materials and minimizing waste disposal. Industrial symbiosis fosters collaboration, reduces costs, and creates a closed-loop system that optimizes resource utilization.

*c. Waste minimization: rethinking consumption and production*

Waste minimization focuses on reducing waste generation at the source. This can be achieved through strategies such as source reduction, product redesign, and process optimization. By adopting practices like waste segregation, recycling, composting, and energy recovery, businesses can divert waste from landfills and create value from otherwise discarded materials. Waste minimization requires a shift in mindset, emphasizing the importance of waste prevention rather than relying solely on end-of-pipe solutions.

*d. Advanced recycling technologies: closing the loop*

Advanced recycling technologies play a vital role in enabling efficient resource utilization and waste reduction. Innovations such as chemical recycling, pyrolysis, and biological processes can transform complex waste streams into high-quality raw materials or energy sources. These technologies have the potential to address challenging waste streams, such as plastics and electronic waste, which are difficult to recycle using traditional methods. By closing the loop through advanced recycling, we can unlock the value of waste and minimize the need for virgin resources.

*e. Data analytics and digitalization: optimizing resource management*

Digitalization and data analytics offer powerful tools for optimizing resource management. By collecting and analyzing data on resource consumption, production processes, and supply chains, businesses can identify inefficiencies, pinpoint opportunities for improvement, and make data-driven decisions. Real-time monitoring, predictive analytics, and smart systems enable the optimization of resource use, energy efficiency, and waste reduction. Digital technologies empower businesses to streamline operations, reduce costs, and enhance sustainability performance.

*f. Circular business models: from ownership to service*

Circular business models, such as product-as-a-service, leasing, and sharing platforms, encourage a shift from ownership to service-based consumption. Instead of buying and discarding products, consumers access goods and services when needed, reducing the demand for resource-intensive production. Circular business models incentivize manufacturers to design products for durability, reparability, and reusability, fostering a circular economy mindset among consumers and businesses.

*g. Stakeholder collaboration and policy support*

Achieving efficient resource utilization and waste reduction requires collaboration among stakeholders and supportive policy frameworks. Collaboration between businesses, governments, research institutions, and civil society organizations can drive innovation, share best practices, and promote knowledge exchange. Governments can play a crucial role by implementing policies that incentivize resource efficiency, establish waste reduction targets, and create a favorable environment for circular economy practices.

### 8.4.4. Facilitating circular supply chain collaborations

*a. The need for circular supply chain collaborations*

Circular supply chain collaborations are essential for achieving a circular economy. By fostering partnerships and cooperation among suppliers, manufacturers, distributors, retailers, and consumers, the entire supply chain can work collectively to optimize resource utilization, minimize waste, and create a closed-loop system. Circular collaborations enhance transparency, innovation, and value creation, while also driving economic and environmental benefits.

*b. Mapping supply chain networks*

A crucial step in facilitating circular supply chain collaborations is mapping the existing supply chain networks. Understanding the flow of materials, resources, and information across the supply chain enables the identification of opportunities for circular interventions. Mapping also helps identify potential collaboration partners and enables the assessment of each actor's impact on resource utilization and waste generation.

*c. Sharing data and information*

Effective circular supply chain collaborations rely on the exchange of data and information. Transparent sharing of data on material inputs, product life cycles, waste generation, and recycling processes enhances supply chain visibility and fosters trust among stakeholders. Technologies such as blockchain, the Internet of Things (IoT), and cloud computing can facilitate secure and real-time data sharing, enabling more efficient decision-making and collaboration.

*d. Co-designing products and packaging*

Collaborative product and packaging design is a key aspect of circular supply chain collaborations. By involving stakeholders from different stages of the supply chain in the design process, products can be optimized for reusability, recyclability, and durability. Co-design efforts ensure that products are designed with the end of their life cycle in mind, facilitating easier disassembly, material recovery, and remanufacturing.

*e. Resource sharing and reverse logistics*

Circular collaborations can be enhanced through resource sharing and reverse logistics initiatives. By sharing underutilized resources, such as manufacturing capacity or transportation infrastructure, supply chain partners can optimize resource utilization and reduce costs. Reverse logistics systems can enable the collection, sorting, and redistribution of products and materials, extending their life cycles and reducing waste.

*f. Circular procurement and supplier engagement*

Circular supply chain collaborations can be strengthened through circular procurement practices and active engagement with suppliers. By integrating circular criteria into procurement processes, organizations can incentivize suppliers to provide products and services with circular attributes. Engaging suppliers in the circularity journey through capacity building, knowledge sharing, and performance evaluation fosters a collective commitment to sustainability across the supply chain.

*g. Monitoring and evaluating circular performance*

To ensure the effectiveness of circular supply chain collaborations, monitoring and evaluation mechanisms are crucial. Key performance indicators (KPIs) can be established to assess resource efficiency, waste reduction, and circularity metrics. Regular monitoring allows stakeholders to identify areas for improvement, track progress, and recognize successful circular initiatives. Sharing performance data and best practices among collaborators encourages continuous learning and innovation.

*h. Policy support and industry standards*

Enabling circular supply chain collaborations requires supportive policy frameworks and industry standards. Governments can incentivize circular practices through regulations, tax incentives, and public procurement policies that prioritize circularity. Industry associations and organizations can develop standards and certifications that guide circular supply chain collaborations, ensuring consistency and accountability across different sectors.

### 8.4.5. Promoting sustainable practices and closing the loop in agriculture

*a. Sustainable agriculture: an overview*

Sustainable agriculture focuses on meeting the needs of the present generation while preserving the ability of future generations to meet their own needs. It encompasses practices that optimize resource use, protect the environment, promote social well-being, and ensure economic viability. Sustainable agriculture emphasizes soil health, water conservation, biodiversity preservation, and the reduction of chemical inputs.

*b. Precision agriculture and digital technologies*

Precision agriculture and digital technologies offer innovative tools for promoting sustainable practices in agriculture. Remote sensing, geographic information systems (GIS), and satellite imagery enable farmers to monitor crop health, optimize inputs, and reduce resource waste. Internet of Things (IoT) devices, drones, and data analytics facilitate real-time decision-making, enhancing efficiency, and minimizing environmental impacts.

*c. Organic farming and agroecology*

Organic farming and agroecology are sustainable farming approaches that prioritize soil health, biodiversity, and natural processes. Organic farming avoids the use of synthetic fertilizers and pesticides, relying instead on organic inputs and ecological pest management. Agroecology integrates ecological principles into farming systems, emphasizing biodiversity, nutrient cycling, and ecosystem services. These approaches promote sustainable land management and reduce chemical pollution.

*d. Closing the loop: nutrient cycling and waste management*

Closing the loop in agriculture involves efficient nutrient cycling and waste management. By implementing practices such as composting, cover cropping, and crop rotation, farmers can recycle organic matter, reduce the need for synthetic fertilizers, and improve soil fertility. Integrated farming systems, such as agroforestry and livestock integration, maximize resource utilization, minimize waste, and enhance ecosystem resilience.

*e. Water conservation and irrigation management*

Water scarcity is a significant challenge in agriculture, particularly in arid and semi-arid regions. Sustainable irrigation practices, such as drip irrigation, precision irrigation, and water-efficient technologies, can minimize water waste and optimize water use. Implementing water-saving strategies, such as rainwater harvesting and water recycling, promotes water conservation and enhances the resilience of agricultural systems.

*f. Sustainable pest and disease management*

Conventional pest and disease management often rely heavily on synthetic pesticides, which can have detrimental effects on the environment and human health. Sustainable pest and disease management approaches prioritize integrated pest management (IPM), biological control, crop rotation, and resistant varieties. These practices reduce pesticide use, preserve beneficial insects, and minimize the development of pesticide resistance.

*g. Circular economy in agriculture: from waste to resources*

Embracing the principles of the circular economy in agriculture involves transforming waste into valuable resources. Agricultural byproducts, such as crop residues and animal manure, can be utilized for energy generation, compost production, or bio-based products. Anaerobic digestion, bioenergy production, and bio-refineries enable the conversion of agricultural waste into renewable energy and high-value bio-based materials.

*h. Education, policy, and stakeholder collaboration*

Promoting sustainable practices and closing the loop in agriculture requires education, policy support, and collaboration among stakeholders. Education and knowledge-sharing initiatives can raise awareness about sustainable farming practices and empower farmers with the necessary skills and information. Governments can develop policies that incentivize sustainable agriculture, provide financial support, and promote research and innovation. Collaboration among farmers, researchers, policymakers, and consumers fosters a collective commitment to sustainability and drives the adoption of sustainable practices (Mangla, Kazançoğlu, Yıldızbaşı, Öztürk, & Çalık, 2022).

Chapter 9

# Case studies and real-world examples

*Unveiling the Blockchain Harvest: Lessons from Real-world Agricultural Transformations* - Unknown

**Abstract**

This chapter provides an in-depth analysis of real-world use cases and successful implementations of blockchain technology in agriculture. It highlights the outcomes, challenges faced, and lessons learned from each case study, offering valuable insights into the practical applications and potential benefits of blockchain in the agricultural industry. It begins by presenting a comprehensive analysis of various real-world use cases of blockchain in agriculture. It explores a range of applications, including supply chain management, traceability, smart contracts, and financial transactions. Moreover, it delves into the outcomes achieved by each case study, showcasing the positive impacts of implementing blockchain. It discusses how blockchain has improved transparency, efficiency, and trust within the agricultural ecosystem and addresses the challenges encountered during the implementation process. Furthermore, It explores the lessons learned from each case study. It emphasizes the key takeaways, best practices, and strategies that have contributed to the success of blockchain implementations.

**Keywords**: Blockchain technology, Agriculture, Real-world use cases, Supply chain management, Traceability, Smart contracts, Financial transactions, Outcomes, Challenges, Lessons learned

**Chapter contents**



***

## 9.1. Successful blockchain implementations in the agriculture industry

### 9.1.1. Overview of successful blockchain applications in agriculture

*a. Supply chain management*

One of the key areas where blockchain has demonstrated immense potential is supply chain management in agriculture. Blockchain enables secure and transparent tracking of products from farm to fork, ensuring that consumers have access to accurate information about the origin, quality, and handling of the produce they consume. By leveraging blockchain technology, companies have successfully eliminated information asymmetry, reduced fraud, and improved efficiency throughout the supply chain (Ali, Chung, Kumar, Zailani, & Tan, 2021; Martínez-Castañeda & Feijoo, 2023; Tian, 2016).

*Case Study: Farm-to-table Traceability*

In this application, blockchain is used to create an immutable record of each stage in the supply chain, including planting, harvesting, processing, and distribution. Consumers can scan a QR code on the product packaging to access detailed information about its journey, such as the farm where it was grown, the date of harvest, transportation conditions, and even the farmer responsible for its cultivation. This transparency enhances trust and allows consumers to make informed choices about the products they purchase.

*b. Smart contracts and payments*

Blockchain technology has enabled the development of smart contracts, which are self-executing agreements with predefined rules stored on the blockchain. These contracts streamline transactions and facilitate secure and transparent payments within the agricultural ecosystem. By automating payment processes and eliminating intermediaries, blockchain-powered smart contracts reduce transaction costs, minimize disputes, and ensure prompt payments to farmers and other stakeholders (Caro et al., 2018; Martínez-Castañeda & Feijoo, 2023).

*Case Study: Fair Trade and Direct Payments*

Through blockchain-based smart contracts, farmers can directly connect with buyers and negotiate fair prices for their produce. Once the agreed-upon conditions are met, the smart contract automatically triggers payment, ensuring that farmers receive fair compensation for their efforts. This eliminates the need for intermediaries and empowers farmers by providing them with greater control over their earnings.

*c. Quality assurance and certification*

Maintaining quality standards and ensuring the authenticity of agricultural products is crucial for both consumers and producers. Blockchain technology offers a robust solution by providing a tamper-proof and verifiable record of certifications and quality assurance processes. This enhances consumer trust, supports fair trade practices, and prevents counterfeit products from entering the market (Antonucci et al., 2019; Detwiler, 2020; Martínez-Castañeda & Feijoo, 2023).

*Case Study: Organic Certification*

Blockchain enables the creation of an immutable ledger that records each step of the organic certification process, from farm inspections to compliance with organic standards. Certifying agencies can securely store and share data on the blockchain, allowing consumers to verify the authenticity of organic products. This transparency ensures that consumers receive genuinely organic products and supports the growth of the organic farming industry.

### 9.1.2. Improved traceability and transparency in the food supply chain

*a. Enhanced traceability*

Traceability is vital in the food industry to identify the origin and journey of a product from farm to table. Traditional traceability systems often rely on paper-based records or centralized databases, which are susceptible to errors, tampering, and limited accessibility. Blockchain technology provides an immutable and decentralized ledger, enabling real-time tracking of food products at every stage of the supply chain (Feng et al., 2020; Martínez-Castañeda & Feijoo, 2023; Saurabh & Dey, 2021).

*Case Study: Seafood Traceability*

Blockchain can revolutionize the seafood industry by ensuring the traceability of fish from catch to consumption. Each fishing vessel, processing facility, and distribution center can record critical information, such as catch location, species, fishing methods, and processing techniques on the blockchain. Consumers can then scan a QR code or use a mobile app to access this information, ensuring that the seafood they consume is sustainably sourced and meets their quality and ethical preferences.

*b. Preventing food fraud*

Food fraud, including the substitution, adulteration, or misrepresentation of food products, poses significant risks to consumer health and trust. Blockchain technology helps combat food fraud by creating an unalterable record of each transaction and product movement within the supply chain. This transparency enables quick identification of fraudulent activities and facilitates targeted interventions.

*Case Study: Olive Oil Authentication*

The olive oil industry has struggled with counterfeit products and mislabeling. Blockchain can address this issue by recording every step of the production process, including harvesting, extraction, and bottling. By scanning a QR code on the bottle, consumers can verify the authenticity of the olive oil, ensuring that it meets quality standards and is sourced from legitimate producers. This authentication process builds consumer confidence and supports fair trade practices.

*c. Supply chain transparency*

Consumers today are increasingly concerned about the social and environmental impact of their food choices. Blockchain technology enables supply chain transparency by capturing and storing data related to sustainability practices, fair trade certifications, and ethical sourcing. This information empowers consumers to make informed decisions and supports companies that prioritize responsible and sustainable practices.

*Case Study: Coffee Supply Chain*

Blockchain can be utilized to create a transparent coffee supply chain, connecting farmers, roasters, distributors, and consumers. Each participant can record relevant data, such as the location of the coffee plantation, farming methods, labor practices, and certifications, on the blockchain. By accessing this information, consumers can choose coffee that aligns with their values, supporting producers who adhere to sustainable and ethical standards.

### 9.1.3. Streamlined logistics and reduced transaction costs

*a. Supply chain optimization*

Blockchain technology offers a decentralized and transparent platform for optimizing supply chain logistics. By providing real-time visibility into inventory levels, transportation routes, and demand patterns, blockchain enables stakeholders to make data-driven decisions, reducing delays, and optimizing resource allocation.

*Case Study: Fresh Produce Logistics*

The transportation of perishable goods, such as fresh produce, requires precise planning and coordination. Blockchain-based platforms allow farmers, distributors, and retailers to track the condition and location of the produce in real time. Temperature sensors and IoT devices can be integrated with the blockchain, ensuring that proper storage conditions are maintained throughout the journey. This transparency and automated monitoring result in reduced spoilage, improved shelf life, and minimized transportation costs.

*b. Eliminating intermediaries*

Traditional transactions often involve multiple intermediaries, such as banks, brokers, and agents, which add complexity and increase costs. Blockchain technology enables peer-to-peer transactions, eliminating the need for intermediaries and reducing transaction fees. Smart contracts, powered by blockchain, automate and enforce the terms of the agreement, ensuring secure and efficient transactions (Bolt, 2019; Martínez-Castañeda & Feijoo, 2023; Stranieri et al., 2021).

*Case Study: Grain Trading*

In the grain trading industry, blockchain can simplify and streamline the buying and selling process. Farmers and buyers can directly connect on a blockchain-based platform, negotiate terms, and execute smart contracts for the purchase of grains. The smart contract ensures that the payment is made to the farmer upon successful delivery and quality confirmation. By eliminating intermediaries, blockchain reduces transaction costs, minimizes delays, and enhances trust between the parties involved.

*c. Cross-border trade and payments*

International trade often faces challenges related to complex documentation, regulatory compliance, and cross-border payments. Blockchain technology provides a secure and efficient platform for international trade by digitizing and automating these processes. Smart contracts on the blockchain can facilitate seamless cross-border payments, reducing transaction costs and eliminating the need for multiple currencies and intermediaries.

*Case Study: Exporting Agricultural Products*

Blockchain simplifies the export process for agricultural products by creating an immutable record of the product's journey, including inspections, certifications, and customs clearance. This transparent and tamper-proof documentation ensures compliance with regulations and reduces the time and costs associated with manual paperwork. Smart contracts can also facilitate instant cross-border payments, ensuring timely compensation for farmers and exporters.

### 9.1.4. Increased trust and consumer confidence in agricultural products

*a. Enhanced food safety and quality assurance*

Foodborne illnesses and contamination incidents can have severe consequences for public health and consumer trust. Blockchain technology provides a transparent and immutable ledger that enables the traceability of agricultural products throughout the supply chain. This traceability ensures that potential issues can be quickly identified, isolated, and resolved, thereby enhancing food safety.

*Case Study: Contaminated Produce Tracing*

In the event of a foodborne illness outbreak linked to specific produce, blockchain enables rapid traceback investigations. By accessing the blockchain, authorities can identify the exact source of contamination, enabling targeted recalls and minimizing the impact on public health. This transparent and efficient process instills consumer confidence and demonstrates the commitment of industry stakeholders to food safety.

*b. Product authenticity and counterfeit prevention*

The prevalence of counterfeit agricultural products, such as organic certifications, high-value crops, and premium goods, poses a significant threat to consumer trust. Blockchain technology enables the creation of an immutable record that verifies the authenticity and origin of products, preventing counterfeiting and misrepresentation.

*Case Study: Organic Product Certification*

Blockchain can be utilized to ensure the integrity of organic product certifications. Certifying agencies can record certification data on the blockchain, including farm inspections, compliance with organic standards, and labeling requirements. Consumers can access this information, providing assurance that the product they purchase is genuinely organic and meets the desired quality standards. This transparency builds trust and encourages consumers to make informed choices.

*c. Ethical sourcing and sustainability*

Consumers are increasingly concerned about the social and environmental impact of the products they purchase. Blockchain technology allows for the transparent recording of information related to ethical sourcing, fair trade certifications, labor practices, and sustainable production methods. This information enables consumers to support producers and brands that align with their values (Chinaka, 2016; Martínez-Castañeda & Feijoo, 2023).

*Case Study: Sustainable Palm Oil Production*

Blockchain can play a vital role in ensuring sustainable palm oil production. By recording each step of the supply chain, from plantation to processing and distribution, blockchain provides transparency into the practices employed. Consumers can verify that the palm oil they purchase is sourced responsibly, without contributing to deforestation or human rights violations. This transparency fosters consumer confidence and encourages sustainable practices throughout the industry.

### 9.1.5. Enhancing sustainability and social responsibility in agriculture

*a. Sustainable farming practices*

Blockchain technology can play a transformative role in promoting sustainable farming practices. By utilizing sensors, Internet of Things (IoT) devices, and data analytics, blockchain enables real-time monitoring and optimization of resource use, including water, energy, and fertilizers. This data-driven approach helps farmers make informed decisions, reduce waste, and mitigate the environmental impact of their operations.

*Case Study: Precision Agriculture*

Blockchain integrated with IoT devices and sensor networks can enable precision agriculture. Farmers can collect data on soil moisture levels, weather conditions, and crop health in real time. This information is securely recorded on the blockchain, allowing farmers to make precise decisions regarding irrigation, fertilization, and pest control. Precision agriculture optimizes resource allocation, minimizes chemical inputs, and reduces the ecological footprint of farming practices.

*b. Supply chain transparency and sustainability*

Blockchain technology provides end-to-end transparency in the agricultural supply chain, enabling the tracking of products from farm to consumer. This transparency supports sustainability efforts by promoting responsible sourcing, reducing food waste, and ensuring fair trade practices. Stakeholders can monitor and verify the environmental and social impact of each stage in the supply chain, fostering greater accountability and encouraging sustainable choices (Khan et al., 2022; Martínez-Castañeda & Feijoo, 2023; Rana et al., 2021).

*Case Study: Sustainable Coffee Supply Chain*

Blockchain facilitates transparency and traceability in the coffee supply chain. From seed to cup, information about coffee plantations, processing methods, certifications, and fair trade practices can be recorded on the blockchain. Consumers can access this data, making informed choices and supporting coffee producers who prioritize sustainability and social responsibility. The transparency provided by blockchain empowers consumers to drive positive change in the industry.

*c. Community engagement and fair compensation*

Blockchain technology can empower farmers and agricultural workers by ensuring fair compensation and enabling direct engagement with consumers. Smart contracts on the blockchain automate payment processes and eliminate intermediaries, ensuring timely and transparent transactions. This fosters trust between producers and consumers, empowering farmers and agricultural workers to receive fair compensation for their efforts.

*Case Study: Direct Farmer-to-consumer Model*

Blockchain-powered platforms enable direct interaction between farmers and consumers, cutting out middlemen and promoting fair compensation for farmers. Farmers can showcase their products, share information about their cultivation practices, and negotiate prices directly with consumers. Smart contracts ensure secure and instant payments, providing financial stability to farmers and fostering a sense of community and collaboration.

## 9.2. Case studies of blockchain adoption by farmers, cooperatives, and organizations

### 9.2.1. Case study 1: Blockchain-based coffee traceability for smallholder farmers

*a. The challenges faced by smallholder coffee farmers*

Small holder coffee farmers, who often operate in remote regions, face numerous challenges in the traditional coffee supply chain. Lack of market access, information asymmetry, and limited bargaining power result in low prices for their coffee beans. Additionally, the complexity of the supply chain makes it difficult to trace the origin and quality of coffee, leading to doubts among buyers and consumers.

*b. Implementing blockchain-based traceability*

By implementing blockchain-based traceability systems, smallholder coffee farmers can overcome these challenges and gain a competitive edge. Each step of the coffee supply chain, from cultivation to export, can be recorded on the blockchain, creating an immutable and transparent ledger accessible to all stakeholders (Cottrill et al., 2018; Mavilia & Pisani, 2022; Tribis, Bouchti, & Bouayad, 2021; van Hilten et al., 2020).

- Farm-level Data: Smallholder farmers can record essential data on the blockchain, such as cultivation practices, harvesting methods, and post-harvest processing. This information helps buyers and consumers understand the quality and sustainability of the coffee produced.
- Cooperative and Certification Data: Cooperatives or farmer associations can record information about the farmer groups involved, certifications obtained (e.g., Fair Trade, Organic), and compliance with social and environmental standards. This data adds credibility and authenticity to the coffee's origin and ethical practices.
- Processing and Logistics Data: Information about the processing facilities, transportation routes, and storage conditions can be recorded on the blockchain. This ensures that the coffee remains traceable and the integrity of the product is maintained throughout the supply chain.

*c. Benefits for smallholder farmers*

Implementing blockchain-based coffee traceability offers several benefits for smallholder farmers:

- Increased Market Access: Transparent supply chains facilitated by blockchain technology provide smallholder farmers with access to broader markets. Buyers and consumers can trace the coffee's journey and verify its quality, resulting in increased demand and fairer prices.
- Trust and Credibility: The immutable nature of blockchain ensures the authenticity and integrity of the coffee's origin, cultivation practices, and certifications. This builds trust among buyers, promoting long-term relationships and market stability.
- Fair Compensation: Through blockchain-based smart contracts, smallholder farmers can receive fair compensation for their coffee beans. Smart contracts automatically execute agreed-upon terms, ensuring prompt and transparent payments to the farmers.

- Empowering Communities: Blockchain-based traceability platforms can foster collaboration and knowledge-sharing among smallholder farmers. Farmers can learn from each other's best practices, access resources, and collectively address challenges, strengthening their community and enhancing their resilience.

### 9.2.2. Case study 2: Using blockchain to combat counterfeit fertilizers in agriculture

*a. The impact of counterfeit fertilizers*

Counterfeit fertilizers are a pressing concern in agriculture, as they undermine the effectiveness of fertilization practices and jeopardize crop health. Farmers unknowingly purchase counterfeit fertilizers that often contain ineffective or harmful substances, resulting in reduced yields, poor crop quality, and economic losses. Traditional methods of verification and quality control often fall short, making it challenging to detect and prevent the distribution of counterfeit products.

*b. Implementing blockchain-based verification and traceability*

Blockchain technology offers a promising solution to combat counterfeit fertilizers by providing a decentralized and transparent platform for verification and traceability. By integrating blockchain into the fertilizer supply chain, each stage of production, distribution, and sale can be recorded on the blockchain, ensuring the integrity and authenticity of the fertilizers.

- Manufacturer and Batch Information: Fertilizer manufacturers can record detailed information about their products on the blockchain, including batch numbers, composition, and production processes. This information serves as a digital fingerprint for each fertilizer batch, enabling easy verification throughout the supply chain.
- Supply Chain Traceability: From manufacturer to retailer, every transaction and movement of fertilizers can be recorded on the blockchain. This creates a transparent and immutable record of the fertilizer's journey, including distribution channels, storage conditions, and handling procedures. Farmers can verify the legitimacy of the product they purchase, ensuring they receive authentic and quality fertilizers.
- Quality Testing and Certifications: Certifications and quality testing results can be securely stored on the blockchain, providing an additional layer of trust and assurance. Farmers can access this information, confirming that the purchased fertilizer meets regulatory standards and specifications.

*c. Benefits for farmers and the industry*

Implementing blockchain technology to combat counterfeit fertilizers offers several benefits for farmers and the agricultural industry:

- Ensuring Authenticity and Quality: Blockchain-based verification and traceability enable farmers to confidently purchase fertilizers that are authentic and of high quality. This improves crop health, maximizes yields, and protects farmers' investments.
- Improved Supply Chain Efficiency: Blockchain streamlines the supply chain by reducing the time and effort required for verification and quality control. With access to transparent information, distributors, retailers, and farmers can make informed decisions and address issues promptly, minimizing delays and disruptions.
- Consumer Confidence: Blockchain's transparency and traceability can help restore consumer confidence in the agricultural industry. Consumers can have assurance that the food they consume is produced using genuine and safe fertilizers, fostering trust and loyalty.
- Industry Collaboration: Blockchain-based platforms can facilitate collaboration among stakeholders, including fertilizer manufacturers, regulatory bodies, and farmers' associations. By sharing data and insights, the industry can collectively combat counterfeit fertilizers and work towards ensuring a safer and more sustainable agricultural ecosystem.

### 9.2.3. Case study 3: Implementing blockchain for fair trade and organic certifications

*a. The challenges in fair trade and organic certifications*

Fairtrade and organic certifications aim to guarantee that agricultural products meet specific social and environmental standards. However, the traditional certification process relies on complex paperwork, manual audits, and limited visibility, leading to inefficiencies, delays, and potential fraudulent practices. Consumers and buyers may have difficulty verifying the authenticity and credibility of certifications.

*b. Implementing blockchain-based certifications*

Blockchain technology can revolutionize the fair trade and organic certification process by providing a decentralized, secure, and transparent platform for certification data. The immutable nature of blockchain ensures the integrity of certification records, enabling easy verification and fostering trust among stakeholders (Cottrill et al., 2018; FAO, 2019; Mavilia & Pisani, 2022; Tribis et al., 2021; Tripoli & Schmidhuber, 2018).

- Certification Data on the Blockchain: Fairtrade and organic certification bodies can record certification data on the blockchain, including farmer information, compliance with standards, audit reports, and inspection records. This creates a tamper-proof digital record accessible to all stakeholders.
- Transparent Supply Chain Tracking: Blockchain enables the recording of supply chain data, including the origin, transportation, processing, and distribution of certified products. Each step of the supply chain can be verified, ensuring that the product maintains its certified status throughout the journey.
- Consumer Access to Certification Information: Through blockchain-based platforms or mobile applications, consumers can access detailed information about fair trade and organic certifications. They can verify the authenticity of certifications, learn about the specific standards met by the product, and make informed purchasing decisions.

*c. Benefits for producers, consumers, and the industry*

Implementing blockchain for fair trade and organic certifications offers numerous benefits for producers, consumers, and the agricultural industry as a whole:

- Enhanced Transparency and Trust: Blockchain's transparency and immutability instill confidence in fair trade and organic certifications. Stakeholders can easily access certification data, ensuring that the products they purchase align with their ethical and sustainability preferences.
- Streamlined Certification Process: Blockchain streamlines the certification process by automating verification, reducing paperwork, and eliminating the need for manual audits. This saves time, reduces costs, and increases efficiency for certification bodies and producers.
- Improved Market Access and Premium Prices: Blockchain-based certifications can open doors to new markets and attract buyers who prioritize sustainability and ethical practices. Producers with transparent and credible certifications are more likely to command premium prices for their products.
- Consumer Empowerment: Blockchain empowers consumers to make informed choices, supporting fair trade and organic practices. By accessing certification information, consumers can support producers who adhere to sustainable and ethical standards, driving positive change in the industry.

### 9.2.4. Case study 4: Blockchain-enabled supply chain management in livestock farming

*a. The challenges in livestock supply chain management*

Livestock supply chains face challenges such as information asymmetry, lack of transparency, and difficulties in tracking and verifying the origin, health, and treatment of animals. Traditional methods of record-keeping and data management often result in delays, errors, and limited visibility, hindering effective supply chain management.

*b. Implementing blockchain-based supply chain management*

Blockchain technology offers a transformative solution for managing the livestock supply chain, providing a decentralized and secure platform for recording and verifying data at each stage of the process (Alshehri, 2023).

- Animal Identity and Health Records: Livestock farmers can record essential information, such as the animal's identity, breed, health records, vaccination history, and feeding practices, on the blockchain. This ensures that accurate and up-to-date information is available throughout the supply chain.
- Traceability and Transparency: Blockchain enables the recording of each transaction, movement, and event in the livestock supply chain. This includes breeding, transportation, feed procurement, and veterinary treatments. Each entry on the blockchain is immutable, creating a transparent and auditable trail of the animal's journey.
- Quality Assurance and Certifications: Certifications, such as organic or animal welfare standards, can be securely stored on the blockchain. This enables easy verification and ensures that consumers have access to accurate and reliable information about the animal's production practices.

*c. Benefits for stakeholders*

Implementing blockchain-enabled supply chain management in livestock farming offers several benefits for the stakeholders involved:

- Consumer Trust and Safety: Blockchain ensures transparency and traceability in the livestock supply chain, giving consumers confidence in the safety, quality, and ethical standards of the meat or dairy products they consume.
- Efficient Recall Management: In the event of a food safety concern or product recall, blockchain enables quick and precise identification of affected animals, minimizing the impact on public health and reducing financial losses for farmers and businesses.

- Regulatory Compliance: Blockchain facilitates compliance with regulations and standards by providing an immutable record of the animal's health, treatment, and origin. This streamlines the auditing process and simplifies reporting for regulatory authorities.
- Supply Chain Efficiency: Blockchain streamlines data management and eliminates manual record-keeping, reducing administrative burdens and enhancing the efficiency of the supply chain. Stakeholders can access real-time information, make informed decisions, and address issues promptly.

### 9.2.5. Case study 5: Decentralized marketplace for direct farmer-consumer transactions

*a. The challenges in traditional agricultural marketplaces*

Traditional agricultural marketplaces are plagued by challenges such as price fluctuations, lack of transparency, and limited access for small-scale farmers. Intermediaries often take a significant portion of the profit, leaving farmers with lower returns, while consumers may have limited information about the origin and quality of the products they purchase.

*b. Implementing a decentralized marketplace*

A decentralized marketplace powered by blockchain technology can address these challenges by facilitating direct interactions and transactions between farmers and consumers.

- Farmer Profiles and Product Listings: Farmers can create profiles on the blockchain-based marketplace, providing information about their farms, cultivation practices, and products. They can list their produce, including detailed descriptions, quantities, and prices.
- Transparent and Immutable Transactions: Blockchain enables transparent and secure transactions, where buyers and farmers can engage directly without the need for intermediaries. Each transaction is recorded on the blockchain, providing an immutable record of the exchange and ensuring trust between the parties involved.
- Smart Contracts and Escrow Services: Smart contracts, programmable agreements on the blockchain, can automate various aspects of the transaction process. They define the terms of the transaction, including pricing, quantity, and delivery conditions. Additionally, escrow services can hold funds in a secure manner until the buyer confirms the satisfactory receipt of the products, ensuring fair and transparent transactions.

- Consumer Reviews and Ratings: Blockchain-based marketplaces can incorporate consumer reviews and ratings for products and farmers. This information helps build trust among consumers and allows farmers to showcase their credibility and quality.

*c. Benefits for farmers and consumers*

Implementing a decentralized marketplace for direct farmer-consumer transactions offers numerous benefits for both parties involved:

- Fairer Prices for Farmers: Eliminating intermediaries enables farmers to receive higher prices for their products, as they have direct access to consumers and can negotiate their own terms.
- Increased Transparency: Blockchain provides transparency throughout the transaction process, allowing consumers to trace the origin, cultivation practices, and journey of the products they purchase. This fosters trust and empowers consumers to make informed choices.
- Support for Small-scale Farmers: Decentralized marketplaces level the playing field for small-scale farmers, providing them with equal opportunities to showcase their products and compete in the market. Direct access to consumers enables them to gain visibility and grow their customer base.
- Enhanced Traceability and Quality Assurance: Consumers can have confidence in the authenticity and quality of the products they purchase. The transparent nature of blockchain allows them to verify the claims made by farmers and make conscious decisions based on reliable information.

## 9.3. Lessons learned and best practices for implementing blockchain in agriculture

### 9.3.1. Identifying suitable use cases for blockchain technology

*a. Criteria for identifying suitable use cases*

To identify suitable use cases for blockchain technology, it is essential to consider the following criteria: the level of decentralization required, the necessity for immutable record-keeping, and the potential for enhancing transparency and trust. By assessing these factors, organizations can pinpoint scenarios where blockchain offers significant advantages over traditional systems. Furthermore, understanding these criteria helps in determining the feasibility and effectiveness of implementing blockchain solutions in various industries (Blokdyk, 2021; Motta et al., 2020):

- Trust and Transparency: Blockchain excels in scenarios where trust and transparency are paramount. Use cases that involve multiple stakeholders, complex transactions, or a need for auditability can benefit from the transparent and immutable nature of blockchain.
- Intermediary Elimination: Blockchain is particularly valuable when there is a desire to eliminate intermediaries or streamline complex processes. If there are unnecessary intermediaries causing inefficiencies, costs, or delays, blockchain can potentially provide a decentralized and direct solution.
- Data Sharing and Collaboration: Use cases that involve multiple entities sharing and collaborating on sensitive data can benefit from blockchain's ability to securely share information while maintaining privacy and control.
- Data Integrity and Security: When data integrity and security are critical, blockchain's immutability and cryptographic properties offer robust protection against tampering, fraud, and unauthorized access.
- Traceability and Supply Chain Management: Blockchain's ability to provide end-to-end traceability and visibility in supply chains is well-suited for use cases where tracking the origin, movement, and authenticity of goods is essential.
- Disintermediation of Trust: Blockchain can be valuable in scenarios where trust traditionally relies on centralized authorities. By decentralizing trust through consensus mechanisms, blockchain enables peer-to-peer interactions and reduces reliance on central entities.

*b. Assessing feasibility and impact*

Once potential use cases are identified, a feasibility and impact assessment should be conducted to evaluate the practicality and potential benefits of implementing blockchain technology.

- Technological Feasibility: Consider whether the required infrastructure, scalability, privacy, and consensus mechanisms can be effectively implemented using existing blockchain platforms or if custom solutions are needed.
- Cost-benefit Analysis: Assess the financial implications of implementing blockchain, including development costs, infrastructure requirements, ongoing maintenance, and potential savings or revenue generation resulting from blockchain adoption.
- User Adoption and Experience: Evaluate the willingness and ability of relevant stakeholders to adopt and interact with blockchain-based systems.

Consider the learning curve, user experience, and potential resistance to change.

- Potential Impact: Determine the potential impact of implementing blockchain technology on the targeted use case. Assess how blockchain can address existing challenges, improve efficiency, enhance security, or create new value propositions.

*c. Iterative approach and collaboration*

Identifying suitable use cases for blockchain technology requires an iterative and collaborative approach. Engage with stakeholders, industry experts, and blockchain professionals to gather insights, brainstorm ideas, and validate assumptions. It is crucial to continuously assess the evolving landscape of blockchain technology, keeping up with advancements and potential solutions that may emerge (Yli-Huumo, Ko, Choi, Park, & Smolander, 2016).

### 9.3.2. Collaboration and partnerships for successful implementation

*a. Importance of collaboration in blockchain implementation*

Collaboration plays a crucial role in blockchain implementation for several reasons:

- Diverse Expertise: Blockchain projects involve a multidisciplinary approach that requires expertise in technology, industry-specific knowledge, regulatory compliance, and business processes. Collaborating with stakeholders who possess diverse skills and experiences enhances the chances of a successful implementation.
- Network Effect: Blockchain's true potential is unlocked when multiple participants join the network. Collaborative efforts enable a broader network of participants, which increases the value and effectiveness of the blockchain solution.
- Shared Resources: Collaborative partnerships allow organizations to pool resources, including funding, technology infrastructure, human capital, and industry connections. Sharing resources mitigates the financial burden and accelerates the development and deployment of blockchain projects.
- Trust and Adoption: Collaboration builds trust among stakeholders, which is essential for blockchain adoption. By working together, organizations can collectively address concerns, align interests, and drive acceptance of blockchain technology within the industry.

*b. Considerations for forming partnerships*

- Identify Complementary Skills and Expertise: When forming partnerships, consider stakeholders who bring complementary skills, expertise, and perspectives to the table. Look for partners with a deep understanding of the industry, technical proficiency, regulatory knowledge, or access to key networks.
- Establish Clear Objectives and Roles: Clearly define the objectives, roles, and responsibilities of each partner in the collaboration. Establishing mutual understanding ensures alignment and effective coordination throughout the blockchain implementation journey.
- Trust and Transparency: Collaboration requires trust and transparency among partners. Establish open lines of communication, share information, and foster an environment that encourages open dialogue and collaboration.
- Governance and Decision-making: Define the governance structure and decision-making processes within the collaboration. Establish mechanisms for resolving conflicts, making consensus-based decisions, and ensuring accountability among partners.
- Legal and Regulatory Considerations: Consider the legal and regulatory implications of the collaboration. Ensure compliance with relevant laws, data privacy regulations, intellectual property rights, and any industry-specific requirements.
- Long-term Sustainability: Evaluate the long-term sustainability of the collaboration. Consider factors such as funding models, ongoing maintenance, scalability, and adaptability to evolving technology and industry trends.

*c. Building a collaborative ecosystem*

- Industry Consortia and Standards Bodies: Engage with industry consortia and standards bodies related to blockchain technology. These organizations provide platforms for collaboration, knowledge sharing, and the establishment of industry-wide standards and best practices.
- Research and Academic Institutions: Collaborate with research institutions and universities with expertise in blockchain technology. Academic partnerships can provide valuable insights, conduct research, and contribute to the development and refinement of blockchain solutions.
- Government and Regulatory Agencies: Engage with government entities and regulatory agencies to ensure compliance and navigate any legal or regulatory hurdles. Collaboration with these stakeholders can foster a

supportive regulatory environment for blockchain implementation (Kshetri, 2022a).

- Technology Providers and Integrators: Collaborate with technology providers and system integrators with expertise in blockchain implementation. These partners can offer technical guidance, support, and integration services to ensure a smooth implementation process.

### 9.3.3. Addressing technical and regulatory challenges

*a. Technical challenges*

Blockchain technology encounters various challenges that warrant attention and resolution. These encompass issues such as scalability, which pertains to the system's ability to handle increasing transaction volumes efficiently, interoperability concerns regarding the seamless interaction between different blockchain platforms and networks, as well as challenges related to privacy and confidentiality, ensuring that sensitive data remains secure and inaccessible to unauthorized parties. Additionally, the significant energy consumption associated with blockchain operations presents a pressing environmental concern that requires mitigation strategies (Zhao et al., 2019).

- Scalability: Blockchain systems face scalability limitations in terms of transaction throughput and network capacity. As the number of transactions increases, it can strain the network and cause delays. Implementing solutions like sharding, off-chain transactions, or layer-two protocols can help address scalability challenges.
- Interoperability: Integrating blockchain systems with existing legacy systems or other blockchain networks can be challenging due to differences in protocols, data formats, and consensus mechanisms. Adopting interoperability standards and leveraging middleware solutions can facilitate seamless data exchange between different systems.
- Privacy and Confidentiality: Public blockchains, by design, offer transparency and immutability, but they lack privacy features. In certain use cases, privacy and confidentiality of sensitive data are crucial. Implementing privacy-enhancing technologies, such as zero-knowledge proofs or secure multiparty computation, can ensure data privacy while leveraging the benefits of blockchain.
- Energy Consumption: Proof-of-work (PoW) consensus algorithms used by some blockchains require significant computational power, resulting in high energy consumption. Exploring alternative consensus mechanisms, such as Proof-of-stake (PoS) or energy-efficient protocols, can mitigate environmental concerns associated with blockchain technology.

*b. Regulatory challenges*

Blockchain technology encounters a myriad of regulatory challenges that demand careful consideration and navigation. These include navigating complex legal frameworks that may vary across jurisdictions, ensuring compliance with stringent data privacy and protection regulations to safeguard sensitive information, addressing issues related to identity verification and digital signatures in a decentralized context, and grappling with jurisdictional challenges that arise due to the global nature of blockchain networks (Queiroz & Wamba, 2019).

- Legal Frameworks: The regulatory landscape for blockchain technology varies across jurisdictions. Lack of clarity or conflicting regulations can pose challenges for organizations implementing blockchain projects. Engaging with regulatory authorities, industry associations, and legal experts can help navigate regulatory complexities and ensure compliance.
- Data Privacy and Protection: Blockchain's immutability and transparency can clash with data privacy regulations, such as the General Data Protection Regulation (GDPR). Implementing privacy-oriented blockchain solutions, pseudonymization techniques, or permissioned networks can help address privacy concerns while adhering to data protection regulations.
- Identity and Digital Signatures: Verifying identities on blockchain platforms and ensuring the validity of digital signatures can be complex. Adhering to identity verification standards and leveraging digital identity solutions can enhance the trustworthiness of blockchain-based transactions.
- Jurisdictional Challenges: Cross-border transactions and the global nature of blockchain systems can raise jurisdictional challenges. Organizations must navigate the legal requirements of different countries, including taxation, cross-border data transfers, and regulatory compliance.

*c. Strategies to address challenges*

- Collaboration and Standards: Collaborate with industry consortia, standardization bodies, and regulatory authorities to establish common standards and best practices. Participating in collaborative efforts facilitates knowledge sharing, addresses technical challenges, and influences regulatory developments.
- Regulatory Compliance: Stay informed about relevant regulations and engage with legal experts to ensure compliance. Proactively design blockchain solutions with privacy and data protection in mind, aligning them with existing regulations and anticipating future regulatory developments.

- Education and Awareness: Educate stakeholders, including decision-makers, users, and the general public, about the benefits and challenges of blockchain technology. Foster awareness about the technical and regulatory considerations involved in blockchain implementation to promote understanding and support.
- Pilot Projects and Proof of Concepts: Conduct pilot projects and proof of concepts to validate the technical feasibility and regulatory compliance of blockchain solutions. Pilot projects enable organizations to identify and address challenges in a controlled environment before scaling up.

### 9.3.4. Ensuring data privacy and security in blockchain applications

*a. Challenges in data privacy and security*

Challenges of blockchain in data privacy and security include: immutable and transparent nature, pseudonymity vs. anonymity, smart contract vulnerabilities, insider threats (Risso et al., 2023).

- Immutable and Transparent Nature: Blockchain's immutability and transparency pose challenges to data privacy. Once data is recorded on the blockchain, it becomes permanently stored and visible to all participants. This feature conflicts with the need to protect sensitive information.
- Pseudonymity vs. Anonymity: Blockchain platforms often employ pseudonymous identifiers rather than real-world identities. While pseudonymity offers privacy benefits, it can be challenging to strike a balance between privacy and accountability, especially in scenarios where legal or regulatory compliance is required.
- Smart Contract Vulnerabilities: Smart contracts, which automate transactions on the blockchain, can be vulnerable to security flaws. Exploiting vulnerabilities in smart contracts can lead to unauthorized access, data breaches, or manipulation of transactions.
- Insider Threats: Organizations implementing blockchain must consider the risk of insider threats, where malicious actors with authorized access misuse or manipulate data on the blockchain. Ensuring appropriate access controls and monitoring mechanisms is crucial.

*b. Strategies for data privacy in blockchain applications*

- Permissioned Blockchains: Implementing permissioned blockchains allows organizations to control access to the network. Only trusted entities or participants with proper authorization can join and validate transactions, enhancing privacy and data confidentiality.

- Off-chain Data Storage: To protect sensitive data, organizations can store it off-chain, outside the blockchain. By storing data off-chain and referencing it through hash values on the blockchain, privacy can be preserved while still maintaining the integrity and transparency of the blockchain.
- Encryption and Data Masking: Employing strong encryption techniques to encrypt sensitive data before storing it on the blockchain provides an additional layer of privacy and security. Data masking, such as replacing sensitive data with pseudonyms, can further protect personal or confidential information.
- Zero-knowledge Proofs: Zero-knowledge proofs enable the verification of information without revealing the underlying data. By using zero-knowledge proofs, blockchain applications can prove the validity of certain data or transactions without disclosing the actual data, ensuring privacy.

*c. Strategies for data security in blockchain applications*

- Secure Smart Contract Development: Implement rigorous security practices during smart contract development to mitigate vulnerabilities. Conduct thorough code audits, adhere to best practices, and implement formal verification techniques to identify and address security flaws.
- Multi-factor Authentication: Implement multi-factor authentication mechanisms to enhance access control and protect against unauthorized access. Multi-factor authentication requires users to provide multiple credentials, such as passwords, biometrics, or hardware tokens, to gain access to blockchain applications.
- Regular Security Audits: Conduct regular security audits of blockchain networks and applications to identify vulnerabilities and address them promptly. Engage independent security experts or third-party auditors to assess the security posture of the blockchain infrastructure and recommend necessary improvements.
- Incident Response and Recovery: Develop robust incident response and disaster recovery plans specific to blockchain applications. This includes protocols for detecting and responding to security breaches, as well as backup and recovery mechanisms to restore data in case of incidents.

*d. Compliance with data protection regulations*

- GDPR Compliance: If blockchain applications involve the personal data of individuals in the European Union, compliance with the General Data Protection Regulation (GDPR) is essential. Implement privacy-oriented features, such as data minimization and user consent management, to meet GDPR requirements.

- Data Retention Policies: Define data retention policies that align with applicable regulations and business requirements. Consider the nature of the data stored on the blockchain and ensure compliance with data protection regulations regarding data storage and retention periods.
- Privacy by Design: Incorporate privacy by design principles into the development and deployment of blockchain applications. Implement privacy features, user-centric controls, and privacy-enhancing technologies from the inception of the project to protect user data.

### 9.3.5. Evaluating the scalability and sustainability of blockchain solutions

*a. Scalability challenges*

- Transaction Throughput: Blockchain networks face limitations in terms of transaction processing speed and capacity. As the number of transactions increases, network congestion and delays can occur. Evaluating the scalability of a blockchain solution involves assessing its ability to handle a higher volume of transactions without compromising performance.
- Network Size and Consensus Mechanisms: The scalability of a blockchain can be affected by the number of participants in the network and the consensus mechanism used. Consensus mechanisms like Proof-of-work (PoW) can limit scalability due to computational requirements, while alternative mechanisms like Proof-of-stake (PoS) or delegated consensus can offer higher scalability.
- Data Storage and Replication: Storing and replicating data across a distributed network can pose scalability challenges. As the size of the blockchain increases, so do the storage and bandwidth requirements. Evaluating the scalability involves considering data management techniques, such as sharding or off-chain storage, to optimize resource utilization.

*b. Strategies for scalability*

Enhancing scalability in blockchain systems necessitates the implementation of various strategic approaches. These encompass the adoption of layered or off-chain solutions to alleviate congestion on the main blockchain, implementing sharding techniques to partition the network for improved throughput, optimizing consensus protocols to enhance transaction processing efficiency, undertaking network upgrades and infrastructure enhancements to bolster overall performance, prioritizing energy efficiency to mitigate environmental footprints, ensuring economic viability for sustainable growth, and fostering robust governance frameworks coupled with active community engagement (Tomu, 2020).

- Layered or Off-chain Solutions: Implementing layer-two solutions, such as payment channels or sidechains, can help alleviate scalability limitations by reducing the burden on the main blockchain. Off-chain solutions enable faster and more efficient transactions without congesting the main blockchain network.
- Sharding: Sharding involves dividing the blockchain network into smaller partitions or shards, each capable of processing a subset of transactions independently. Sharding improves scalability by allowing multiple parallel transactions to occur simultaneously.
- Consensus Protocol Optimization: Evaluating and selecting the appropriate consensus protocol is crucial for scalability. Consensus mechanisms like PoS, delegated proof-of-stake (DPoS), or practical Byzantine fault tolerance (PBFT) offer higher transaction throughput and scalability compared to PoW.
- Network Upgrades and Infrastructure Improvements: Regular upgrades to the blockchain network infrastructure, including improvements in network protocols, bandwidth capacity, and storage capabilities, can enhance scalability.
- Energy Efficiency: Blockchain networks, especially those using PoW consensus, consume significant amounts of energy. Evaluating the sustainability of a blockchain solution involves exploring energy-efficient consensus mechanisms, such as PoS or improved PoW algorithms, to reduce environmental impact.
- Environmental Footprint: Assessing the environmental impact of blockchain solutions involves considering factors such as carbon emissions, electronic waste, and the use of non-renewable resources. Implementing sustainable practices, such as using renewable energy sources or offsetting carbon emissions, can enhance the sustainability of blockchain applications.
- Economic Viability: Sustainable blockchain solutions should be economically viable in the long run. Evaluating the cost-effectiveness of maintaining and operating the blockchain network, including transaction fees, resource requirements, and incentive structures, is crucial for long-term sustainability.
- Governance and Community Engagement: Establishing effective governance mechanisms and engaging the blockchain community in decision-making processes contribute to the sustainability of blockchain solutions. Transparent governance, community participation, and consensus-building foster trust and longevity.

*c. Evaluation framework*

- Scalability Metrics: Define scalability metrics, such as transaction per second (TPS), network latency, or block confirmation time, to measure the scalability of a blockchain solution accurately. Conduct benchmark tests and simulations to assess performance under different transaction loads.
- Sustainability Assessment: Evaluate the sustainability of a blockchain solution by analyzing its environmental impact, energy consumption, economic viability, and governance model. Consider industry standards, best practices, and the specific requirements of the application domain.
- Future-proofing: Assess the potential for future growth and advancements in technology. Evaluate the flexibility and adaptability of the blockchain solution to accommodate scalability improvements and sustainability measures as the technology evolves.

# Chapter 10

# Future trends and challenges

*Forging Ahead: Navigating the Future Intersection of Blockchain and Agriculture* - Unknown

**Abstract**

This chapter explores the future trends and potential developments of blockchain technology in agriculture. It discusses the emerging possibilities and advancements that may shape the adoption and implementation of blockchain in the agricultural industry. Additionally, the chapter examines the challenges and obstacles that stakeholders may encounter in the widespread adoption of blockchain and provides insights on how to overcome these challenges to shape a promising future for blockchain in agriculture. The chapter begins by discussing the future trends and potential developments of blockchain technology in agriculture. Next, the chapter explores the challenges and obstacles that may hinder the widespread adoption of blockchain in agriculture. It addresses concerns related to scalability, energy consumption, regulatory frameworks, data privacy, and interoperability between different systems. Furthermore, the chapter offers insights into how stakeholders can overcome these challenges and shape the future of blockchain in agriculture.

**Keywords**: Blockchain technology, Agriculture, Future trends, Potential developments, Adoption, Implementation, Challenges, Scalability, Energy consumption, Data privacy

**Chapter contents**



***

## 10.1. Emerging trends in blockchain technology and agriculture

The evolving landscape of blockchain technology in agriculture is characterized by several emerging trends, including the Integration of blockchain with emerging technologies, Blockchain interoperability and standardization, Tokenization and fractional ownership of agricultural assets, Blockchain-based data marketplaces for agriculture, and the Integration of sustainability metrics into blockchain systems (Aldrighetti, Canavari, & Hingley, 2021; Zhao et al., 2019). These trends signify a transformative shift towards more efficient, transparent, and sustainable agricultural practices, offering promising opportunities for stakeholders across the industry.

### 10.1.1. Integration of blockchain with emerging technologies

*a. Integration with AI*

- Data Integrity and Provenance: Blockchain can ensure the integrity and immutability of data used in AI systems. By recording data transactions on the blockchain, organizations can track the origin, ownership, and changes made to the data, enabling transparent and auditable data provenance.
- Federated Learning and Data Sharing: Blockchain facilitates secure and decentralized data sharing among AI models and participants. Federated learning, where models are trained locally and only aggregated updates are shared, can be enhanced with blockchain to ensure privacy, data authenticity, and fair reward distribution.
- AI Model Marketplace: Blockchain enables the creation of decentralized marketplaces for AI models, where developers can share, sell, and license their models. Smart contracts and tokenization can automate transactions, intellectual property rights management, and incentivize model development.
- Trust and Explainability: Blockchain's transparency and auditability can enhance trust in AI systems. By recording AI model training data, algorithms, and outcomes on the blockchain, stakeholders can verify the fairness, bias, and explainability of AI decisions.

*b. Integration with IoT*

Blockchain integration with IoT offers a myriad of benefits. These encompass secure data exchange, device identity and authentication, supply chain traceability, and the establishment of decentralized IoT networks (Detwiler, 2020; Galvez et al., 2018; Lezoche, Hernandez, Díaz, Panetto, & Kacprzyk, 2020; Rejeb, Keogh, Zailani, Treiblmaier, & Rejeb, 2020; Sharma, Tripathi, & Mittal, 2022; Torky & Hassanein, 2020).

- Secure Data Exchange: Blockchain can provide a secure and decentralized infrastructure for IoT devices to exchange data and execute transactions autonomously. Smart contracts on the blockchain can enable secure and trustless interactions between IoT devices, ensuring data integrity and privacy.

- Device Identity and Authentication: Blockchain-based identity management solutions can authenticate and validate the identity of IoT devices, preventing unauthorized access and ensuring the integrity of IoT networks. Immutable records on the blockchain can track device ownership, firmware updates, and device lifecycle events.
- Supply Chain Traceability: Blockchain combined with IoT enables end-to-end traceability in supply chains. IoT sensors can collect real-time data on the movement, storage conditions, and quality of goods, while blockchain ensures the transparency and integrity of this data throughout the supply chain.
- Decentralized IoT Networks: Blockchain can support the creation of decentralized IoT networks, where devices can autonomously negotiate and execute transactions. Blockchain's consensus mechanisms can ensure trust and integrity in a distributed IoT ecosystem.

*c. Integration with other emerging technologies*

Integrating blockchain with other emerging technologies holds promise for generating a multitude of benefits. These emerging technologies encompass edge computing, augmented reality (AR) and virtual reality, and quantum computing, as discussed by Akram, Malik, Singh, Anita, and Tanwar (2020) and Alfa, Alhassan, Olaniyi, and Olalere (2021). Such integration heralds the potential for innovative solutions and transformative advancements across various domains.

- Edge Computing: Integrating blockchain with edge computing allows for decentralized data storage and processing at the network edge. Blockchain's tamper-proof and distributed ledger ensure the integrity of data processed and shared across edge devices.
- Augmented Reality (AR) and Virtual Reality (VR): Blockchain can enhance the trust and ownership of digital assets in AR and VR environments. By recording ownership and transaction history on the blockchain, users can prove the authenticity and uniqueness of digital objects.
- Quantum Computing: The integration of blockchain with quantum-resistant cryptography can safeguard blockchain networks against potential threats posed by quantum computing. Quantum-resistant algorithms and protocols can ensure the security of blockchain systems in the future.

*d. Challenges and considerations*

- Scalability: Integrating blockchain with emerging technologies may exacerbate scalability challenges. It is important to assess the performance and scalability of the integrated system, considering the transaction volume, network size, and computational requirements.
- Interoperability: Integration requires seamless interoperability between blockchain and other technologies. Standards and protocols that enable interoperability should be considered to ensure efficient data exchange and compatibility between different systems.
- Data Privacy and Compliance: Integrating blockchain with emerging technologies necessitates careful consideration of data privacy regulations. Organizations must implement privacy-preserving measures and ensure compliance with relevant data protection laws.
- Technical Complexity: Integration may introduce technical complexities, requiring expertise in both blockchain and the respective emerging technology. Collaboration between domain experts and technology specialists is essential to navigate these complexities effectively.

### 10.1.2. Blockchain interoperability and standardization

*a. Understanding blockchain interoperability*

- Definition: Blockchain interoperability refers to the ability of different blockchain networks to communicate, exchange data, and interact with each other seamlessly. It enables the transfer of assets, smart contracts, and information across heterogeneous blockchain platforms.
- Benefits of Interoperability: Interoperability enhances the efficiency, scalability, and usability of blockchain technology. It enables cross-chain transactions, data sharing, and collaboration, fostering innovation and expanding the reach of blockchain applications.
- Interoperability Approaches: Various approaches are being explored to achieve blockchain interoperability, including cross-chain bridges, interoperability protocols, sidechains, atomic swaps, and decentralized exchanges. These mechanisms aim to establish trust and enable secure communication between different blockchain networks (Xiong et al., 2020)

*b. Importance of standardization*

- Definition: Standardization involves the development and adoption of common protocols, formats, and interfaces that enable seamless interoperability between blockchain networks. It establishes a shared language and ensures compatibility across different systems.

- Benefits of Standardization: Standardization simplifies integration efforts, reduces complexity, and facilitates widespread adoption of blockchain technology. It enables interoperability between diverse blockchain networks, fostering collaboration, and unlocking the potential for cross-industry applications.
- Standardization Challenges: Standardizing blockchain technology poses challenges due to the decentralized and evolving nature of the technology. Technical, governance, and regulatory aspects need to be considered to establish robust standards that cater to the diverse needs of blockchain networks (Kamilaris et al., 2019).

*c. Approaches to blockchain interoperability*

- Cross-chain Bridges: Cross-chain bridges facilitate asset transfer and data exchange between different blockchain networks. These bridges act as connectors, allowing tokens and information to move between otherwise isolated chains.
- Interoperability Protocols: Interoperability protocols define a set of rules and standards that enable communication and data exchange between different blockchains. Examples include Polkadot, Cosmos, and Interledger Protocol (ILP), which facilitate interoperability at the protocol level.
- Sidechains: Sidechains are blockchain networks that are connected to a mainchain, enabling the transfer of assets and data between the mainchain and sidechains. This approach allows for scalability and specialized use cases while maintaining interoperability with the main chain.
- Atomic Swaps: Atomic swaps enable peer-to-peer exchange of digital assets across different blockchains without the need for intermediaries. They ensure trustless and secure transactions between participants on separate blockchain networks.
- Decentralized Exchanges (DEXs): DEXs facilitate the trading of digital assets directly between users on different blockchain networks. These exchanges utilize smart contracts and cryptographic mechanisms to enable peer-to-peer transactions.

*d. Standardization efforts*

- Technical Standards: Developing technical standards, such as data formats, cryptographic algorithms, consensus mechanisms, and interoperability protocols, is crucial for seamless blockchain interoperability.
- Governance and Regulatory Standards: Establishing governance models and regulatory frameworks that encourage cooperation, interoperability, and compliance across blockchain networks ensures the effective implementation of standardized practices.
- Industry Collaborations: Industry collaborations, consortiums, and open-source initiatives play a vital role in driving blockchain standardization efforts. Collaborative platforms foster consensus-building, knowledge-sharing, and the development of interoperability frameworks.

### 10.1.3. Tokenization and fractional ownership of agricultural assets

*a. Understanding tokenization*

- Definition: Tokenization involves representing real-world assets, such as land, crops, livestock, or agricultural machinery, as digital tokens on a blockchain. Each token represents a fraction of ownership or investment in the underlying asset.
- Benefits of Tokenization:

  Tokenization presents a plethora of advantages, spanning from fractional ownership and increased accessibility to enhanced liquidity, and transparency and trust, as underscored by Lin et al. (2017). By leveraging tokenization, individuals and entities stand to unlock new avenues for asset ownership, foster greater market participation, bolster liquidity, and instill confidence through transparent, trustworthy transactions.

+ Fractional Ownership: Tokenization enables the division of assets into smaller units, allowing multiple individuals or entities to own fractions of an asset. This fractional ownership model enhances liquidity and opens investment opportunities to a broader range of investors.
+ Increased Accessibility: Tokenization democratizes access to agricultural assets, breaking down traditional barriers to entry. Investors of all sizes can participate in the agricultural sector, diversifying their portfolios and supporting the growth of the industry.
+ Enhanced Liquidity: Tokenization introduces liquidity to traditionally illiquid agricultural assets. Token holders can trade their ownership stakes on secondary markets, providing a mechanism for buying, selling, and transferring ownership more efficiently.

+ Transparency and Trust: Blockchain's transparent and immutable nature ensures the integrity of ownership records and transaction history. Token holders can verify ownership rights, track asset performance, and gain trust in the agricultural investment ecosystem.

*b. Tokenization in agriculture*

- Agricultural Asset Classes: Various agricultural assets can be tokenized, including farmland, plantations, livestock, machinery, and even agricultural produce. Each asset class presents unique opportunities and considerations for tokenization.
- Real Estate Crowdfunding: Tokenization enables fractional ownership of farmland or agricultural real estate, allowing investors to participate in the agricultural sector without the need for large capital investments. Crowdfunding platforms and marketplaces facilitate the issuance and trading of agricultural tokens.
- Commodity Trading: Tokenizing agricultural commodities, such as grains, coffee, or cotton, streamlines trading, enhances supply chain transparency, and enables fractional ownership of these commodities. Tokens represent ownership rights to physical or digital inventories.
- Livestock and Animal Products: Tokenizing livestock allows for fractional ownership and investment in animals, such as cattle or poultry. It provides opportunities for farmers, investors, and consumers to participate in livestock production and gain from the associated value chain.

*c. Challenges and implications*

- Regulatory Frameworks: Tokenization and fractional ownership of agricultural assets require careful consideration of regulatory frameworks. Compliance with securities laws, property rights, and agricultural regulations is crucial to ensure legal and regulatory compliance.
- Asset Valuation: Determining the value of tokenized agricultural assets can be challenging. Factors such as land quality, productivity, market conditions, and demand-supply dynamics must be carefully evaluated to arrive at accurate asset valuations.
- Market Liquidity: The liquidity of secondary markets for tokenized agricultural assets depends on factors such as market size, investor demand, and trading platforms. Developing robust marketplaces and establishing investor confidence contribute to the liquidity of tokenized assets.

- Technological Infrastructure: Tokenization requires a reliable technological infrastructure, including blockchain networks, smart contracts, and secure storage solutions for digital assets. Ensuring scalability, security, and interoperability is vital for the smooth functioning of tokenized agricultural markets.

*d. Future perspectives*

- Fractional Farming and Cooperative Models: Tokenization can facilitate cooperative farming models, where multiple stakeholders collectively own and manage agricultural assets. Fractional ownership and tokenization enable collaboration, risk-sharing, and collective decision-making in the agricultural sector.
- Sustainable and Impact Investing: Tokenization can promote sustainable and impact investing in agriculture. Investors can choose to support environmentally friendly farming practices, fair trade initiatives, or organic agriculture by investing in tokenized assets aligned with their values.
- Data-driven Decision Making: Tokenization provides a wealth of data on agricultural assets, including performance metrics, environmental impact, and market trends. Analyzing this data can empower stakeholders with valuable insights for informed decision-making and efficient resource allocation.

### 10.1.4. Blockchain-based data marketplaces for agriculture

*a. Understanding data marketplaces*

- Definition: A data marketplace is a platform that facilitates the buying and selling of data assets between data providers and data consumers. It acts as an intermediary, ensuring secure and transparent transactions while preserving data privacy and ownership rights.
- Benefits of Data Marketplaces:

+ Access to Diverse Data: Data marketplaces provide a centralized hub for aggregating and accessing diverse agricultural data from various sources, such as farms, weather stations, satellite imagery, and IoT devices. This enables data consumers to access comprehensive and varied datasets for analysis and decision-making.
+ Data Monetisation: Data providers, including farmers, agricultural organizations, and research institutions, can monetize their data assets by

offering them for sale on data marketplaces. This incentivizes data sharing and promotes collaboration within the agricultural ecosystem.

+ Improved Decision-making: Data consumers, such as farmers, agri-businesses, and researchers, can leverage data marketplaces to access high-quality, real-time data for informed decision-making. By analyzing and integrating data from multiple sources, stakeholders can optimize farming practices, mitigate risks, and enhance productivity.
+ Data Security and Privacy: Blockchain-based data marketplaces ensure the integrity and security of data transactions. Smart contracts and encryption mechanisms preserve data privacy, enabling data providers to maintain control over their data while allowing selective sharing with trusted consumers.

*b. Blockchain technology in data marketplaces*

Blockchain assumes a pivotal role within data marketplaces, offering a suite of functionalities that encompass transparent and immutable transactions, data ownership and licensing, data quality and verification, and data access control and privacy, as articulated by (Dutta, Choi, et al., 2020) and Upadhyay (2020). Through blockchain integration, data marketplaces can uphold integrity, streamline ownership protocols, ensure data veracity, and fortify access controls, thereby fostering trust and efficiency within the data economy.

- Transparent and Immutable Transactions: Blockchain technology ensures transparency and immutability of data transactions within the marketplace. All transactions, including data purchases, usage, and ownership transfers, are recorded on the blockchain, providing a trusted and auditable history of data transactions.
- Data Ownership and Licensing: Smart contracts on the blockchain enable data providers to define ownership rights, licensing terms, and usage restrictions for their data assets. This allows for automated and transparent royalty payments and ensures fair compensation for data providers.
- Data Quality and Verification: Blockchain-based data marketplaces can implement mechanisms to verify the quality and authenticity of data. Smart contracts can incorporate validation processes, such as data certification, reputation systems, and consensus mechanisms, to establish trust in the data being transacted.
- Data Access Control and Privacy: Blockchain enables granular control over data access and privacy. Data providers can selectively share data with specific consumers while retaining control over their data assets. Encryption and access control mechanisms can be implemented to safeguard sensitive data.

*c. Potential applications of blockchain-based data marketplaces in agriculture*

The potential applications of blockchain-based data marketplaces in agriculture are highly promising, encompassing vital domains such as weather and climate data, soil and crop data, supply chain traceability, and market and price data (Al-Amin, Sharkar, Kaiser, & Biswas, 2021; Xu, Li, et al., 2022). By leveraging blockchain technology, the agricultural sector stands to benefit from enhanced data reliability, transparency in supply chains, and informed decision-making processes, ultimately fostering sustainability and efficiency across the industry.

- Weather and Climate Data: Farmers can access localized weather and climate data from various sources, enabling better crop management, irrigation planning, and pest control.
- Soil and Crop Data: Data marketplaces can facilitate the sharing of soil health information, nutrient levels, and crop performance data. This supports precision agriculture practices, optimizing fertilizer usage and crop yield.
- Supply Chain Traceability: Blockchain-based data marketplaces can enhance supply chain traceability by enabling the exchange of data related to product origin, quality certifications, and transportation records. This promotes transparency and trust in the agricultural supply chain.
- Market and Price Data: Farmers and agribusinesses can access real-time market data, including commodity prices, demand trends, and trading volumes. This empowers them to make informed marketing and pricing decisions.

*d. Challenges and considerations*

- Data Quality and Standardization: Ensuring the quality, accuracy, and standardization of data is crucial for building trust and facilitating meaningful data transactions within the marketplace. Establishing data quality protocols and incentivizing data providers can address this challenge.
- Governance and Regulatory Compliance: Blockchain-based data marketplaces must navigate regulatory frameworks, data protection laws, and intellectual property rights. Collaboration with regulatory bodies and adherence to privacy and security standards are essential.
- Interoperability and Integration: Integrating diverse data sources and systems into a unified marketplace poses technical challenges. Standards, APIs, and interoperability frameworks need to be established to ensure seamless integration and data compatibility.
- Data Privacy and Security: Implementing robust privacy and security measures to protect sensitive data from unauthorized access or tampering

is critical. Encryption, data anonymization techniques, and secure data storage solutions are essential components of a secure data marketplace.

### 10.1.5. Integration of sustainability metrics blockchain systems

*a. Understanding sustainability metrics*

- Definition: Sustainability metrics are quantitative or qualitative measures that assess the environmental, social, and economic impacts of agricultural practices. These metrics evaluate factors such as carbon emissions, water usage, biodiversity conservation, fair labor practices, and resource efficiency.
- Importance of Sustainability Metrics: Sustainability metrics provide a standardized framework for evaluating and monitoring sustainability performance. They enable stakeholders to track progress, identify areas for improvement, and make data-driven decisions to enhance sustainable practices in agriculture.

*b. Integration of sustainability metrics into blockchain systems*

- Enhanced Transparency: Blockchain technology provides a transparent and immutable ledger, making it ideal for recording and verifying sustainability metrics. By integrating sustainability data into blockchain systems, stakeholders gain access to accurate and tamper-proof sustainability records (Menon & Jain, 2021).
- Immutable Audit Trail: Sustainability metrics recorded on the blockchain create an immutable audit trail that can be accessed and verified by relevant stakeholders. This enhances trust and accountability, as sustainability claims can be verified through transparent and traceable data.
- Smart Contracts for Compliance: Smart contracts can be used to automate compliance with sustainability standards and regulations. These contracts can enforce rules related to emissions, water usage, waste management, and other sustainability metrics, ensuring adherence to predefined benchmarks.
- Supply Chain Transparency: Integrating sustainability metrics into blockchain systems allows for comprehensive supply chain transparency. Stakeholders can trace the entire lifecycle of agricultural products, verifying sustainability claims at each stage of the supply chain, from production to consumption.
- Consumer Engagement: Blockchain-based systems can empower consumers by providing access to detailed sustainability information. With transparency and verifiable metrics, consumers can make informed purchasing decisions

aligned with their sustainability preferences, supporting environmentally conscious agricultural practices.

*c. Benefits of integrating sustainability metrics into blockchain systems*

- Data Integrity and Trust: Blockchain ensures the integrity and immutability of sustainability data, eliminating concerns of data manipulation or falsification. This fosters trust among stakeholders, facilitating collaborative efforts towards sustainable agriculture.
- Efficient Data Sharing: Blockchain-based systems enable efficient and secure sharing of sustainability metrics across multiple entities in the agricultural ecosystem. This streamlines data collection, aggregation, and reporting processes, eliminating the need for manual reconciliation and enhancing data accuracy.
- Performance Monitoring and Benchmarking: Integrating sustainability metrics into blockchain systems facilitates real-time monitoring of sustainability performance. This enables stakeholders to track progress, compare performance against benchmarks, and identify areas for improvement.
- Incentivising Sustainable Practices: Blockchain systems can introduce token-based incentives to reward sustainable agricultural practices. Tokens can be issued to farmers or organizations that achieve specific sustainability targets, fostering a culture of sustainability and promoting positive behavior change.

*d. Challenges and considerations*

- Data Standardization: Standardizing sustainability metrics is crucial to ensure consistent and comparable data across different agricultural operations. Developing industry-wide standards for sustainability metrics promotes data compatibility and facilitates meaningful comparisons.
- Data Collection and Verification: Reliable data collection mechanisms and verification processes are essential to maintain the accuracy and credibility of sustainability metrics. IoT devices, sensors, satellite imagery, and third-party audits can aid in data collection and verification.
- Privacy and Data Confidentiality: Integrating sustainability metrics into blockchain systems requires careful consideration of data privacy and confidentiality. Implementing privacy-preserving techniques, such as selective data disclosure and encryption, helps protect sensitive information while ensuring transparency.

- Scalability and Interoperability: As sustainability data volume increases, ensuring the scalability and interoperability of blockchain systems becomes crucial. Scalable blockchain architectures and interoperability protocols enable seamless integration with existing data management systems.

### 10.2. Potential challenges and barriers to blockchain adoption

The adoption of blockchain technology may encounter a spectrum of challenges and barriers. These encompass technical hurdles and scalability issues, regulatory and legal complexities, data privacy and security apprehensions, adoption obstacles particularly affecting small-scale farmers and stakeholders, and resistance to change within existing frameworks (FAO, 2019; Sahebi, Masoomi, & Ghorbani, 2020). Addressing these multifaceted challenges calls for a holistic approach, integrating technological innovation with regulatory clarity, stakeholder engagement, and capacity-building initiatives to foster widespread acceptance and utilization of blockchain solutions across diverse sectors.

#### 10.2.1. Technical challenges and scalability issues

*a. Blockchain technical challenges*

Blockchain technology also encounters various technical challenges, spanning scalability, performance, energy consumption, interoperability, and data privacy and security (Prewett, Prescott, & Phillips, 2020). Addressing these hurdles is crucial for advancing the widespread adoption and effective implementation of blockchain solutions across diverse industries.

- Scalability: Scalability is one of the most significant challenges facing blockchain technology. As the number of transactions increases, blockchain networks may face limitations in processing capacity and transaction throughput. This can result in delays, increased transaction costs, and reduced overall system efficiency.
- Performance: Blockchain networks may experience performance issues due to the computational overhead required for consensus mechanisms and cryptographic operations. Slow transaction confirmation times and network congestion can hinder the smooth operation of blockchain applications in agriculture.
- Energy Consumption: The energy consumption of blockchain networks, particularly proof-of-work (PoW) consensus mechanisms, has raised concerns about the environmental impact. Energy-intensive mining operations can undermine the sustainability goals of the agricultural industry.

- Interoperability: Achieving interoperability between different blockchain networks and legacy systems is a complex challenge. The lack of standardized protocols and formats can hinder data exchange and integration between disparate systems, limiting the seamless flow of information across the agricultural ecosystem.
- Data Privacy and Security: While blockchain technology provides inherent security through immutability and consensus mechanisms, protecting sensitive data and ensuring privacy can be challenging. Striking a balance between transparency and privacy is crucial, especially when dealing with personal or confidential agricultural data.

*b. Scalability issues in agriculture*

- Data Volume: The agricultural sector generates vast amounts of data, ranging from crop monitoring information to supply chain data. Scaling blockchain systems to handle and process this massive volume of data in real-time poses a significant challenge.
- Integration with IoT: The Internet of Things (IoT) plays a crucial role in agriculture by collecting data from sensors, drones, and other devices. Integrating blockchain with IoT technologies and scaling the infrastructure to handle the continuous influx of IoT data presents scalability challenges.
- Supply Chain Complexity: Agricultural supply chains are complex, involving multiple stakeholders, processes, and data sources. Scaling blockchain systems to accommodate the diverse requirements and interactions within the supply chain can be demanding.
- Transaction Speed: In time-sensitive agricultural processes, such as real-time monitoring and decision-making, fast transaction speeds are essential. Scaling blockchain networks to improve transaction-processing times without compromising security is a challenge that needs to be addressed.

*c. Potential solutions and mitigation strategies*

- Layer 2 Scaling Solutions: Layer 2 solutions, such as state channels and sidechains, offer scalability enhancements by processing transactions off-chain while leveraging the security of the underlying blockchain. These solutions can alleviate the burden on the main blockchain network and improve transaction throughput.
- Consensus Algorithm Optimization: Exploring alternative consensus mechanisms, such as proof-of-stake (PoS), can reduce energy consumption and improve scalability. PoS-based blockchains can achieve higher transaction speeds and scalability while maintaining security.

- Interoperability Protocols: Developing standardized interoperability protocols and frameworks can facilitate seamless data exchange and integration between different blockchain networks and legacy systems. This allows for efficient collaboration and data sharing across the agricultural ecosystem.
- Off-chain Data Storage: Offloading non-essential or bulky data to off-chain storage solutions, such as distributed file systems or cloud storage, can improve scalability by reducing the storage requirements on the main blockchain network.
- Hybrid Approaches: Combining blockchain technology with other technologies, such as cloud computing or edge computing, can address scalability issues by leveraging the strengths of each approach. This hybrid model allows for efficient data processing and storage while maintaining the security and immutability of the blockchain.

### 10.2.2. Regulatory and legal considerations

*a. Regulatory frameworks*

Regulatory frameworks encompass a spectrum of considerations, comprising jurisdictional compliance, data protection and privacy, anti-money laundering, know your customer regulations, and intellectual property rights Sternberg, Hofmann, and Roeck (2021). These frameworks serve as essential pillars for ensuring legal adherence, safeguarding user privacy, combating illicit activities, and protecting intellectual assets within the realm of blockchain technology and its applications.

- Jurisdictional Compliance: Different countries and regions have varying regulations concerning data privacy, security, intellectual property, and financial transactions. Understanding and complying with jurisdiction-specific regulations is crucial when implementing blockchain solutions in agriculture.
- Data Protection and Privacy: Blockchain's transparency can conflict with data protection laws and privacy regulations. Striking a balance between the immutable nature of blockchain and the need to comply with data protection requirements is essential. Implementing privacy-enhancing technologies and adhering to data privacy regulations can help address these concerns.
- Anti-money Laundering (AML) and Know Your Customer (KYC) Regulations: Blockchain-based solutions involving financial transactions, tokenization, or cryptocurrency usage must adhere to AML and KYC regulations.

Implementing robust identity verification processes and transaction monitoring mechanisms can help ensure compliance with these regulations.

- Intellectual Property Rights: Blockchain technology can facilitate the transfer and management of intellectual property rights in agriculture. Understanding the legal frameworks surrounding patent, copyright, and trademark laws is crucial to protect intellectual property assets and prevent infringement.

*b. Smart contracts and legal contracts*

- Legal Enforceability: Smart contracts, while self-executing and tamper-resistant, must align with existing legal frameworks to ensure their enforceability. Legal contracts based on smart contracts should be drafted carefully, taking into account applicable laws and regulations, to ensure their validity and enforceability in case of disputes.
- Contractual Clarity: Smart contracts should clearly define the rights, obligations, and conditions of the parties involved. Ambiguities and loopholes should be addressed to prevent misunderstandings and potential legal issues. Legal expertise is essential in drafting smart contracts that accurately reflect the intent of the parties and comply with legal requirements.
- Dispute Resolution: In the event of disputes arising from smart contracts or blockchain-based transactions, mechanisms for dispute resolution must be established. Exploring alternative dispute resolution methods, such as arbitration or mediation, can provide efficient and cost-effective resolution processes.

*c. Governance and liability*

- Decentralized Governance: Decentralized blockchain networks may lack a central authority, raising questions about governance and decision-making processes. Establishing governance frameworks and consensus mechanisms that align with legal requirements can address these concerns.
- Liability and Accountability: Blockchain-based systems involve multiple stakeholders, and determining liability and accountability can be complex. Clear delineation of roles, responsibilities, and liabilities among stakeholders is crucial to address potential legal issues and ensure accountability for actions taken on the blockchain.
- Regulatory Reporting and Compliance: Blockchain-based solutions that involve financial transactions, supply chain management, or certification processes may require regulatory reporting and compliance. Establishing

mechanisms to generate accurate reports, maintain records, and adhere to regulatory requirements is essential.

*d. International considerations*

- Cross-border Transactions: Blockchain technology enables seamless cross-border transactions and collaborations. However, varying legal frameworks, trade regulations, and data protection laws across jurisdictions can complicate international operations. Understanding the legal requirements of different countries and implementing mechanisms to ensure compliance is crucial.
- Export Controls and Sanctions: Blockchain applications in agriculture should consider export controls and sanctions imposed by different countries. Compliance with trade regulations and restrictions on the transfer of certain agricultural products or technologies is necessary to avoid legal implications.

### 10.2.3. Data privacy and security concerns

*a. Data privacy challenges*

Data privacy presents a series of intricate challenges within the blockchain ecosystem, encompassing the immutable nature of blockchain, the delicate balance between pseudonymity and anonymity, and the protection of personally identifiable information (Mirabelli & Solina, 2020). Addressing these challenges requires innovative approaches that balance the transparency and security inherent to blockchain technology while upholding individuals' rights to privacy and data protection.

- Immutable Nature of Blockchain: Blockchain's immutability poses challenges to data privacy. Once data is recorded on the blockchain, it becomes permanent and cannot be altered or erased. Ensuring that sensitive agricultural data, such as personal information, trade secrets, or proprietary data, is not publicly accessible is critical.
- Pseudonymity vs. Anonymity: Blockchain transactions are often pseudonymous, meaning that users are identified by their cryptographic addresses rather than their real identities. However, the potential for data linkage and analysis can compromise privacy. Balancing pseudonymity with the need for transparency and accountability is a challenge.
- Personally Identifiable Information (PII): Agricultural systems may involve the collection and storage of personally identifiable information, such as farmer data, supply chain details, or consumer preferences. Safeguarding

this information and ensuring compliance with data protection regulations, such as the General Data Protection Regulation (GDPR), is crucial.

*b. Security considerations*

- Distributed Nature of Blockchain: Blockchain's distributed architecture offers inherent security benefits by eliminating single points of failure. However, it also introduces vulnerabilities, such as the potential for 51% attacks, Sybil attacks, or consensus algorithm vulnerabilities. Implementing robust security measures to protect against these threats is essential.
- Private Key Management: Blockchain relies on cryptographic keys for access and authentication. Safeguarding private keys is critical to prevent unauthorized access or theft of digital assets. Secure key management practices, such as hardware wallets or multi-signature schemes, must be implemented to protect against private key compromise.
- Smart Contract Vulnerabilities: Smart contracts, although designed to be tamper-resistant, can contain coding errors or vulnerabilities. Exploiting these vulnerabilities can lead to financial losses or unauthorized access to data. Conducting rigorous code audits, implementing security best practices, and utilizing formal verification methods can mitigate these risks.
- Malicious Actors and Insider Threats: The agricultural sector is not immune to malicious actors seeking to exploit vulnerabilities or insider threats. Vigilance in identifying and addressing potential threats, implementing access controls, and conducting regular security audits can mitigate risks posed by these actors.

*c. Mitigation strategies*

- Privacy-preserving Techniques: Implementing privacy-preserving techniques, such as zero-knowledge proofs, ring signatures, or homomorphic encryption, can enhance data privacy on the blockchain. These techniques allow for selective data disclosure while maintaining confidentiality.
- Consent and Data Minimisation: Obtaining informed consent from individuals and stakeholders before collecting and using their data is crucial. Implementing data minimization strategies by collecting only necessary information and anonymizing data whenever possible helps protect privacy.
- Secure Identity Management: Implementing robust identity management systems, including strong authentication mechanisms and user verification processes, helps prevent unauthorized access and protects sensitive data.

- Continuous Security Monitoring: Regular security assessments, penetration testing, and monitoring of blockchain networks and applications are essential to detect and mitigate potential security vulnerabilities or attacks promptly.
- Collaboration with Security Experts: Engaging cybersecurity experts and collaborating with industry peers, regulatory bodies, and academia can help address emerging security threats, share best practices, and stay abreast of evolving security measures.

### 10.2.4. Adoption barriers for small-scale farmers and stakeholders

*a. Limited technological infrastructure*

The adoption of blockchain and its applications may face impediments due to limited technological infrastructure. Such constraints encompass factors like access to stable internet connectivity, specific hardware infrastructure requirements, and the need for technical expertise (Jardim, Pranto, Ruivo, & Oliveira, 2021). Overcoming these barriers necessitates investments in infrastructure development and educational initiatives to ensure broader accessibility and proficiency in utilizing blockchain technology effectively.

- Connectivity: Access to stable Internet connectivity is crucial for utilizing blockchain applications. However, in many rural areas, small-scale farmers face limited or unreliable Internet access, hindering their ability to connect and participate in blockchain networks.
- Hardware Requirements: Blockchain systems may require specific hardware infrastructure, such as computers or smartphones, to access and interact with the technology. Small-scale farmers with limited access to such devices may face challenges in adopting blockchain solutions.
- Technical Expertise: Blockchain technology can be complex, requiring technical knowledge and skills for implementation and operation. Small-scale farmers may lack the necessary technical expertise, making it difficult for them to adopt and utilize blockchain solutions effectively.

*b. Financial constraints*

Access to blockchain and its applications may be impeded by financial constraints, which encompass factors such as the cost of implementation and lack of capital (Khan, Haleem, Husain, Samson, & Pathak, 2023; Rijanto, 2021). Overcoming these challenges requires innovative financing models, collaborative efforts, and strategic resource allocation to ensure that cost barriers do not limit the potential benefits of blockchain technology for individuals and organizations alike.

- Cost of Implementation: Implementing blockchain solutions often involves significant upfront costs, including infrastructure setup, system integration, and ongoing maintenance. Small-scale farmers with limited financial resources may find it challenging to invest in the necessary infrastructure and technology.
- Lack of Capital: Access to capital is essential for small-scale farmers to adopt and leverage blockchain solutions effectively. Limited financial resources and difficulty in securing loans or investments can pose significant barriers to adoption.

*c. Limited awareness and education*

- Knowledge Gap: Many small-scale farmers may lack awareness and understanding of blockchain technology and its potential benefits. Bridging the knowledge gap through education and awareness programs is crucial for encouraging adoption and empowering farmers with the necessary information.
- Training and Support: Lack of training and technical support can hinder the successful adoption of blockchain solutions. Providing training programs and ongoing support to small-scale farmers can help them navigate the technology and address any challenges they encounter.

*d. Trust and perception*

- Trust in Technology: Blockchain is still a relatively new technology, and there may be skepticism or lack of trust among small-scale farmers regarding its reliability, security, and long-term viability. Building trust in blockchain solutions through transparent communication and successful use cases is essential.
- Perception of Complexity: Small-scale farmers may perceive blockchain technology as complex and difficult to understand. Simplifying the user experience and providing user-friendly interfaces can help overcome this perception and encourage adoption.

*e. Interoperability and standardization*

- Integration with Existing Systems: Small-scale farmers often operate within complex agricultural ecosystems involving multiple stakeholders and legacy systems. Integrating blockchain solutions with existing systems and ensuring interoperability can be challenging, potentially hindering adoption.
- Lack of Standardization: The absence of standardized protocols and formats for data exchange and smart contracts can create compatibility issues and

limit the seamless integration of blockchain technology into existing agricultural processes.

*f. Regulatory and legal considerations*

- Compliance Challenges: Small-scale farmers may face challenges in navigating the regulatory and legal frameworks associated with blockchain technology. Understanding and complying with data protection, financial, and agricultural regulations are crucial for ensuring regulatory compliance.
- Intellectual Property Rights: Concerns around intellectual property rights and ownership of data can arise with blockchain implementation. Addressing these concerns and providing clarity on ownership rights is important to build trust and encourage adoption.

### 10.2.5. Overcoming resistance to change and fostering collaboration

*a. Building awareness and demonstrating benefits*

- Education and Communication: Educating stakeholders about the potential benefits of blockchain technology is crucial in overcoming resistance. Clear and concise communication about how blockchain can improve efficiency, transparency, and trust in agricultural systems can help alleviate concerns and generate enthusiasm for adoption.
- Use Case Demonstrations: Showcasing successful use cases and real-world examples of blockchain implementations in agriculture can be persuasive in overcoming skepticism. Demonstrating tangible benefits, such as improved traceability, reduced fraud, or streamlined processes, can help stakeholders visualize the value of blockchain technology.

*b. Stakeholder engagement and collaboration*

- Inclusive Decision-making: Involving all relevant stakeholders in the decision-making process fosters a sense of ownership and encourages collaboration. Engaging farmers, industry associations, regulators, technology providers, and researchers ensures that diverse perspectives are considered, making the implementation process more inclusive and successful.
- Public-private Partnerships: Collaboration between the public and private sectors is essential to drive the adoption of blockchain technology in agriculture. Governments can provide support through policy frameworks, funding initiatives, and creating an enabling environment, while private sector entities can bring technological expertise and resources to implement blockchain solutions.

- Industry Standards and Consortia: Establishing industry standards and consortia dedicated to blockchain technology in agriculture promotes collaboration and knowledge sharing. These platforms enable stakeholders to collaborate, develop common protocols, and address industry-wide challenges collectively, fostering a supportive ecosystem for blockchain adoption.

*c. Addressing concerns and mitigating risks*

Ensuring trust in blockchain technology necessitates the proactive addressing of concerns such as data privacy and security, as well as the prudent mitigation of risks such as gradual implementation and pilots to circumvent the pitfalls associated with large-scale deployment (Rugeviciute & Mehrpouya, 2019). By systematically managing these challenges, stakeholders can instill confidence in the reliability and integrity of blockchain solutions, paving the way for broader adoption and sustained innovation in various sectors.

- Data Privacy and Security: Addressing concerns around data privacy and security is crucial for fostering trust in blockchain technology. Implementing robust data protection measures, encryption techniques, and privacy-enhancing technologies can help alleviate these concerns and demonstrate a commitment to safeguarding stakeholders' data.
- Interoperability: Ensuring interoperability among different blockchain platforms and systems promotes collaboration and data sharing across the agricultural ecosystem. Developing common data formats, application programming interfaces (APIs), and interoperability frameworks enables seamless integration and data exchange between stakeholders.
- Gradual Implementation and Pilots: Starting with small-scale pilot projects allows stakeholders to experience the benefits of blockchain technology firsthand, mitigating the risk associated with large-scale implementation. Gradually expanding the implementation based on the success of these pilots builds confidence and reduces resistance to change.

*d. Capacity building and support*

- Training and Skill Development: Providing training programs and workshops on blockchain technology equips stakeholders with the necessary knowledge and skills to adopt and utilize the technology effectively. Technical training, awareness sessions, and mentorship programs can empower individuals and organizations to embrace blockchain solutions.
- Technical Support and Assistance: Offering technical support and assistance during the implementation phase helps address challenges and ensures the smooth adoption of blockchain technology. Access to dedicated support

channels, forums, and documentation can assist stakeholders in resolving issues and maximizing the value of blockchain solutions.

- Incentives and Rewards: Providing incentives, such as grants, subsidies, or recognition, can motivate stakeholders to adopt blockchain technology. Rewarding early adopters and showcasing their success stories can inspire others to overcome resistance and actively participate in the blockchain revolution.

*10.3. Predictions for the future of blockchain in agriculture*

Forecasts regarding the future trajectory of blockchain in agriculture are diverse and expansive. These encompass increased adoption of blockchain in global food supply chains, the proliferation of blockchain-enabled smart contracts facilitating autonomous transactions, the broadening of blockchain applications to encompass emerging markets, the evolution of blockchain-based certification and labeling systems, and the revolutionizing of agricultural finance and investment through blockchain technology (Friedman & Ormiston, 2022; Kopyto, Lechler, von der Gracht, & Hartmann, 2020). These prognostications underscore the transformative potential of blockchain in revolutionizing various facets of the agricultural sector, paving the way for enhanced efficiency, transparency, and sustainability.

### 10.3.1. Increased adoption of blockchain in global food supply chains

*a. Drivers for adoption*

The drivers propelling the adoption of blockchain in global food supply chains are diverse and multifaceted. These encompass consumer demand for transparency, the imperative for food safety and quality assurance, the growing emphasis on sustainability and ethical sourcing, and the pursuit of supply chain efficiency and cost reduction (Cottrill et al., 2018; Singh & Sharma, 2023). Recognizing these driving forces underscores the critical role blockchain technology plays in reshaping and enhancing the integrity, sustainability, and efficiency of food supply chains worldwide.

- Consumer Demand for Transparency: Consumers are increasingly concerned about the origins and quality of their food. Blockchain technology provides a transparent and immutable record of every step in the supply chain, addressing consumer demand for traceability and information about the products they purchase.
- Food Safety and Quality Assurance: Blockchain enables real-time monitoring and verification of food safety and quality parameters. Rapid identification

of potential issues, such as contamination or spoilage, allows for timely interventions, reducing the risk of foodborne illnesses and product recalls.

- Sustainability and Ethical Sourcing: Blockchain can facilitate the tracking of sustainable and ethically sourced ingredients or materials throughout the supply chain. Consumers are more conscious of sustainability practices, and blockchain offers a mechanism to verify and validate such claims, fostering trust and supporting responsible sourcing initiatives.
- Supply Chain Efficiency and Cost Reduction: Blockchain streamlines supply chain operations by automating processes, reducing paperwork, and minimizing manual interventions. Improved efficiency leads to cost savings, optimized inventory management, and faster response times, benefiting all participants in the supply chain.

*b. Benefits of blockchain adoption*

- Enhanced Traceability and Transparency: Blockchain creates an immutable and auditable record of every transaction and event within the supply chain. This transparency enables stakeholders to trace the journey of products, verify authenticity, and identify bottlenecks or inefficiencies.
- Improved Food Safety and Quality: By capturing data at each stage of the supply chain, blockchain enables real-time monitoring of critical parameters, such as temperature, humidity, or storage conditions. Rapid identification of deviations ensures timely corrective actions, reducing the risk of compromised food safety and quality.
- Efficient Recall Management: In the event of a product recall, blockchain enables rapid and precise identification of affected batches. By pinpointing the exact origin and destination of the products, recalls can be limited to the affected lots, reducing waste and minimizing the impact on consumers and businesses.
- Supply Chain Optimization: Blockchain facilitates better coordination and collaboration among supply chain participants. Smart contracts, automated workflows, and real-time data sharing enable seamless coordination, reducing delays, errors, and costs associated with manual processes.
- Trust and Brand Reputation: Blockchain's transparency and traceability instill trust in consumers, fostering brand loyalty and reputation. Companies can leverage blockchain to showcase their commitment to quality, sustainability, and ethical practices, differentiating themselves in the market.

*c. Challenges and considerations*

- Scalability: As global food supply chains involve numerous transactions and participants, scalability becomes a challenge for blockchain technology. Ensuring that blockchain networks can handle the volume and velocity of data generated by these supply chains is crucial for successful adoption.
- Interoperability and Standardization: Interoperability among different blockchain platforms and standardization of data formats are essential for seamless data exchange and collaboration across diverse supply chain networks. Establishing common protocols and industry standards is a key consideration for widespread adoption.
- Data Privacy and Security: Protecting sensitive data while maintaining transparency is a delicate balance. Robust data privacy measures, encryption techniques, and access controls are necessary to ensure the security of data shared on the blockchain.
- Collaboration and Governance: Effective collaboration among supply chain stakeholders, including producers, manufacturers, distributors, and retailers, is essential for blockchain implementation. Establishing governance frameworks and defining roles, responsibilities, and data-sharing agreements foster trust and cooperation.

### 10.3.2. Blockchain-enabled smart contracts for autonomous transactions

*a. Understanding smart contracts*

- Definition and Characteristics: Smart contracts are programmable contracts that execute automatically when predefined conditions are met. They are stored on a blockchain and enforced by its underlying consensus mechanism. Smart contracts are immutable, transparent, and self-verifiable, providing security and eliminating the need for intermediaries.
- Benefits of Smart Contracts: Smart contracts offer several advantages, including increased efficiency, reduced costs, enhanced security, and improved trust. They eliminate the need for manual verification and execution of contracts, minimizing errors, delays, and disputes.
- Turing Completeness: Smart contracts built on blockchain platforms such as Ethereum are Turing complete, meaning they can perform any computation that a regular computer can. This feature enables the execution of complex business logic and the automation of various types of transactions.

*b. Autonomous transactions enabled by smart contracts*

- Payment and Financial Transactions: Smart contracts can facilitate autonomous payment and financial transactions without the need for intermediaries. By automatically verifying and executing payment conditions, smart contracts enable secure and instant transfers of digital assets, cryptocurrencies, and traditional currencies.
- Supply Chain Management: Smart contracts can automate supply chain transactions, including procurement, inventory management, and shipping. By defining conditions and triggering actions based on real-time data inputs, smart contracts enable transparent and efficient supply chain operations.
- Intellectual Property and Royalty Management: Smart contracts can automate the management and distribution of intellectual property rights and royalties. By codifying the terms of agreements, tracking usage, and executing automatic payments, smart contracts ensure fair and transparent compensation for creators and rights holders.
- Decentralized Exchanges and Marketplaces: Smart contracts enable the creation of decentralized exchanges and marketplaces, where buyers and sellers can interact directly without intermediaries. By automating transaction settlements and enforcing predefined rules, smart contracts enhance trust and efficiency in peer-to-peer exchanges.

*c. Considerations and challenges*

- Code Security and Auditing: Smart contracts are vulnerable to coding errors and security breaches. Comprehensive testing, code auditing, and adherence to best practices are essential to minimize risks and ensure the reliability of smart contracts.
- Scalability and Performance: As blockchain networks grow, scalability becomes a challenge for executing smart contracts. Ensuring that blockchain platforms can handle a large volume of transactions and compute complex computations efficiently is critical for autonomous transactions at scale.
- Legal and Regulatory Considerations: Smart contracts raise legal and regulatory considerations, as traditional legal frameworks may not fully accommodate their unique characteristics. Clarifying the legal status of smart contracts and addressing issues such as liability and dispute resolution are important for their broader adoption.
- User Experience and Adoption: User-friendly interfaces and simplified interaction models are crucial for the widespread adoption of blockchain-

enabled smart contracts. Improving the user experience and abstracting the complexity of smart contract execution can encourage broader participation from individuals and businesses.

### 10.3.3. Expansion of blockchain applications to emerging markets

*a. Bridging financial inclusion*

- Cross-border Payments: Blockchain technology can facilitate faster, more secure, and cost-effective cross-border payments, especially in regions with limited access to traditional banking services. By leveraging blockchain's decentralized nature, emerging markets can benefit from efficient remittance systems and reduced transaction fees.
- Financial Services for the Unbanked: Blockchain-based platforms enable the creation of decentralized financial services, such as digital wallets, lending, and microinsurance, that can reach the unbanked populations in emerging markets. These solutions provide access to financial services, promote economic empowerment, and foster entrepreneurship.

*b. Enhancing supply chain transparency*

- Agricultural Value Chains: Blockchain technology can enhance transparency and traceability in agricultural value chains in emerging markets. From farm to fork, blockchain-based solutions enable small-scale farmers to record and verify their products' origin, quality, and certifications, creating trust and expanding market opportunities.
- Fair Trade and Ethical Sourcing: Blockchain can enable transparent verification of fair trade and ethical sourcing practices in emerging markets. By recording and immutably documenting supply chain data, blockchain allows consumers to make informed choices and supports sustainable and socially responsible production.

*c. Strengthening public services*

- Identity Management: Blockchain-based identity management systems can provide individuals in emerging markets with secure and verifiable digital identities. This empowers citizens to access government services, financial inclusion, healthcare, and educational opportunities while minimizing identity fraud.
- Public Procurement: Blockchain can improve transparency and efficiency in public procurement processes in emerging markets. By recording procurement transactions on an immutable ledger, blockchain ensures

fairness, reduces corruption, and enhances accountability in government spending.

*d. Enabling trustworthy digital transactions*

- E-commerce and Marketplace Platforms: Blockchain can foster trust in E-commerce and online marketplace transactions in emerging markets. By leveraging smart contracts and decentralized escrow services, blockchain ensures secure transactions, dispute resolution, and protection against fraud.
- Intellectual Property Protection: Emerging markets often face challenges related to intellectual property protection. Blockchain can provide a decentralized and transparent platform for creators to register, protect, and track their intellectual property rights, fostering innovation and creativity.

*e. Overcoming challenges and considerations*

- Infrastructure and Connectivity: The adoption of blockchain in emerging markets may require investments in digital infrastructure and connectivity. Addressing challenges related to Internet access, bandwidth, and power supply is crucial for the widespread adoption and usability of blockchain technology.
- Education and Awareness: Promoting education and awareness about blockchain technology among stakeholders in emerging markets is essential. Capacity-building initiatives, training programs, and partnerships with educational institutions can empower individuals and organizations to leverage blockchain effectively.
- Regulatory Environment: Establishing clear and supportive regulatory frameworks is crucial for the expansion of blockchain applications in emerging markets. Governments need to foster an enabling environment that encourages innovation while addressing concerns related to privacy, security, and legal compliance.

### 10.3.4. Evolution of blockchain-based certification and labeling systems

*a. Traditional certification and labeling challenges*

- Trust and Verification: Traditional certification and labeling systems often rely on centralized authorities, making it challenging for consumers to verify the accuracy and authenticity of claims. Lack of transparency and trust can lead to skepticism and confusion among consumers.

- Complexity and Paperwork: The manual and paper-based nature of traditional certification and labeling processes can be time-consuming, error-prone, and costly for businesses. The complexity of compliance requirements and the need for multiple audits and documentation create administrative burdens.
- Counterfeiting and Fraud: Traditional certification and labeling systems are susceptible to counterfeiting and fraudulent practices. Without robust mechanisms for verifying the authenticity of certifications and labels, consumers and businesses face the risk of purchasing counterfeit or misleading products.

*b. The rise of blockchain-based certification and labeling systems*

- Immutable and Transparent Records: Blockchain technology provides a decentralized and immutable ledger where certification and labeling data can be recorded transparently. Every transaction and modification is securely documented, enhancing the integrity and reliability of certification information.
- Verification and Trustworthiness: Blockchain enables stakeholders to verify the authenticity of certifications and labels by tracing them back to their source. Immutable records on the blockchain provide a tamper-proof audit trail, building trust and confidence among consumers.
- Supply Chain Traceability: Blockchain-based certification and labeling systems enable end-to-end traceability of products throughout the supply chain. By recording each stage of production, processing, and distribution on the blockchain, stakeholders can ensure compliance with standards and sustainability practices.
- Consumer Empowerment: Blockchain-based systems empower consumers to make informed choices by providing access to detailed information about product origins, manufacturing processes, and sustainability practices. This transparency enables conscious consumerism and supports ethical purchasing decisions.

*c. Application across industries*

- Food and Agriculture: Blockchain-based certification and labeling systems can enhance the traceability of food products, ensuring safety, quality, and adherence to organic or fair trade standards. Consumers can verify the origin of ingredients, production methods, and ethical sourcing practices.
- Fashion and Textiles: Blockchain enables the tracking of the entire supply chain in the fashion and textiles industry, ensuring transparency and

ethical sourcing of materials. Certification and labeling systems can verify fair labor practices, sustainable manufacturing, and supply chain transparency.

- Renewable Energy: Blockchain-based certification systems can validate the generation and consumption of renewable energy, ensuring accurate reporting and supporting carbon credit trading. Transparency in energy production and usage promotes sustainability and incentivizes renewable energy adoption.
- Healthcare and Pharmaceuticals: Blockchain can enhance the integrity of certifications and labels in the healthcare and pharmaceutical industries. Patients and consumers can verify the authenticity of drugs, medical devices, and certifications related to quality standards and safety.

*d. Collaboration and standardization*

- Industry Collaboration: Blockchain-based certification and labeling systems require collaboration among industry stakeholders, including producers, manufacturers, certification bodies, and regulators. Establishing industry-wide partnerships and data-sharing agreements is crucial for standardizing practices and ensuring interoperability.
- Standards and Protocols: Developing common standards and protocols for blockchain-based certification and labeling systems is essential for seamless integration and interoperability. Standardization facilitates data exchange, supports interoperability among different systems, and ensures trust and transparency.

### 10.3.5. Transformation of agricultural finance and investment through blockchain

*a. Challenges in agricultural finance*

- Limited Access to Credit: Small-scale farmers often face challenges in accessing affordable credit due to a lack of collateral, credit history, and documentation. Traditional financial systems may overlook these farmers, hindering their ability to invest in their farms and improve productivity.
- Lack of Transparency: The lack of transparency in agricultural finance can lead to information asymmetry, making it difficult for investors to assess risks and farmers to access fair financing. This opacity hampers the flow of capital and limits investment opportunities in the agricultural sector.
- Inefficient Processes: Traditional financial processes in agricultural finance can be slow, cumbersome, and costly. Manual paperwork, intermediaries,

and complex verification procedures contribute to delays, inefficiencies, and increased transaction costs.

*b. Blockchain solutions in agricultural finance*

- Decentralized Financing Platforms: Blockchain-based platforms enable decentralized financing solutions, such as peer-to-peer lending and crowdfunding, connecting farmers directly with lenders and investors. Smart contracts facilitate automated loan disbursements, repayment schedules, and transparent monitoring of funds utilization.
- Tokenization of Agricultural Assets: Blockchain enables the tokenization of agricultural assets, such as land, crops, and livestock. By representing these assets as digital tokens, fractional ownership and investment opportunities are created, allowing investors to diversify their portfolios and farmers to unlock liquidity.
- Supply Chain Financing: Blockchain can facilitate supply chain financing by providing transparent and immutable records of transactions throughout the agricultural value chain. Smart contracts automate payment settlements, reducing the risk for financiers and enabling faster access to funds for farmers.
- Data-driven Credit Scoring: Blockchain's ability to securely store and verify data opens up possibilities for data-driven credit scoring models. By leveraging farmers' transactional and production data recorded on the blockchain, financial institutions can assess creditworthiness and offer tailored financial products.

*c. Benefits and impact*

- Increased Access to Financing: Blockchain-based agricultural finance solutions can provide small-scale farmers with access to affordable credit and investment opportunities. By leveraging alternative data and bypassing traditional barriers, blockchain expands financial inclusion in the agricultural sector.
- Improved Transparency and Trust: Blockchain enhances transparency by providing a decentralized and immutable ledger for financial transactions. Investors can verify the authenticity of agricultural projects, while farmers can showcase their track records and achievements, fostering trust between stakeholders.
- Efficiency and Cost Reduction: Blockchain streamlines and automates processes, reducing paperwork, eliminating intermediaries, and minimizing transaction costs. This efficiency leads to faster financing decisions, quicker

disbursements, and lower administrative burdens for farmers and financial institutions.

- Risk Mitigation: Blockchain's transparent and traceable nature helps mitigate risks in agricultural finance. Immutable records on the blockchain enable better risk assessment, fraud detection, and prevention of double financing, contributing to more secure and reliable financial transactions.

*d. Regulatory and adoption considerations*

- Regulatory Frameworks: The adoption of blockchain in agricultural finance requires supportive regulatory frameworks that address legal, privacy, and security concerns. Regulators need to provide clarity on the use of smart contracts, digital assets, and investor protection to foster a conducive environment.
- Collaborative Ecosystems: Collaboration among financial institutions, technology providers, governments, and agricultural stakeholders is crucial for successful adoption. Establishing partnerships and building ecosystems that leverage blockchain technology can accelerate its integration into existing financial systems.
- Education and Capacity Building: Educating farmers, investors, and financial institutions about blockchain's benefits and functionalities is essential for widespread adoption. Capacity-building programs can facilitate understanding, promote best practices, and ensure efficient utilization of blockchain-based agricultural finance solutions.

# References


Agrawal, T. K., Kumar, V., Pal, R., Wang, L., & Chen, Y. (2021). Blockchain-based Framework for Supply Chain Traceability: A Case Example of Textile and Clothing Industry. *Computers & Industrial Engineering, 154,* 107130. doi: https://doi.org/10.1016/j.cie.2021.107130

Ahmed, R. A., Hemdan, E. E. D., El-Shafai, W., Ahmed, Z. A., El-Rabaie, E. S. M., & Abd El-Samie, F. E. (2022). Climate-smart Agriculture Using Intelligent Techniques, Blockchain and Internet of Things: Concepts, Challenges, and Opportunities. *Transactions on Emerging Telecommunications Technologies, 33*(11), e4607. doi:https://doi.org/10.1002/ett.4607

Akanfe, O., Lawong, D., & Rao, H. R. (2024). Blockchain Technology and Privacy Regulation: Reviewing Frictions and Synthesizing Opportunities. *International Journal of Information Management, 76,* 102753. doi:https://doi.org/10.1016/j.ijinfomgt.2024.102753

Akram, S. V., Malik, P. K., Singh, R., Anita, G., & Tanwar, S. (2020). Adoption of Blockchain Technology in Various Realms: Opportunities and Challenges. *Security and Privacy, 3*(5), e109. doi:https://doi.org/10.1016/j.tifs.2019.07.034

Al-Amin, S., Sharkar, S. R., Kaiser, M. S., & Biswas, M. (2021). *Towards A Blockchain-based Supply Chain Management for E-agro Business System.* Paper presented at the Proceedings of International Conference on Trends in Computational and Cognitive Engineering: Proceedings of TCCE 2020.

Al Sadawi, A., Madani, B., Saboor, S., Ndiaye, M., & Abu-Lebdeh, G. (2021). A Comprehensive Hierarchical Blockchain System for Carbon Emission Trading Utilizing Blockchain of Things and Smart Contract. *Technological Forecasting and Social Change, 173,* 121124. doi:https://doi.org/10.1016/j.techfore.2021.121124

Aldrighetti, A., Canavari, M., & Hingley, M. K. (2021). A Delphi Study on Blockchain Application to Food Traceability. *International Journal on Food System Dynamics, 12*(1), 6-18. doi:https://doi.org/10.18461/ijfsd.v12i1.72

Alfa, A. A., Alhassan, J. K., Olaniyi, O. M., & Olalere, M. (2021). Blockchain Technology in IoT Systems: Current Trends, Methodology, Problems, Applications, and Future Directions. *Journal of Reliable Intelligent Environments, 7*(2), 115-143. doi:https://doi.org/10.1007/s40860-020-00116-z

Ali, M. H., Chung, L., Kumar, A., Zailani, S., & Tan, K. H. (2021). A Sustainable Blockchain Framework for The Halal Food Supply Chain: Lessons from Malaysia. *Technological Forecasting and Social Change, 170,* 120870. doi:https://doi.org/10.1016/j.techfore.2021.120870

Alshehri, M. (2023). Blockchain-assisted Internet of Things Framework in Smart Livestock Farming. *Internet of Things, 22,* 100739. doi:https://doi.org/10.1016/j.iot.2023.100739

Antonucci, F., Figorilli, S., Costa, C., Pallottino, F., Raso, L., & Menesatti, P. (2019). A Review on Blockchain Applications In The Agri-food Sector. *Journal of the Science of Food and Agriculture, 99*(14), 6129-6138. doi:https://doi.org/10.1002/jsfa.9912

Astill, J., Dara, R. A., Campbell, M., Farber, J. M., Fraser, E. D., Sharif, S., & Yada, R. Y. (2019). Transparency in Food Supply Chains: A Review of Enabling Technology Solutions. *Trends in Food Science & Technology, 91*, 240-247. doi:https://doi.org/10.1016/j.tifs.2019.07.024

Awazi, N. P., Tchamba, M. N., & Avana, T. M.-L. (2019). Climate Change Resiliency Choices of Small-scale Farmers in Cameroon: Determinants and Policy Implications. *Journal of Environmental Management, 250*, 109560. doi:https://doi.org/10.1016/j.jenvman.2019.109560

Awotunde, J. B., Misra, S., Ayoade, O. B., Ogundokun, R. O., & Abiodun, M. K. (2022). Blockchain-based Framework for Secure Medical Information in Internet of Things System *Blockchain Applications in the Smart Era* (pp. 147-169): Springer.

Awwad, M., Kalluru, S. R., Airpulli, V. K., Zambre, M. S., Marathe, A., & Jain, P. (2018). *Blockchain Technology for Efficient Management of Supply Chain.* Paper presented at the Proceedings of the International Conference on Industrial Engineering and Operations Management.

Babbar, S., & Prasad, S. (1998). International Purchasing, Inventory Management and Logistics Research. *International Journal of Operations & Production Management, 18*(1), 6-36. doi:10.1108/01443579810192763

Badea, L., & Mungiu-Pupăzan, M. C. (2021). The Economic and Environmental Impact of Bitcoin. *IEEE Access, 9*, 48091-48104. doi:10.1109/ACCESS.2021.3068636

Balzarova, M. A. (2021). Blockchain Technology-A New Era of Ecolabeling Schemes? *Corporate Governance: The International Journal of Business in Society, 21*(1), 159-174. doi:https://doi.org/10.1108/CG-08-2020-0328

Baralla, G., Ibba, S., Marchesi, M., Tonelli, R., & Missineo, S. (2019). *A Blockchain Based System to Ensure Transparency and Reliability in Food Supply Chain.* Paper presented at the Euro-Par 2018: Parallel Processing Workshops: Euro-Par 2018 International Workshops, Turin, Italy, August 27-28, 2018, Revised Selected Papers 24.

Barrett, C. B. (2010). Smallholder Market Participation: Concepts and Evidence from Eastern and Southern Africa *Food security in Africa*: Edward Elgar Publishing.

Behnke, K., & Janssen, M. (2020). Boundary Conditions for Traceability in Food Supply Chains Using Blockchain Technology. *International Journal of Information Management, 52*, 101969. doi:https://doi.org/10.1016/j.ijinfomgt.2019.05.025

Belen-Saglam, R., Altuncu, E., Lu, Y., & Li, S. (2023). A Systematic Literature Review of The Tension Between the GDPR and Public Blockchain Systems. *Blockchain: Research and Applications, 4*(2), 100129. doi:https://doi.org/10.1016/j.bcra.2023.100129

Ben-Daya, M., Hassini, E., Bahroun, Z., & Banimfreg, B. H. (2020). The Role of Internet of Things in Food Supply Chain Quality Management: A Review. *Quality management journal, 28*(1), 17-40. doi:https://doi.org/10.1080/10686967.2020.1838978

Bermeo-Almeida, O., Cardenas-Rodriguez, M., Samaniego-Cobo, T., Ferruzola-Gómez, E., Cabezas-Cabezas, R., & Bazán-Vera, W. (2018). *Blockchain in Agriculture: A Systematic Literature Review.* Paper presented at the Technologies and Innovation: 4th International Conference, CITI 2018, Guayaquil, Ecuador, November 6-9, 2018, Proceedings 4.

Bhardwaj, A., Shah, S. B. H., Shankar, A., Alazab, M., Kumar, M., & Gadekallu, T. R. (2021). Penetration Testing Framework for Smart Contract Blockchain. *Peer-to-Peer Networking and Applications, 14*, 2635-2650. doi:https://doi.org/10.1007/s12083-020-00991-6

Bhat, S. A., Huang, N.-F., Sofi, I. B., & Sultan, M. (2022). Agriculture-Food Supply Chain Management Based on Blockchain and IoT: A Narrative on Enterprise Blockchain Interoperability. *Agriculture, 12*(1), 1-40. doi:https://doi.org/10.3390/agriculture12010040

Blokdyk, G. (2021). *Blockchain In Agriculture A Complete Guide – 2019*: 5STARCooks.

Bolt, J. (2019). *Financial Resilience of Kenyan Smallholders Affected by Climate Change, and The Potential for Blockchain Technology*. Retrieved from https://edepot.wur.nl/472583

Borah, M. D., Naik, V. B., Patgiri, R., Bhargav, A., Phukan, B., & Basani, S. G. (2020). Supply Chain Management in Agriculture Using Blockchain and IoT. *Advanced applications of blockchain technology*, 227-242. doi:https://doi.org/10.1007/978-981-13-8775-3_11

Borselli, A. (2020). *Smart Contracts in Insurance: A Law and Futurology Perspective* (Vol. 1): Springer.

Bumblauskas, D., Mann, A., Dugan, B., & Rittmer, J. (2020). A Blockchain Use Case in Food Distribution: Do You Know Where Your Food Has Been? *International Journal of Information Management, 52*, 102008. doi:https://doi.org/10.1016/j.ijinfomgt.2019.09.004

Buterin, V. (2014). A Next-generation Smart Contract And Decentralized Application Platform. *white paper, 3*(37), 2-1.

Buterin, V. (2023). Ethereum Whitepaper. Retrieved from https://ethereum.org/en/whitepaper/

Cachin, C. (2016). *Architecture of The Hyperledger Blockchain Fabric*. Paper presented at the Workshop on Distributed Cryptocurrencies And Consensus Ledgers.

Cao, S., Powell, W., Foth, M., Natanelov, V., Miller, T., & Dulleck, U. (2021). Strengthening Consumer Trust in Beef Supply Chain Traceability with A Blockchain-based Human-machine Reconcile Mechanism. *Computers and Electronics in Agriculture, 180*, 105886. doi:https://doi.org/10.1016/j.compag.2020.105886

Cappiello, C., Comuzzi, M., Daniel, F., & Meroni, G. (2019). *Data Quality Control in Blockchain Applications*. Paper presented at the Business Process Management: Blockchain and Central and Eastern Europe Forum: BPM 2019 Blockchain and CEE Forum, Vienna, Austria, September 1–6, 2019, Proceedings.

Caro, M. P., Ali, M. S., Vecchio, M., & Giaffreda, R. (2018). *Blockchain-based Traceability in Agri-food Supply Chain Management: A Practical Implementation*. Paper presented at the 2018 IoT Vertical and Topical Summit on Agriculture-Tuscany (IOT Tuscany).

Casino, F., Kanakaris, V., Dasaklis, T. K., Moschuris, S., & Rachaniotis, N. P. (2019). Modeling Food Supply Chain Traceability Based on Blockchain Technology. *Ifac-Papersonline, 52*(13), 2728-2733. doi:https://doi.org/10.1016/j.ifacol.2019.11.620

Casino, F., Kanakaris, V., Dasaklis, T. K., Moschuris, S., Stachtiaris, S., Pagoni, M., & Rachaniotis, N. P. (2021). Blockchain-based Food Supply Chain Traceability: A Case Study in The Dairy Sector. *International Journal of Production Research, 59*(19), 5758-5770. doi:https://doi.org/10.1080/00207543.2020.1789238

Chang, S. E., Chen, Y.-C., & Lu, M.-F. (2019). Supply Chain Re-engineering Using Blockchain Technology: A Case of Smart Contract Based Tracking Process. *Technological Forecasting and Social Change, 144*, 1-11. doi:https://doi.org/10.1016/j.techfore.2019.03.015

Chauhan, A., Malviya, O. P., Verma, M., & Mor, T. S. (2018, 16-20 July 2018). *Blockchain and Scalability.* Paper presented at the 2018 IEEE International Conference on Software Quality, Reliability and Security Companion (QRS-C).

Chauhan, C., Parida, V., & Dhir, A. (2022). Linking Circular Economy and Digitalisation Technologies: A Systematic Literature Review of Past Achievements and Future Promises. *Technological Forecasting and Social Change, 177*, 121508. doi:https://doi.org/10.1016/j.techfore.2022.121508

Chen, H., Chen, Z., Lin, F., & Zhuang, P. (2021). Effective Management for Blockchain-based Agri-food Supply Chains Using Deep Reinforcement Learning. *IEEE Access, 9*, 36008-36018. doi:10.1109/ACCESS.2019.2918000

Chinaka, M. (2016). *Blockchain Technology-Applications in Improving Financial Inclusion in Developing Economies: Case Study for Small Scale Agriculture in Africa.* Massachusetts Institute of Technology. Retrieved from http://hdl.handle.net/1721.1/104542

Chun-Ting, P., Meng-Ju, L., Nen-Fu, H., Jhong-Ting, L., & Jia-Jung, S. (2020). *Agriculture Blockchain Service Platform for Farm-to-fork Traceability with IoT Sensors.* Paper presented at the 2020 international conference on information networking (ICOIN).

Clohessy, T., Acton, T., & Rogers, N. (2019). Blockchain Adoption: Technological, Organisational and Environmental Considerations. In H. Treiblmaier & R. Beck (Eds.), *Business Transformation through Blockchain: Volume I* (pp. 47-76). Cham: Springer International Publishing.

Coble, K. H., Mishra, A. K., Ferrell, S., & Griffin, T. (2018). Big Data in Agriculture: A Challenge for the Future. *Applied Economic Perspectives and Policy, 40*(1), 79-96. doi:https://doi.org/10.1093/aepp/ppx056

Cottrill, K., DeCovny, S., & Harris, P. (2018). *Blockchain and the Future of Food: Driving Efficiency, Transparency and Trust in Food Supply Chains*: Chain Publish Insights.

Creydt, M., & Fischer, M. (2019). Blockchain and More-algorithm Driven Food Traceability. *Food control, 105*, 45-51. doi:https://doi.org/10.1016/j.foodcont.2019.05.019

Cumming, D. J., Johan, S., & Pant, A. (2019). Regulation of the Crypto-Economy: Managing Risks, Challenges, and Regulatory Uncertainty. *Journal of Risk and Financial Management, 12*(3). Retrieved from doi:10.3390/jrfm12030126

Dakshayini, M., & Balaji Prabhu, B. (2020). *An Effective Big Data and Blockchain (BD-BC) Based Decision Support Model for Sustainable Agriculture System.* Paper presented at the EAI International Conference on Big Data Innovation for Sustainable Cognitive Computing: BDCC 2018.

Detwiler, D. (2020). *Building The Future of Food Safety Technology: Blockchain and Beyond*: Academic Press.

Dey, K., & Shekhawat, U. (2021). Blockchain for Sustainable E-agriculture: Literature Review, Architecture for Data Management, and Implications. *Journal of cleaner production, 316*, 128254. doi:https://doi.org/10.1016/j.jclepro.2021.128254

Driscoll, K., Hall, B., Sivencrona, H., & Zumsteg, P. (2003, 2003//). *Byzantine Fault Tolerance, from Theory to Reality.* Paper presented at the Computer Safety, Reliability, and Security, Berlin, Heidelberg.

Dutta, P., Choi, T.-M., Somani, S., & Butala, R. (2020). Blockchain Technology in Supply Chain Operations: Applications, Challenges and Research Opportunities. *Transportation research part e: Logistics and transportation review, 142*, 102067. doi:https://doi.org/10.1016/j.tre.2020.102067

Dutta, R., Das, A., Dey, A., & Bhattacharya, S. (2020, 2020//). *Blockchain vs GDPR in Collaborative Data Governance.* Paper presented at the Cooperative Design, Visualization, and Engineering, Cham.

El Ioini, N., & Pahl, C. (2018). *A Review of Distributed Ledger Technologies.* Paper presented at the On the Move to Meaningful Internet Systems. OTM 2018 Conferences: Confederated International Conferences: CoopIS, C&TC, and ODBASE 2018, Valletta, Malta, October 22-26, 2018, Proceedings, Part II.

Esmaeilian, B., Sarkis, J., Lewis, K., & Behdad, S. (2020). Blockchain for The Future of Sustainable Supply Chain Management in Industry 4.0. *Resources, Conservation and Recycling, 163*, 105064. doi:https://doi.org/10.1016/j.resconrec.2020.105064

FAO. (2019). *E-agriculture in Action: Blockchain for Agriculture: Challenges and Opportunities.*

Feng, H., Wang, X., Duan, Y., Zhang, J., & Zhang, X. (2020). Applying Blockchain Technology to Improve Agri-food Traceability: A Review of Development Methods, Benefits and Challenges. *Journal of cleaner production, 260*, 121031. doi:https://doi.org/10.1016/j.jclepro.2020.121031

Fernando, Y., Rozuar, N. H. M., & Mergeresa, F. (2021). The Blockchain-enabled Technology and Carbon Performance: Insights From Early Adopters. *Technology in Society, 64*, 101507. doi:https://doi.org/10.1016/j.techsoc.2020.101507

Ferrández-Pastor, F.-J., Mora-Pascual, J., & Díaz-Lajara, D. (2022). Agricultural Traceability Model Based on IoT and Blockchain: Application in Industrial Hemp Production. *Journal of Industrial Information Integration, 29*, 100381. doi:https://doi.org/10.1016/j.jii.2022.100381

Finck, M. (2018). Blockchains: Regulating the Unknown. *German Law Journal, 19*(4), 665-692. doi:10.1017/S2071832200022847

Frank, D. (2005). Logistics, Inventory Control, and Supply Chain Management. *Choices, 20*(4), 287-291.

Friedman, N., & Ormiston, J. (2022). Blockchain as A Sustainability-oriented Innovation?: Opportunities for and Resistance to Blockchain Technology as A Driver of Sustainability in Global Food Supply Chains. *Technological Forecasting and Social Change, 175*, 121403. doi:https://doi.org/10.1016/j.techfore.2021.121403

Galvez, J. F., Mejuto, J. C., & Simal-Gandara, J. (2018). Future Challenges on The Use of Blockchain for Food Traceability Analysis. *TrAC Trends in Analytical Chemistry, 107*, 222-232. doi:https://doi.org/10.1016/j.trac.2018.08.011

Gatteschi, V., Lamberti, F., Demartini, C., Pranteda, C., & Santamaría, V. (2018). Blockchain and Smart Contracts for Insurance: Is The Technology Mature Enough? *Future internet, 10*(2), 20. doi:https://doi.org/10.3390/fi10020020

Gayvoronskaya, T., & Meinel, C. (2021). *Blockchain*: Springer.

Ghiro, L., Restuccia, F., D'Oro, S., Basagni, S., Melodia, T., Maccari, L., & Cigno, R. L. (2021). What is a Blockchain? A Definition to Clarify the Role of the Blockchain in the Internet of Things. *arXiv preprint arXiv:2102.03750.* doi: https://doi.org/10.48550/arXiv.2102.03750

Giancaspro, M. (2017). Is A 'Smart Contract' Really A Smart Idea? Insights from A Legal Perspective. *Computer law & security review, 33*(6), 825-835. doi: https://doi.org/10.1016/j.clsr.2017.05.007

Gouët, C., & Van Paassen, A. (2012). Smallholder Marketing Cooperatives and Smallholders' Market Access: Lessons Learned from the Actors Involved. *The Journal of Agricultural Education and Extension, 18*(4), 369-385. doi:10.1080/1389224X.2012.691784

Greenow, L., & Muñiz, V. (1988). Market trade in Decentralized Development: The Case of Cajamarca, Peru. *The Professional Geographer, 40*(4), 416-427. doi:https://doi.org/10.1111/j.0033-0124.1988.00416.x

Gu, J., Zhao, L., Yue, X., Arshad, N. I., & Mohamad, U. H. (2023). Multistage Quality Control In Manufacturing Process Using Blockchain with Machine Learning Technique. *Information Processing & Management, 60*(4), 103341. doi:https://doi.org/10.1016/j.ipm.2023.103341

Guggisberg, S. (2022). Transparency In the Activities of The Food and Agriculture Organization for Sustainable Fisheries. *Marine Policy, 136*, 104498. doi: https://doi.org/10.1016/j.marpol.2021.104498

Gupta, R., & Shankar, R. (2023). Managing Food Security Using Blockchain-enabled Traceability System. *Benchmarking: An International Journal.* doi: https://doi.org/10.1108/BIJ-01-2022-0029

Hamilton, M. (2020). Blockchain Distributed Ledger Technology: An Introduction and Focus on Smart Contracts. *Journal of Corporate accounting & finance, 31*(2), 7-12. doi:https://doi.org/10.1002/jcaf.22421

Hang, L., Ullah, I., & Kim, D.-H. (2020). A Secure Fish Farm Platform Based on Blockchain for Agriculture Data Integrity. *Computers and Electronics in Agriculture, 170*, 105251. doi:https://doi.org/10.1016/j.compag.2020.105251

Heard-Lauréote, K. (2007). A Transparency Gap?: The Case of European Agricultural Committee Governance. *Public Policy and Administration, 22*(2), 239-258. doi:10.1177/0952076707075905

Hong, W., Mao, J., Wu, L., & Pu, X. (2021). Public Cognition of The Application of Blockchain in Food Safety Management-Data From China's Zhihu Platform. *Journal of cleaner production, 303*, 127044. doi:https://doi.org/10.1016/j.jclepro.2021.127044

Hu, S., Huang, S., Huang, J., & Su, J. (2021). Blockchain and Edge Computing Technology Enabling Organic Agricultural Supply Chain: A Framework Solution To Trust Crisis. *Computers & Industrial Engineering, 153*, 107079. doi:https://doi.org/10.1016/j.cie.2020.107079

Hu, S., Huang, S., & Qin, X. (2022). Exploring Blockchain-supported Authentication Based on Online and Offline Business in Organic Agricultural Supply Chain.

*Computers & Industrial Engineering, 173*, 108738. doi:https://doi.org/10.1016/j.cie.2022.108738

Huang, H., Zhou, X., & Liu, J. (2019, 2019//). *Food Supply Chain Traceability Scheme Based on Blockchain and EPC Technology.* Paper presented at the Smart Blockchain, Cham.

Huq, F. A., & Stevenson, M. (2020). Implementing Socially Sustainable Practices in Challenging Institutional Contexts: Building Theory from Seven Developing Country Supplier Cases. *Journal of business ethics, 161*(2), 415-442. doi:10.1007/s10551-018-3951-x

Iansiti, M., & Lakhani, K. R. (2017). The Truth about Blockchain. *Harvard business review, 95*(1), 118-127.

Iftekhar, A., Cui, X., Hassan, M., & Afzal, W. (2020). Application of Blockchain and Internet of Things To Ensure Tamper-proof Data Availability for Food Safety. *Journal of Food Quality, 2020*, 1-14. doi:https://doi.org/10.1155/2020/5385207

Iqbal, R., & Butt, T. A. (2020). Safe Farming as A Service of Blockchain-based Supply Chain Management for Improved Transparency. *Cluster Computing, 23*(3), 2139-2150. doi:10.1007/s10586-020-03092-4

Jabbar, S., Lloyd, H., Hammoudeh, M., Adebisi, B., & Raza, U. (2021). Blockchain-enabled Supply Chain: Analysis, Challenges, and Future Directions. *Multimedia Systems, 27*, 787-806. doi:https://doi.org/10.1007/s00530-020-00687-0

Jannat, M. U., Ahamed, R., Mamun, A., Ferdaus, J., Costa, R., & Biswas, M. (2021). *Organic Food Supply Chain Traceability Using Blockchain Technology.* Paper presented at the 2021 International Conference on Science & Contemporary Technologies (ICSCT), Dhaka, Bangladesh.

Jardim, L., Pranto, S., Ruivo, P., & Oliveira, T. (2021). What Are The Main Drivers of Blockchain Adoption within Supply Chain?–An Exploratory Research. *Procedia Computer Science, 181*, 495-502. doi:https://doi.org/10.1016/j.procs.2021.01.195

Jha, N., Prashar, D., Khalaf, O. I., Alotaibi, Y., Alsufyani, A., & Alghamdi, S. (2021). Blockchain Based Crop Insurance: A Decentralized Insurance System for Modernization of Indian Farmers. *Sustainability, 13*(16), 8921. doi:https://doi.org/10.3390/su13168921

Jin, C., Bouzembrak, Y., Zhou, J., Liang, Q., Van Den Bulk, L. M., Gavai, A., . . . Marvin, H. J. (2020). Big Data in Food Safety-A review. *Current Opinion in Food Science, 36*, 24-32. doi:https://doi.org/10.1016/j.cofs.2020.11.006

Kamath, R. (2018). Food Traceability on Blockchain: Walmart's Pork and Mango Pilots with IBM. *The Journal of the British Blockchain Association, 1*(1). doi:http://dx.doi.org/10.31585/jbba-1-1-(10)2018

Kamble, S. S., Gunasekaran, A., & Sharma, R. (2020). Modeling The Blockchain Enabled Traceability in Agriculture Supply Chain. *International Journal of Information Management, 52*, 101967. doi:https://doi.org/10.1016/j.ijinfomgt.2019.05.023

Kamilaris, A., Fonts, A., & Prenafeta-Boldú, F. X. (2019). The Rise of Blockchain Technology in Agriculture and Food Supply Chains. *Trends in Food Science & Technology, 91*, 640-652. doi:https://www.sciencedirect.com/science/article/pii/S0924224418303686

Katsikouli, P., Wilde, A. S., Dragoni, N., & Høgh-Jensen, H. (2021). On The Benefits and Challenges of Blockchains for Managing Food Supply Chains. *Journal of the Science of Food and Agriculture, 101*(6), 2175-2181. doi:https://doi.org/10.1002/jsfa.10883

Kaur, G., & Gandhi, C. (2020). Chapter 15 - Scalability in Blockchain: Challenges and Solutions. In S. Krishnan, V. E. Balas, E. G. Julie, Y. H. Robinson, S. Balaji, & R. Kumar (Eds.), *Handbook of Research on Blockchain Technology* (pp. 373-406): Academic Press.

Kayikci, Y., Subramanian, N., Dora, M., & Bhatia, M. S. (2022). Food Supply Chain in The Era of Industry 4.0: Blockchain Technology Implementation Opportunities and Impediments from The Perspective of People, Process, Performance, and Technology. *Production planning & control, 33*(2-3), 301-321. doi:https://doi.org/10.1080/09537287.2020.1810757

Kelly, V., Adesina, A. A., & Gordon, A. (2003). Expanding Access to Agricultural Inputs in Africa: A Review of Recent Market Development Experience. *Food Policy, 28*(4), 379-404. doi:https://doi.org/10.1016/j.foodpol.2003.08.006

Khan, D., Jung, L. T., & Hashmani, M. A. (2021). Systematic Literature Review of Challenges in Blockchain Scalability. *Applied Sciences, 11*(20). Retrieved from doi:10.3390/app11209372

Khan, H. H., Malik, M. N., Konečná, Z., Chofreh, A. G., Goni, F. A., & Klemeš, J. J. (2022). Blockchain Technology for Agricultural Supply Chains during The COVID-19 Pandemic: Benefits and Cleaner Solutions. *Journal of cleaner production, 347*, 131268. doi:https://doi.org/10.1016/j.jclepro.2022.131268

Khan, S., Haleem, A., Husain, Z., Samson, D., & Pathak, R. (2023). Barriers to Blockchain Technology Adoption in Supply Chains: The Case of India. *Operations Management Research*, 1-16. doi:https://doi.org/10.1007/s12063-023-00358-z

Khan, S. N., Loukil, F., Ghedira-Guegan, C., Benkhelifa, E., & Bani-Hani, A. (2021). Blockchain Smart Contracts: Applications, Challenges, and Future Trends. *Peer-to-Peer Networking and Applications, 14*, 2901-2925. doi:https://doi.org/10.1007/s12083-021-01127-0

Kim, H. M., & Laskowski, M. (2018). Agriculture on The Blockchain: Sustainable Solutions for Food, Farmers, and Financing. *Supply Chain Revolution, Barrow Books*. doi:https://dx.doi.org/10.2139/ssrn.3028164

Kim, M., Hilton, B., Burks, Z., & Reyes, J. (2018). *Integrating Blockchain, Smart Contract-tokens, and IoT To Design A Food Traceability Solution*. Paper presented at the 2018 IEEE 9th annual information technology, electronics and mobile communication conference (IEMCON).

Kopyto, M., Lechler, S., von der Gracht, H. A., & Hartmann, E. (2020). Potentials of Blockchain Technology in Supply Chain Management: Long-Term Judgments of An International Expert Panel. *Technological Forecasting and Social Change, 161*, 120330. doi:https://doi.org/10.1016/j.techfore.2020.120330

Kouhizadeh, M., Saberi, S., & Sarkis, J. (2021). Blockchain Technology and The Sustainable Supply Chain: Theoretically Exploring Adoption Barriers. *International Journal of Production Economics, 231*, 107831. doi:https://doi.org/10.1016/j.ijpe.2020.107831

Kouhizadeh, M., Zhu, Q., & Sarkis, J. (2020). Blockchain and The Circular Economy: Potential Tensions and Critical Reflections from Practice. *Production planning & control, 31*(11-12), 950-966. doi:https://doi.org/10.1080/09537287.2019.1695925

Krzyzanowski Guerra, K., & Boys, K. A. (2022). A New Food Chain: Adoption and Policy Implications to Blockchain Use In Agri-food Industries. *Applied Economic Perspectives and Policy, 44*(1), 324-349. doi:https://doi.org/10.1002/aepp.13163

Kshetri, N. (2017). Can Blockchain Strengthen The Internet of Things? *IT professional, 19*(4), 68-72. doi:https://doi.org/10.1109/MITP.2017.3051335

Kshetri, N. (2019). Blockchain and The Economics of Food Safety. *IT professional, 21*(3), 63-66. doi:http://dx.doi.org/10.1109/MITP.2019.2906761

Kshetri, N. (2021a). Blockchain-based Smart Contracts To Provide Crop Insurance for Smallholder Farmers in Developing Countries. *IT professional, 23*(6), 58-61. doi:http://dx.doi.org/10.1109/MITP.2021.3123416

Kshetri, N. (2021b). Blockchain and Sustainable Supply Chain Management in Developing Countries. *International Journal of Information Management, 60,* 102376. doi:https://doi.org/10.1016/j.ijinfomgt.2021.102376

Kshetri, N. (2022a). Blockchain as A Tool To Facilitate Property Rights Protection in The Global South: Lessons from India's Andhra Pradesh State. *Third World Quarterly, 43*(2), 371-392. doi:https://doi.org/10.1080/01436597.2021.2013116

Kshetri, N. (2022b). Blockchain Systems and Ethical Sourcing in The Mineral and Metal Industry: A Multiple Case Study. *The International Journal of Logistics Management, 33*(1), 1-27. doi:https://doi.org/10.1108/IJLM-02-2021-0108

Kumar, G., Saha, R., Buchanan, W. J., Geetha, G., Thomas, R., Rai, M. K., . . . Alazab, M. (2020). Decentralized Accessibility of E-commerce Products through Blockchain Technology. *Sustainable Cities and Society, 62,* 102361. doi:https://doi.org/10.1016/j.scs.2020.102361

Kumar, S., Raut, R. D., Agrawal, N., Cheikhrouhou, N., Sharma, M., & Daim, T. (2022). Integrated Blockchain and Internet of Things in The Food Supply Chain: Adoption Barriers. *Technovation, 118,* 102589. doi:https://doi.org/10.1016/j.technovation.2022.102589

Kuo, T.-T., Kim, H.-E., & Ohno-Machado, L. (2017). Blockchain Distributed Ledger Technologies for Biomedical and Health Care Applications. *Journal of the American Medical Informatics Association, 24*(6), 1211-1220. doi:https://doi.org/10.1093/jamia/ocx068

Kushwaha, S. S., Joshi, S., Singh, D., Kaur, M., & Lee, H. N. (2022). Systematic Review of Security Vulnerabilities in Ethereum Blockchain Smart Contract. *IEEE Access, 10,* 6605-6621. doi:10.1109/ACCESS.2021.3140091

Lafourcade, P., & Lombard-Platet, M. (2020). About Blockchain Interoperability. *Information Processing Letters, 161,* 105976. doi:https://doi.org/10.1016/j.ipl.2020.105976

Lambert, D. M., & Stock, J. R. (1993). *Strategic Logistics Management* (Vol. 69): Irwin Homewood, IL.

Lashkari, B., & Musilek, P. (2021). A Comprehensive Review of Blockchain Consensus Mechanisms. *IEEE Access, 9,* 43620-43652. doi:https://doi.org/10.1109/ACCESS.2021.3065880

Leduc, G., Kubler, S., & Georges, J.-P. (2021). Innovative Blockchain-based Farming Marketplace and Smart Contract Performance Evaluation. *Journal of cleaner production, 306,* 127055. doi:https://doi.org/10.1016/j.jclepro.2021.127055

Lee, J. H. (2019). *Systematic approach to analyzing security and vulnerabilities of blockchain systems.* Massachusetts Institute of Technology. Retrieved from https://hdl.handle.net/1721.1/121793

Leire, C., & Mont, O. (2010). The Implementation of Socially Responsible Purchasing. *Corporate Social Responsibility and Environmental Management, 17*(1), 27-39. doi:https://doi.org/10.1002/csr.198

Lezoche, M., Hernandez, J. E., Díaz, M. d. M. E. A., Panetto, H., & Kacprzyk, J. (2020). Agri-food 4.0: A Survey of The Supply Chains and Technologies for The Future Agriculture. *Computers in industry, 117,* 103187. doi:https://doi.org/10.1016/j.compind.2020.103187

Li, C., Li, P., Zhou, D., Yang, Z., Wu, M., Yang, G., . . . Yao, A. C.-C. (2020). *A Decentralized Blockchain with High Throughput and Fast Confirmation.* Paper presented at the 2020 {USENIX} Annual Technical Conference ({USENIX} {ATC} 20).

Li, H., Luo, J., & Zhou, G. (2023). Application of Blockchain Technology Based on Big Data Analysis in Sustainable Agriculture. *Agronomy Journal, 115*(1), 81-95. doi:https://doi.org/10.1002/agj2.21105

Li, J., & Kassem, M. (2021). Applications of Distributed Ledger Technology (DLT) and Blockchain-enabled Smart Contracts in Construction. *Automation in Construction, 132,* 103955. doi:https://doi.org/10.1016/j.autcon.2021.103955

Li, X., Jiang, P., Chen, T., Luo, X., & Wen, Q. (2020). A Survey on The Security of Blockchain Systems. *Future Generation Computer Systems, 107,* 841-853. doi:https://doi.org/10.1016/j.future.2017.08.020

Li, X., Wang, D., & Li, M. (2020). Convenience Analysis of Sustainable E-agriculture Based on Blockchain Technology. *Journal of cleaner production, 271,* 122503. doi:https://doi.org/10.1016/j.jclepro.2020.122503

Lin, J., Shen, Z., Zhang, A., & Chai, Y. (2018). *Blockchain and IoI Based Food Traceability for Smart Agriculture.* Paper presented at the Proceedings of the 3rd international conference on crowd science and engineering.

Lin, W., Huang, X., Fang, H., Wang, V., Hua, Y., Wang, J., . . . Yau, L. (2020). Blockchain Technology in Current Agricultural Systems: From Techniques to Applications. *IEEE Access, 8,* 143920-143937. doi:https://doi.org/10.1109/ACCESS.2020.3014522

Lin, Y.-P., Petway, J. R., Anthony, J., Mukhtar, H., Liao, S.-W., Chou, C.-F., & Ho, Y.-F. (2017). Blockchain: The Evolutionary Next Step for ICT e-agriculture. *Environments, 4*(3), 50. doi:https://doi.org/10.3390/environments4030050

Litke, A., Anagnostopoulos, D., & Varvarigou, T. (2019). Blockchains for Supply Chain Management: Architectural Elements and Challenges towards A Global Scale Deployment. *Logistics, 3*(1), 5. doi:https://doi.org/10.3390/logistics3010005

Liu, K.-H., Chang, S.-F., Huang, W.-H., & Lu, I.-C. (2019). The Framework of The Integration of Carbon Footprint and Blockchain: Using Blockchain as A Carbon Emission Management Tool. *Technologies and Eco-innovation towards*

*Sustainability I: Eco Design of Products and Services*, 15-22. doi:https://doi.org/10.1007/978-981-13-1181-9_2

Liu, X., Farahani, B., & Firouzi, F. (2020). Distributed Ledger Technology. *Intelligent Internet of Things: From Device to Fog and Cloud*, 393-431. doi:https://doi.org/10.1007/978-3-030-30367-9_8

Lubbe, J. C. (1998). *Basic Methods of Cryptography*: Cambridge University Press.

Luu, L., Narayanan, V., Zheng, C., Baweja, K., Gilbert, S., & Saxena, P. (2016). *A Secure Sharding Protocol for Open Blockchains*. Paper presented at the Proceedings of The 2016 ACM SIGSAC Conference on Computer And Communications Security.

Makkar, H. P., & Costa, C. (2020). Potential Blockchain Applications in Animal Production and Health Sector. *CABI Reviews*(2020). doi:http://dx.doi.org/10.1079/PAVSNNR202015035

Mangla, S. K., Kazançoğlu, Y., Yıldızbaşı, A., Öztürk, C., & Çalık, A. (2022). A Conceptual Framework for Blockchain-based Sustainable Supply Chain and Evaluating Implementation Barriers: A Case of The Tea Supply Chain. *Business Strategy and the Environment, 31*(8), 3693– 3716. doi:https://doi.org/10.1002/bse.3027

Manteghi, Y., Arkat, J., & Mahmoodi, A. (2023). The Competition between Conventional and Organic Food Production in The Presence of The Blockchain Technology. *Trends in Food Science & Technology*. doi:https://doi.org/10.1016/j.tifs.2023.05.003

Manupati, V. K., Schoenherr, T., Ramkumar, M., Wagner, S. M., Pabba, S. K., & Inder Raj Singh, R. (2020). A Blockchain-based Approach for A Multi-echelon Sustainable Supply Chain. *International Journal of Production Research, 58*(7), 2222-2241. doi:https://doi.org/10.1080/00207543.2019.1683248

Martínez-Castañeda, M., & Feijoo, C. (2023). Use of Blockchain in The Agri-food Value Chain: State of The Art in Spain and Some Lessons from The Perspective of Public Support. *Telecommunications Policy*, 102574. doi:https://doi.org/10.1016/j.telpol.2023.102574

Mavilia, R., & Pisani, R. (2022). Blockchain for Agricultural Sector: The Case of South Africa. *African Journal of Science, Technology, Innovation and Development, 14*(3), 845-851. doi:https://doi.org/10.1080/20421338.2021.1908660

Mazzei, D., Baldi, G., Fantoni, G., Montelisciani, G., Pitasi, A., Ricci, L., & Rizzello, L. (2020). A Blockchain Tokenizer for Industrial IOT Trustless Applications. *Future Generation Computer Systems, 105*, 432-445. doi:https://doi.org/10.1016/j.future.2019.12.020

McBratney, A., Whelan, B., Ancev, T., & Bouma, J. (2005). Future Directions of Precision Agriculture. *Precision Agriculture, 6*(1), 7-23. doi:10.1007/s11119-005-0681-8

McIntosh, C. (2018). The Use of Financial Technology in The Agriculture Sector. (872), 32.

Menon, S., & Jain, K. (2021). Blockchain Technology for Transparency in Agri-food Supply Chain: Use Cases, Limitations, and Future Directions. *IEEE Transactions on Engineering Management*. doi:http://dx.doi.org/10.1109/TEM.2021.3110903

Mezquita, Y., Valdeolmillos, D., González-Briones, A., Prieto, J., & Corchado, J. M. (2019, 2019//). *Legal Aspects and Emerging Risks in the Use of Smart Contracts Based on Blockchain.* Paper presented at the Knowledge Management in Organizations, Cham.

Miller, T., Cao, S., Foth, M., Boyen, X., & Powell, W. (2023). An Asset-backed Decentralised Finance Instrument for Food Supply Chains-A Case Study from The Livestock Export Industry. *Computers in industry, 147,* 103863. doi:https://doi.org/10.1016/j.compind.2023.103863

Mirabelli, G., & Solina, V. (2020). Blockchain and Agricultural Supply Chains Traceability: Research Trends and Future Challenges. *Procedia Manufacturing, 42,* 414-421. doi:https://doi.org/10.1016/j.promfg.2020.02.054

Motta, G. A., Tekinerdogan, B., & Athanasiadis, I. N. (2020). Blockchain Applications in The Agri-food Domain: The First Wave. *frontiers in Blockchain, 3,* 6. doi:http://dx.doi.org/10.3389/fbloc.2020.00006

Mukherjee, A. A., Singh, R. K., Mishra, R., & Bag, S. (2021). Application of Blockchain Technology for Sustainability Development in Agricultural Supply Chain: Justification Framework. *Operations Management Research, 15,* 46-61. doi:https://doi.org/10.1007/s12063-021-00180-5

Nayal, K., Raut, R. D., Narkhede, B. E., Priyadarshinee, P., Panchal, G. B., & Gedam, V. V. (2021). Antecedents for Blockchain Technology-enabled Sustainable Agriculture Supply Chain. *Annals of operations research,* 1-45. doi:https://doi.org/10.1007/s10479-021-04423-3

Neethirajan, S., & Kemp, B. (2021). Digital Livestock Farming. *Sensing and Bio-Sensing Research, 32,* 100408. doi:https://doi.org/10.1016/j.sbsr.2021.100408

Nesarani, A., Ramar, R., & Pandian, S. (2020). An Efficient Approach for Rice Prediction from Authenticated Block Chain Node Using Machine Learning Technique. *Environmental Technology & Innovation, 20,* 101064. doi:https://doi.org/10.1016/j.eti.2020.101064

Nguyen, C. T., Hoang, D. T., Nguyen, D. N., Niyato, D., Nguyen, H. T., & Dutkiewicz, E. (2019). Proof-of-stake Consensus Mechanisms for Future Blockchain Networks: Fundamentals, Applications and Opportunities. *IEEE Access, 7,* 85727-85745. doi:https://doi.org/10.1109/ACCESS.2019.2925010

Niknejad, N., Ismail, W., Bahari, M., Hendradi, R., & Salleh, A. Z. (2021). Mapping the Research Trends on Blockchain Technology in Food and Agriculture Industry: A Bibliometric Analysis. *Environmental Technology & Innovation, 21,* 101272. doi:https://doi.org/10.1016/j.eti.2020.101272

O'Leary, D. E. (2017). Configuring Blockchain Architectures for Transaction Information in Blockchain Consortiums: The Case of Accounting and Supply Chain Systems. *Intelligent Systems in Accounting, Finance and Management, 24*(4), 138-147. doi:https://doi.org/10.1002/isaf.1417

OECD. (2020). OECD-FAO Agricultural Outlook 2020–2029. *Oecd.* Retrieved from https://www.oecd-ilibrary.org/agriculture-and-food/oecd-fao-agricultural-outlook-2020-2029_1112c23b-en

Ølnes, S., Ubacht, J., & Janssen, M. (2017). Blockchain in Government: Benefits and Implications of Distributed Ledger Technology for Information Sharing (Vol. 34, pp. 355-364): Elsevier.

Pakseresht, A., Ahmadi Kaliji, S., & Xhakollari, V. (2022). How Blockchain Facilitates the Transition toward Circular Economy in the Food Chain? *Sustainability, 14*(18), 11754. doi:https://doi.org/10.3390/su141811754

Pal, A., Tiwari, C. K., & Behl, A. (2021). Blockchain Technology in Financial Services: A Comprehensive Review of The Literature. *Journal of Global Operations and Strategic Sourcing, 14*(1), 61-80. doi:https://doi.org/10.1108/JGOSS-07-2020-0039

Pandey, V., Pant, M., & Snasel, V. (2022). Blockchain Technology in Food Supply Chains: Review and Bibliometric Analysis. *Technology in Society*, 101954. doi:https://doi.org/10.1016/j.techsoc.2022.101954

Parmentola, A., Petrillo, A., Tutore, I., & De Felice, F. (2022). Is Blockchain Able to Enhance Environmental Sustainability? A Systematic Review and Research Agenda from The Perspective of Sustainable Development Goals (SDGs). *Business Strategy and the Environment, 31*(1), 194-217. doi:https://doi.org/10.1002/bse.2882

Pilkington, M. (2016). Blockchain technology: principles and applications *Research Handbook on Digital Transformations* (pp. 225-253): Edward Elgar Publishing.

Pincheira, M., Vecchio, M., Giaffreda, R., & Kanhere, S. S. (2021). Cost-effective IoI Devices as Trustworthy Data Sources for A Blockchain-based Water Management System in Precision Agriculture. *Computers and Electronics in Agriculture, 180*, 105889. doi:https://doi.org/10.1016/j.compag.2020.105889

Polemis, M. L., & Tsionas, M. G. (2023). The Environmental Consequences of Blockchain Technology: A Bayesian Quantile Cointegration Analysis for Bitcoin. *International journal of finance & economics, 28*(2), 1602-1621. doi:https://doi.org/10.1002/ijfe.2496

Pranto, T. H., Noman, A. A., Mahmud, A., & Haque, A. B. (2021). Blockchain and Smart Contract for IoT Enabled Smart Agriculture. *PeerJ Computer Science, 7*, e407. doi:https://doi.org/10.7717/peerj-cs.407

Prashar, D., Jha, N., Jha, S., Lee, Y., & Joshi, G. P. (2020). Blockchain-based Traceability and Visibility for Agricultural Products: A Decentralized Way of Ensuring Food Safety In India. *Sustainability, 12*(8), 3497. doi:https://doi.org/10.3390/su12083497

Prewett, K. W., Prescott, G. L., & Phillips, K. (2020). Blockchain Adoption Is Inevitable—Barriers and Risks Remain. *Journal of Corporate accounting & finance, 31*(2), 21-28. doi:https://doi.org/10.1002/jcaf.22415

Pufahl, L., Ohlsson, B., Weber, I., Harper, G., & Weston, E. (2021). Enabling Financing in Agricultural Supply Chains Through Blockchain: Interorganizational Process Innovation Through Blockchain *Business Process Management Cases Vol. 2: Digital Transformation-Strategy, Processes and Execution* (Vol. 2, pp. 41-56). Berlin, Heidelberg: Springer.

Queiroz, M. M., & Wamba, S. F. (2019). Blockchain Adoption Challenges in Supply Chain: An Empirical Investigation of The Main Drivers in India and The USA. *International Journal of Information Management, 46*, 70-82. doi:https://doi.org/10.1016/j.ijinfomgt.2018.11.021

Radziwill, N. (2018). Blockchain Revolution: How The Technology Behind Bitcoin Is Changing Money, Business, And The World. The *Quality Management Journal, 25*(1), 64-65.

Ramadoss, T. S., Alam, H., & Seeram, R. (2018). Artificial Intelligence and Internet of Things Enabled Circular Economy. *The International Journal of Engineering and Science, 7*(9), 55-63. doi:http://dx.doi.org/10.9790/1813-0709035563

Rana, R. L., Giungato, P., Tarabella, A., & Tricase, C. (2019). Blockchain Applications and Sustainability Issues. *Amfiteatru Economic, 21*(13), 861-870.

Rana, R. L., Tricase, C., & De Cesare, L. (2021). Blockchain Technology for A Sustainable Agri-food Supply Chain. *British Food Journal, 123*(11), 3471-3485. doi:https://doi.org/10.1108/BFJ-09-2020-0832

Ratta, P., Kaur, A., Sharma, S., Shabaz, M., & Dhiman, G. (2021). Application of Blockchain and Internet of Things in Healthcare and Medical Sector: Applications, Challenges, and Future Perspectives. *Journal of Food Quality, 2021*, 1-20. doi:https://doi.org/10.1155/2021/7608296

Rejeb, A., Keogh, J. G., Zailani, S., Treiblmaier, H., & Rejeb, K. (2020). Blockchain Technology in The Food Industry: A Review of Potentials, Challenges and Future Research Directions. *Logistics, 4*(4), 27. doi:https://doi.org/10.3390/logistics4040027

Rejeb, A., Simske, S., Rejeb, K., Treiblmaier, H., & Zailani, S. (2020). Internet of Things Research in Supply Chain Management and Logistics: A Bibliometric Analysis. *Internet of Things, 12*, 100318. doi:https://doi.org/10.1016/j.iot.2020.100318

Rijanto, A. (2021). Business Financing and Blockchain Technology Adoption in Agroindustry. *Journal of Science and Technology Policy Management, 12*(3), 215-235. doi:https://doi.org/10.1108/JSTPM-03-2020-0065

Risso, L. A., Ganga, G. M. D., Godinho Filho, M., de Santa-Eulalia, L. A., Chikhi, T., & Mosconi, E. (2023). Present and Future Perspectives of Blockchain in Supply Chain Management: A Review of Reviews And Research Agenda. *Computers & Industrial Engineering*, 109195. doi:https://doi.org/10.1016/j.cie.2023.109195

Rugeviciute, A., & Mehrpouya, A. (2019). Blockchain, A Panacea for Development Accountability? A Study of The Barriers and Enablers for Blockchain's Adoption by Development Aid Organizations. *frontiers in Blockchain, 2*, 15. doi:https://doi.org/10.3389/fbloc.2019.00015

Ruml, A., & Qaim, M. (2021). Smallholder Farmers' Dissatisfaction with Contract Schemes in Spite of Economic Benefits: Issues of Mistrust and Lack of Transparency. *The Journal of Development Studies, 57*(7), 1106-1119. doi:10.1080/00220388.2020.1850699

Saberi, S., Kouhizadeh, M., Sarkis, J., & Shen, L. (2019). Blockchain Technology and Its Relationships to Sustainable Supply Chain Management. *International Journal of Production Research, 57*(7), 2117-2135. doi:https://doi.org/10.1080/00207543.2018.1533261

Sahebi, I. G., Masoomi, B., & Ghorbani, S. (2020). Expert Oriented Approach for Analyzing The Blockchain Adoption Barriers in Humanitarian Supply Chain. *Technology in Society, 63*, 101427. doi:https://doi.org/10.1016/j.techsoc.2020.101427

Sajja, G. S., Rane, K. P., Phasinam, K., Kassanuk, T., Okoronkwo, E., & Prabhu, P. (2023). Towards Applicability of Blockchain in Agriculture Sector. *Materials*

*Today: Proceedings, 80*, 3705-3708. doi:https://doi.org/10.1016/j.matpr.2021.07.366

Salah, K., Nizamuddin, N., Jayaraman, R., & Omar, M. (2019). Blockchain-based Soybean Traceability in Agricultural Supply Chain. *IEEE Access, 7*, 73295-73305.

Sander, F., Semeijn, J., & Mahr, D. (2018). The Acceptance of Blockchain Technology in Meat Traceability and Transparency. *British Food Journal*. doi:https://doi.org/10.1108/BFJ-07-2017-0365

Sanka, A. I., & Cheung, R. C. C. (2021). A Systematic Review of Blockchain Scalability: Issues, Solutions, Analysis and Future Research. *Journal of Network and Computer Applications, 195*, 103232. doi:https://doi.org/10.1016/j.jnca.2021.103232

Saurabh, S., & Dey, K. (2021). Blockchain Technology Adoption, Architecture, and Sustainable Agri-food Supply Chains. *Journal of cleaner production, 284*, 124731. doi:https://doi.org/10.1016/j.jclepro.2020.124731

Sayeed, S., Marco-Gisbert, H., & Caira, T. (2020). Smart Contract: Attacks and Protections. *IEEE Access, 8*, 24416-24427. doi:10.1109/ACCESS.2020.2970495

Schneider, L., & Wallenburg, C. M. (2012). Implementing Sustainable Sourcing—Does Purchasing Need to Change? *Journal of Purchasing and Supply Management, 18*(4), 243-257. doi:https://doi.org/10.1016/j.pursup.2012.03.002

Shahab, S., & Allam, Z. (2020). Reducing Transaction Costs of Tradable Permit Schemes Using Blockchain Smart Contracts. *Growth and Change, 51*(1), 302-308. doi:https://doi.org/10.1111/grow.12342

Shakhbulatov, D., Arora, A., Dong, Z., & Rojas-Cessa, R. (2019). *Blockchain Implementation for Analysis of Carbon Footprint Across Food Supply Chain*. Paper presented at the 2019 IEEE International Conference on Blockchain (Blockchain).

Shamieh, F., Wang, X., & Hussein, A. R. (2020). Transaction Throughput Provisioning Technique for Blockchain-based Industrial IoT Networks. *IEEE Transactions on Network Science and Engineering, 7*(4), 3122-3134. doi:10.1109/TNSE.2020.3017389

Sharma, R., Samad, T. A., Jabbour, C. J. C., & de Queiroz, M. J. (2021). Leveraging Blockchain Technology for Circularity in Agricultural Supply Chains: Evidence from A Fast-growing Economy. *Journal of Enterprise Information Management*(ahead-of-print). doi:https://doi.org/10.1108/JEIM-02-2021-0094

Sharma, V., Tripathi, A. K., & Mittal, H. (2022). Technological Revolutions in Smart Farming: Current Trends, Challenges & Future Directions. *Computers and Electronics in Agriculture*, 107217. doi:https://doi.org/10.1016/j.compag.2022.107217

Shi, X., Xiao, H., Liu, W., Lackner, K. S., Buterin, V., & Stocker, T. F. (2023). Confronting the Carbon-footprint Challenge of Blockchain. *Environmental Science & Technology, 57*(3), 1403-1410. doi:10.1021/acs.est.2c05165

Shiferaw, B., Hellin, J., & Muricho, G. (2011). Improving Market Access and Agricultural Productivity Growth in Africa: What Role for Producer Organizations and Collective Action Institutions? *Food Security, 3*(4), 475-489. doi:10.1007/s12571-011-0153-0

Singh, V., & Sharma, S. K. (2023). Application of Blockchain Technology in Shaping The Future of Food Industry Based on Transparency and Consumer

Trust. *Journal of Food Science and Technology, 60*(4), 1237-1254. doi:https://doi.org/10.1007/s13197-022-05360-0

Singhal, B., Dhameja, G., & Panda, P. S. (2018). *Beginning Blockchain: A Beginner's Guide to Building Blockchain Solutions* (Vol. 1): Springer.

Song, K., & Li, C. (2021). Blockchain-enabled Relay-aided Wireless Networks for Sustainable E-agriculture. *Journal of cleaner production, 281*, 124496. doi: https://doi.org/10.1016/j.jclepro.2020.124496

Srivastava, G., Parizi, R. M., & Dehghantanha, A. (2020). The Future of Blockchain Technology in Healthcare Internet of Things Security. *Blockchain cybersecurity, trust and privacy*, 161-184. doi:https://doi.org/10.1007/978-3-030-38181-3_9

Staples, M., Chen, S., Falamaki, S., Ponomarev, A., Rimba, P., Tran, A., . . . Zhu, J. (2017). Risks And Opportunities for Systems Using Blockchain and Smart Contracts. Data61. *CSIRO), Sydney.*

Sternberg, H. S., Hofmann, E., & Roeck, D. (2021). The Struggle Is Real: Insights from A Supply Chain Blockchain Case. *Journal of Business Logistics, 42*(1), 71-87. doi:https://doi.org/10.1111/jbl.12240

Stranieri, S., Riccardi, F., Meuwissen, M. P., & Soregaroli, C. (2021). Exploring The Impact of Blockchain on The Performance of Agri-food Supply Chains. *Food control, 119*, 107495. doi:https://doi.org/10.1016/j.foodcont.2020.107495

Sunny, J., Undralla, N., & Pillai, V. M. (2020). Supply Chain Transparency Through Blockchain-based Traceability: An Overview with Demonstration. *Computers & Industrial Engineering, 150*, 106895. doi:https://doi.org/10.1016/j.cie.2020.106895

Sylvester, G. (2019). *E-agriculture in Action: Blockchain for Agriculture, Opportunities and Challenges.* Rome, Italy: FAO.

Tao, Q., Cui, X., Huang, X., Leigh, A. M., & Gu, H. (2019). Food Safety Supervision System Based on Hierarchical Multi-domain Blockchain Network. *IEEE Access, 7*, 51817-51826. doi:http://dx.doi.org/10.1109/ACCESS.2019.2911265

Tarhini, M. (2021). Application of Asset Tokenization, Smart Contracts and Decentralized Finance in Agriculture. *Revista de Studii Financiare, 6*(10), 152-163.

Tatar, U., Gokce, Y., & Nussbaum, B. (2020). Law Versus Technology: Blockchain, GDPR, and Tough Tradeoffs. *Computer law & security review, 38*, 105454. doi:https://doi.org/10.1016/j.clsr.2020.105454

Tegeltija, S., Dejanović, S., Feng, H., Stankovski, S., Ostojić, G., Kučević, D., & Marjanović, J. (2022). Blockchain Framework for Certification of Organic Agriculture Production. *Sustainability, 14*(19), 11823. doi:https://doi.org/10.3390/su141911823

Tian, F. (2016). *An Agri-food Supply Chain Traceability System for China Based on RFID & Blockchain Technology.* Paper presented at the 2016 13th international conference on service systems and service management (ICSSSM).

Tomu, T. (2020). *The Rise of Blockchain for Agriculture: Exploring the Opportunities, Benefits, Limitations and Risks Associated with Applying Distributed Ledger Technology to Agriculture*: Tendai Tomu.

Torky, M., & Hassanein, A. E. (2020). Integrating Blockchain and The Internet of Things in Precision Agriculture: Analysis, Opportunities, and Challenges. *Computers and Electronics in Agriculture, 178,* 105476. doi:https://doi.org/10.1016/j.compag.2020.105476

Triana Casallas, J. A., Cueva Lovelle, J. M., & Rodríguez Molano, J. I. (2020). Smart Contracts with Blockchain in The Public Sector. *International Journal of Interactive Multimedia and Artificial Intelligence, 10,* 63-72. doi:http://dx.doi.org/10.9781/ijimai.2020.07.005

Tribis, Y., Bouchti, A., & Bouayad, H. (2021). Blockchain Technology-Based Supply Chain: State-of-the-art and Future Prospects. *International Journal of Innovative Technology and Exploring Engineering. 10*(3), 125-136. doi:10.35940/ijitee.C8384.0110321

Tripathi, A. K., Akul Krishnan, K., & Pandey, A. C. (2023). A Novel Blockchain and Internet of Things-Based Food Traceability System for Smart Cities. *Wireless Personal Communications, 129*(3), 2157-2180. doi:10.1007/s11277-023-10230-9

Tripoli, M., & Schmidhuber, J. (2018). Emerging Opportunities for the Application of Blockchain in the Agri-food Industry. *FAO and ICTSD,* 40.

Tsang, Y. P., Choy, K. L., Wu, C. H., Ho, G. T. S., & Lam, H. Y. (2019). Blockchain-driven IoT for Food Traceability with An Integrated Consensus Mechanism. *IEEE Access, 7,* 129000-129017. doi:10.1109/access 2019.2940227

Tse, D., Zhang, B., Yang, Y., Cheng, C., & Mu, H. (2017). *Blockchain Application in Food Supply Information Security.* Paper presented at the 2017 IEEE international conference on industrial engineering and engineering management (IEEM).

Uddin, M., Salah, K., Jayaraman, R., Pesic, S., & Ellahham, S. (2021). Blockchain for Drug Traceability: Architectures and Open Challenges. *Health informatics journal, 27*(2), 14604582211011228. doi:https://doi.org/10.1177/14604582211011228

Upadhyay, N. (2020). Demystifying Blockchain: A Critical Analysis of Challenges, Applications and Opportunities. *International Journal of Information Management, 54,* 102120. doi:https://doi.org/10.1016/j.ijinfomgt.2020.102120

van Hilten, M., Ongena, G., & Ravesteijn, P. (2020). Blockchain for Organic Food Traceability: Case Studies on Drivers and Challenges. *frontiers in Blockchain, 3,* 43. doi:https://doi.org/10.3389/fbloc.2020.567175

van Wassenaer, L., Verdouw, C., Kassahun, A., van Hilten, M., van der Meij, K., & Tekinerdogan, B. (2023). Tokenizing Circularity in Agri-food Systems: A Conceptual Framework and Exploratory Study. *Journal of cleaner production, 413,* 137527. doi:https://doi.org/10.1016/j.jclepro.2023.137527

Vangala, A., Sutrala, A. K., Das, A. K., & Jo, M. (2021). Smart Contract-based Blockchain-envisioned Authentication Scheme for Smart Farming. *IEEE Internet of Things Journal, 8*(13), 10792-10806. doi:http://dx.doi.org/10.1109/JIOT.2021.3050676

Walters, N. (2019). Privacy Law Issues in Blockchains: An Analysis of PIPEDA, The GDPR, and Proposals for Compliance. *Canadian Journal of Law and Technology.*

Wang, M., Wang, B., & Abareshi, A. (2020). Blockchain Technology and Its Role in Enhancing Supply Chain Integration Capability and Reducing Carbon Emission:

A Conceptual Framework. *Sustainability, 12*(24), 10550. doi:https://doi.org/10.3390/su122410550

Wang, W., Hoang, D. T., Hu, P., Xiong, Z., Niyato, D., Wang, P., . . . Kim, D. I. (2019). A Survey on Consensus Mechanisms and Mining Strategy Management in Blockchain Networks. *IEEE Access, 7*, 22328-22370. doi:https://doi.org/10.1109/ACCESS.2019.2896108

Wang, Y., Han, J. H., & Beynon-Davies, P. (2019). Understanding Blockchain Technology for Future Supply Chains: A Systematic Literature Review and Research Agenda. *Supply Chain Management, 24*(1), 62-84. doi:https://doi.org/10.1108/SCM-03-2018-0148

Watson, K., & Achinelli, M. L. (2008). Context and Contingency: The Coffee Crisis for Conventional Small-scale Coffee Farmers in Brazil. *The Geographical Journal, 174*(3), 223-234. doi:https://doi.org/10.1111/j.1475-4959.2008.00277.x

Wendl, M., Doan, M. H., & Sassen, R. (2023). The Environmental Impact of Cryptocurrencies Using Proof of Work and Proof of Stake Consensus Algorithms: A Systematic Review. *Journal of Environmental Management, 326*, 116530. doi:https://doi.org/10.1016/j.jenvman.2022.116530

Wirth, C., & Kolain, M. (2018). Privacy by Blockchain Design: A Blockchain-enabled GDPR-Compliant Approach for Handling Personal Data. doi:http://dx.doi.org/10.18420/blockchain2018_03

Wong, P.-M., RK Sinha, S., & Chui, C.-K. (2021). Blockchain in Manufacturing Quality Control: A Computer Simulation Study. *PLoS one, 16*(3), e0247925. doi:https://doi.org/10.1371/journal.pone.0247925

Woo, J., Fatima, R., Kibert, C. J., Newman, R. E., Tian, Y., & Srinivasan, R. S. (2021). Applying Blockchain Technology for Building Energy Performance Measurement, Reporting, and Verification (MRV) and The Carbon Credit Market: A Review of The Literature. *Building and Environment, 205*, 108199. doi:https://doi.org/10.1016/j.buildenv.2021.108199

Xiong, H., Dalhaus, T., Wang, P., & Huang, J. (2020). Blockchain Technology for Agriculture: Applications and Rationale. *frontiers in Blockchain, 3*, 7. doi: http://dx.doi.org/10.3389/fbloc.2020.00007

Xu, X., Weber, I., Staples, M., Xu, X., Weber, I., & Staples, M. (2019). Case study: AgriDigital: Blockchain Technology in The Trade and Finance of Agriculture Supply Chains. *Architecture for blockchain applications*, 239-255. doi:https://doi.org/10.1007/978-3-030-03035-3_12

Xu, Y., Li, X., Zeng, X., Cao, J., & Jiang, W. (2022). Application of Blockchain Technology in Food Safety Control: Current Trends and Future Prospects. *Critical reviews in food science and nutrition, 62*(10), 2800-2819. doi:https://doi.org/10.1080/10408398.2020.1858752

Xu, Y., Zhang, H., Ji, H., Yang, L., Li, X., & Leung, V. C. M. (2022). Transaction Throughput Optimization for Integrated Blockchain and MEC System in IoT. *IEEE Transactions on Wireless Communications, 21*(2), 1022-1036. doi:10.1109/TWC.2021.3100985

Yadav, S., & Singh, S. P. (2020). Blockchain Critical Success Factors for Sustainable Supply Chain. *Resources, Conservation and Recycling, 152*, 104505. doi: https://doi.org/10.1016/j.resconrec.2019.104505

Yaga, D., Mell, P., Roby, N., & Scarfone, K. (2019). Blockchain Technology Overview. *arXiv preprint arXiv:1906.11078*. doi:https://doi.org/10.48550/arXiv.1906.11078

Yamashita, K., Nomura, Y., Zhou, E., Pi, B., & Jun, S. (2019, 24-24 Feb. 2019). *Potential Risks of Hyperledger Fabric Smart Contracts.* Paper presented at the 2019 IEEE International Workshop on Blockchain Oriented Software Engineering (IWBOSE).

Yiannas, F. (2018). A New Era of Food Transparency Powered by Blockchain. *Innovations: Technology, Governance, Globalization, 12*(1-2), 46-56. doi:https://doi.org/10.1162/inov_a_00266

Yik, M. H.-Y., Wong, V. C.-W. T., Wong, T.-H., & Shaw, P.-C. (2021). HerBchain, a Blockchain-based Informative Platform for Quality Assurance and Quality Control of Herbal Products. *Journal of Traditional and Complementary Medicine, 11*(6), 598-600. doi:https://doi.org/10.1016/j.jtcme.2021.07.005

Yli-Huumo, J., Ko, D., Choi, S., Park, S., & Smolander, K. (2016). Where Is Current Research on Blockchain Technology?—A Systematic Review. *PLoS one, 11*(10), e0163477. doi:https://doi.org/10.1371/journal.pone.0163477

Yousefi, S., & Tosarkani, B. M. (2022). An Analytical Approach for Evaluating The Impact of Blockchain Technology on Sustainable Supply Chain Performance. *International Journal of Production Economics, 246*, 108429. doi:https://doi.org/10.1016/j.ijpe.2022.108429

Zamani, E., He, Y., & Phillips, M. (2020). On the Security Risks of the Blockchain. *Journal of computer information systems, 60*(6), 495-506. doi:10.1080/08874417.2018.1538709

Zhang, C., Wu, C., & Wang, X. (2020). *Overview of Blockchain Consensus Mechanism.* Paper presented at the Proceedings of the 2020 2nd International Conference on Big Data Engineering.

Zhao, G., Liu, S., Lopez, C., Lu, H., Elgueta, S., Chen, H., & Boshkoska, B. M. (2019). Blockchain Technology in Agri-food Value Chain Management: A Synthesis of Applications, Challenges and Future Research Directions. *Computers in industry, 109*, 83-99. doi:https://doi.org/10.1016/j.compind.2019.04.002

Zheng, K., Zheng, L. J., Gauthier, J., Zhou, L., Xu, Y., Behl, A., & Zhang, J. Z. (2022). Blockchain Technology for Enterprise Credit Information Sharing in Supply Chain Finance. *Journal of Innovation & Knowledge, 7*(4), 100256. doi:https://doi.org/10.1016/j.jik.2022.100256

Zheng, Z., Xie, S., Dai, H.-N., Chen, X., & Wang, H. (2018). Blockchain Challenges And Opportunities: A Survey. *International journal of web and grid services, 14*(4), 352-375. doi:https://doi.org/10.1504/IJWGS.2018.095647

Zyskind, G., & Nathan, O. (2015). *Decentralizing Privacy: Using Blockchain To Protect Personal Data.* Paper presented at the 2015 IEEE Security and Privacy Workshops.

# About the Author

Dr. An Hoai, Duong holds a Bachelor's degree and Master's degree in Agricultural Economics, along with a PhD degree in Economics awarded by Griffith University, Australia. Dr. Duong has been actively involved in the educational and research landscape of the agriculture industry for over two decades. From 2008 to 2011 and then again from 2017 to 2023, Dr. Duong held dual roles as a university lecturer and a dedicated researcher at Thai Nguyen University of Agriculture and Forestry in Vietnam. His tenure has been marked by substantial contributions that have enriched both the academic sphere and the agricultural sector. Throughout his career, he has taught various subjects in agriculture and agricultural economics, equipping students with the necessary knowledge and skills to navigate the complexities of the agricultural industry. Beyond his role as an educator, Dr. Duong actively engages in diverse projects within the agriculture sector, effectively bridging the gap between theoretical principles and practical applications in the field. Through these initiatives, he continually strives to drive innovation and catalyze positive transformations within the sector.

Recognizing the transformative potential of blockchain technology, Dr. Duong has been at the forefront of incorporating blockchain applications in agriculture. With more than five years of experience teaching and researching blockchain applied in agriculture, he has developed a deep understanding of the practical implications and possibilities of blockchain in revolutionizing the food supply chain. Dr. Duong's expertise lies in exploring the ways in which blockchain can enhance supply chain management, improve traceability and transparency, and foster sustainable practices in agriculture. Driven by his passion for advancing the agricultural sector, he continues to engage in cutting-edge research and collaborate with industry stakeholders to drive innovation and promote sustainable practices in agriculture through blockchain technology.

Dr. Duong's expertise, experience, and dedication to the field make him a recognized authority in the intersection of agricultural economics and blockchain technology. His book, "Blockchain Applications in Agriculture:

Revolutionising the Food Supply Chain," reflects his deep knowledge and insights, offering readers a comprehensive understanding of the potential of blockchain to transform the agricultural industry.

# Index

## A



## B



## C

## D



## E

## F

## N



## O



## P



## Q



## R

## S

## T

www.ingramcontent.com/pod-product-compliance
Lightning Source LLC
LaVergne TN
LVHW010723210926
848485LV00007B/1431/J
* 9 7 8 1 6 4 8 8 9 9 4 9 2 *